Natural Compounds for Controlling Plant Pathogens

Natural Compounds for Controlling Plant Pathogens

Guest Editors

José Sebastián Dambolena
Virginia Lara Usseglio

MDPI

Basel • Beijing • Wuhan • Barcelona • Belgrade • Novi Sad • Cluj • Manchester

Guest Editors
José Sebastián Dambolena
Instituto Multidisciplinario de Biología Vegetal (IMBIV-CONICET)
Universidad Nacional de Córdoba
Córdoba
Argentina

Virginia Lara Usseglio
Instituto Multidisciplinario de Biología Vegetal (IMBIV-CONICET)
Universidad Nacional de Córdoba
Córdoba
Argentina

Editorial Office
MDPI AG
Grosspeteranlage 5
4052 Basel, Switzerland

This is a reprint of the Special Issue, published open access by the journal *Plants* (ISSN 2223-7747), freely accessible at: https://www.mdpi.com/journal/plants/special_issues/6F8Q1S5H34.

For citation purposes, cite each article independently as indicated on the article page online and as indicated below:

Lastname, A.A.; Lastname, B.B. Article Title. *Journal Name* **Year**, *Volume Number*, Page Range.

ISBN 978-3-7258-7609-9 (Hbk)
ISBN 978-3-7258-7610-5 (PDF)
https://doi.org/10.3390/books978-3-7258-7610-5

Contents

About the Editors

José Sebastián Dambolena

José Sebastián Dambolena is an Associate Professor at the Universidad Nacional de Córdoba (Argentina) and an Independent Researcher at the Consejo Nacional de Investigaciones Científicas y Técnicas (CONICET). His research focuses on phytopathogenic fungi, particularly Fusarium species affecting maize, and on the occurrence, biosynthesis, and molecular and environmental regulation of mycotoxins, such as fumonisins. His work integrates fundamental and applied approaches to understand and control fungal contamination in crops, including the study of natural compounds and their effects on fungal growth and toxin production. In recent years, he has also incorporated the study of mycoviruses, exploring their diversity and their potential as biological control agents. His research contributes to the development of sustainable strategies for plant disease management and food safety.

Virginia Lara Usseglio

Virginia Lara Usseglio is an Assistant Researcher at the Consejo Nacional de Investigaciones Científicas y Técnicas (CONICET) and an Assistant "A" Professor at the Universidad Nacional de Córdoba (Argentina). Her research focuses on identifying natural alternatives for pest control in maize cropping systems by investigating plant–insect, plant–fungus, and insect–plant–fungus interactions. She aims to determine their emergent properties and the role of natural compounds in mediating these relationships, identifying natural compounds with pesticidal potential and elucidating their mechanisms of action. Within this framework, she also examines how climate change influences these interactions and the effectiveness of natural pesticides, contributing to the development of sustainable, environmentally friendly strategies for crop protection and improved agricultural resilience.

Preface

This Reprint addresses sustainable plant pathogen control using natural compounds. It encompasses advances in the discovery of bioactive molecules, evaluation of their efficacy, and elucidation of their mechanisms of action. Motivated by environmental concerns and the rise of resistance to synthetic pesticides, among others, the objective of this Reprint is to promote eco-friendly crop protection strategies. It is intended for researchers, agronomists, and professionals in plant science and sustainable agriculture.

José Sebastián Dambolena and Virginia Lara Usseglio
Guest Editors

plants

MDPI

Editorial

Natural Compounds for Controlling Plant Pathogens

José Sebastián Dambolena * and Virginia Lara Usseglio

Instituto Multidisciplinario de Biología Vegetal (IMBIV-CONICET), Facultad de Ciencias Exactas, Físicas y Naturales, Universidad Nacional de Córdoba, Córdoba X5000HUA, Argentina; virginia.usseglio@unc.edu.ar
* Correspondence: jdambolena@unc.edu.ar

Agricultural production systems across the world are currently facing unprecedented challenges. Rapid environmental changes, increasing climatic variability, and the growing demand for food production have intensified the pressure on agroecosystems [1,2]. Among the most important factors limiting crop productivity are plant diseases caused by fungi, insects, and bacteria. These pathogens can lead to substantial yield losses, deterioration in crop quality, and significant economic damage, representing potentially serious risks for food safety and human health via toxin production [3,4]. Consequently, the effective management of plant pathogens remains a central priority for both agricultural research and global food security.

Pest management has traditionally relied on a combination of strategies that include developing resistant crop varieties and applying synthetic pesticides, among others. Over the past decades, these approaches have played a crucial role in protecting crops and maintaining agricultural productivity. However, the extensive application and misuse of these methods have raised increasing concerns regarding their environmental and ecological impacts. Residues of synthetic agrochemicals may accumulate in soil and water systems, negatively affecting non-target organisms and contributing to biodiversity loss. Moreover, the continuous exposure of pathogens to chemical treatments often accelerates the development of resistance, thereby reducing the long-term effectiveness of these strategies [5–10].

In response to these challenges, developing environmentally sustainable alternatives has become an important focus of research in crop protection. Among the most promising approaches is the use of natural compounds derived from plants, microorganisms, and other biological sources. Natural products represent extraordinarily diverse chemical groups, including phenolics, terpenoids, alkaloids, and flavonoids, many of which exhibit strong pesticidal properties. Compared with synthetic pesticides, natural compounds often present several advantages that make them attractive candidates for sustainable crop protection. Many of these substances are biodegradable and tend to have lower environmental persistence, thereby reducing the risk of long-term environmental contamination. In addition, they frequently display lower toxicity toward non-target organisms and may present improved safety profiles for both agricultural workers and consumers. Another important advantage lies in their structural diversity and complex modes of action, which may reduce the likelihood of pests developing resistance [11–16].

This Special Issue *"Natural Compounds for Controlling Plant Pathogens"* was organized with the aim of bringing together recent advances in this dynamic and interdisciplinary field. The collection brings together original research articles that address different aspects of natural-product-based plant disease management, including discovering new bioactive substances, evaluating their activity against important phytopathogens, and exploring their potential mechanisms of action.

 https://doi.org/10.3390/plants15071084

The nine articles included in this volume illustrate the diversity of natural sources and research approaches currently being explored for plant pathogen control. These are as follows:

- "Paralysis Activity of "Basic Substances" and Rose Extracts on *Meloidogyne incognita* Second-Stage Juveniles"
- "Chemical Composition and Antifungal Activity of *Artemisia sieversiana* Essential Oil Growing in Jilin Against Black Spot on Yanbian Pingguoli Pear in China"
- "Organic Compounds as a Natural Alternative for Pest Control: How Will Climate Change Affect Their Effectiveness?"
- "Antimicrobial Potential of Six Plant Essential Oils Against *Pseudomonas syringae* pv. actinidiae: In Vitro Activity and In Planta Efficacy Do Not Always Align"
- "In Vitro Evaluation of the Antifungal Activity of *Trigonella foenum-graecum* Seed Extract and Its Potential Application in Plant Protection"
- "Antifungal Activity of *Artemisia capillaris* Essential Oil Against Alternaria Species Causing Black Spot on Yanbian Pingguoli Pear in China"
- "Eco-Friendly Crop Protection: *Argyrantemum frutescens*, a Source of Biofungicides"
- "Bioactive Sesquiterpenoids from *Santolina chamaecyparissus* L. Flowers: Chemical Profiling and Antifungal Activity Against Neocosmospora Species"
- "Antifungal Activity of Genistein Against Phytopathogenic Fungi *Valsa mali* Through ROS-Mediated Lipid Peroxidation"

Taken together, the contributions in this volume illustrate the diversity of natural compounds with potential applications in crop protection, highlighting their relevance in developing environmentally friendly strategies. The collective contributions of this Special Issue therefore represent an important step toward advancing knowledge in the field of natural-product-based plant protection. By highlighting both fundamental research and potential practical applications, the articles included in this volume contribute to the development of innovative approaches aimed at improving plant health, reducing environmental impacts, and promoting more sustainable agricultural systems.

Finally, we would like to express our sincere gratitude to all the authors who contributed their valuable research to this Special Issue. Their work reflects the growing global effort to develop environmentally sustainable solutions for plant disease management. We hope that the research presented in this volume will inspire further studies and foster new collaborations aimed at advancing the use of natural compounds in sustainable crop protection.

Conflicts of Interest: The authors declare no conflicts of interest.

References

1. OECD-FAO. *Perspectivas Para El Medio Ambiente. Agricultura Mundial: Hacia Los Años 2015/2030*; FAO: Roma, Italy, 2015.
2. FAO. *GIEWS Crop Prospects and Food Situation #4, December 2019*; FAO: Roma, Italy, 2019.
3. Deutsch, C.A.; Tewksbury, J.J.; Tigchelaar, M.; Battisti, D.S.; Merrill, S.C.; Huey, R.B.; Naylor, R.L. Increase in Crop Losses to Insect Pests in a Warming Climate. *Science* **2018**, *361*, 916–919. [CrossRef] [PubMed]
4. Vila-Donat, P.; Marín, S.; Sanchis, V.; Ramos, A.J. A Review of the Mycotoxin Adsorbing Agents, with an Emphasis on Their Multi-Binding Capacity, for Animal Feed Decontamination. *Food Chem. Toxicol.* **2018**, *114*, 246–259. [CrossRef]
5. Wan, N.F.; Woodcock, B.A.; Scherber, C.; Wyckhuys, K.A.G.; Li, Z.; Qian, X. Leaving Synthetic Pesticides Behind. *Science* **2025**, *388*, 712–713. [CrossRef]
6. Ngegba, P.M.; Cui, G.; Khalid, M.Z.; Zhong, G. Use of Botanical Pesticides in Agriculture as an Alternative to Synthetic Pesticides. *Agriculture* **2022**, *12*, 600. [CrossRef]
7. Brattsten, L.B.; Holyoke, C.W.; Leeper, J.R.; Raffa, K.F. Insecticide Resistance: Challenge to Pest Management and Basic Research. *Science* **1986**, *231*, 1255–1260. [CrossRef] [PubMed]
8. Presello, D.A.; Giménez, F.J.; Ferraguti, F.J. La Producción de Maíz En Argentina. *ACI Av. Cienc. Ing.* **2022**, *14*, 13. [CrossRef]

 https://doi.org/10.3390/plants15071084

9. Blanco, C.A.; Chiaravalle, W.; Dalla-Rizza, M.; Farias, J.R.; García-Degano, M.F.; Gastaminza, G.; Mota-Sánchez, D.; Murúa, M.G.; Omoto, C.; Pieralisi, B.K.; et al. Current Situation of Pests Targeted by Bt Crops in Latin America. *Curr. Opin. Insect Sci.* **2016**, *15*, 131–138. [CrossRef] [PubMed]
10. Fai, P.B.A.; Ncheuveu, N.T.; Tchamba, M.N.; Ngealekeloeh, F. Ecological Risk Assessment of Agricultural Pesticides in the Highly Productive Ndop Flood Plain in Cameroon Using the PRIMET Model. *Environ. Sci. Pollut. Res.* **2019**, *26*, 24885–24899. [CrossRef] [PubMed]
11. Isman, M.B. Plant Essential Oils for Pest and Disease Management. *Crop Prot.* **2000**, *19*, 603–608. [CrossRef]
12. Isman, M.B. Botanical Insecticides in the Twenty-First Century-Fulfilling Their Promise? *Annu. Rev. Entomol.* **2020**, *65*, 233–249. [CrossRef] [PubMed]
13. Regnault-Roger, C.; Vincent, C.; Arnason, J.T. Essential Oils in Insect Control: Low-Risk Products in a High-Stakes World. *Annu. Rev. Entomol.* **2012**, *57*, 405–424. [CrossRef] [PubMed]
14. Sparks, T.C.; Bryant, R.J. Impact of Natural Products on Discovery of, and Innovation in, Crop Protection Compounds. *Pest Manag. Sci.* **2022**, *78*, 399–408. [CrossRef] [PubMed]
15. Sparks, T.C.; Sparks, J.M.; Duke, S.O. Natural Product-Based Crop Protection Compounds—Origins and Future Prospects. *J. Agric. Food Chem.* **2023**, *71*, 2259–2269. [CrossRef]
16. Daraban, G.M.; Hlihor, R.M.; Suteu, D. Pesticides vs. Biopesticides: From Pest Management to Toxicity and Impacts on the Environment and Human Health. *Toxics* **2023**, *11*, 983. [CrossRef] [PubMed]

https://doi.org/10.3390/plants15071084

plants

MDPI

Article

Paralysis Activity of "Basic Substances" and Rose Extracts on *Meloidogyne incognita* Second-Stage Juveniles

Rodanthi Askianaki [1,*], Nikolaos G. Tsiropoulos [1], Kyriakos D. Giannoulis [2] and Nikoletta Ntalli [1,*]

[1] Analytical Chemistry and Pesticides Laboratory, Department of Agriculture Crop Production and Rural Environment, University of Thessaly, 38446 Volos, Greece; ntsirop@uth.gr

[2] Laboratory of Agronomy and Applied Crop Physiology, Department of Agriculture, Crop Production and Rural Environment, University of Thessaly, Fytokou Street, 38446 Volos, Greece; kgiannoulis@uth.gr

* Correspondence: roaskianaki@uth.gr (R.A.); nntalli@uth.gr (N.N.); Tel.: +30-2421093194 (N.N.)

Abstract

To date, searching for bionematicidals is essential. In the absence of nematicides, "Basic Substances" are gaining ground since they are cost-effective, do not mandate an expiration date and have no inherent capacity to cause endocrine-disrupting neurotoxic or immunotoxic effects. Most "Basic Substances" are authorized for the control of phytoparasitic fungi and insects, whereas nematicidals are yet to be available. In this study, we employed "Basic Substances" and in particular, beer, sodium bicarbonate, and sodium chloride, together with rose aromatotherapy by-products, on nematicidal bioassays against *Meloidogyne incognita*. We report that chemical composition analysis of the nematicidal rose extracts correlates with bioactivity. Paralysis-based bioassays were used as primary criteria to assess efficacy, specifically targeting second-stage juveniles of *Meloidogyne incognita*. The evaluated treatments were assessed after one day, two days, and three days of J2 immersion in test solutions. According to our results, the "Basic Substances" demonstrated a significant paralysis effect on J2, thus indicating, for the first time, the considerable significance of their authorization to the root knot nematodes. Similarly, the rose extracts were found to be nematicidal, and since they are foodstuffs, and thus *nonconcern compounds*, "Basic Substances" can be developed as aromatherapy by-products in the frame of a circular economy.

Keywords: rose hydrosol; beer; sodium bicarbonate; sodium chloride; root knot nematodes

1. Introduction

As the criteria for the authorization of Plant Protection Products (PPPs) become increasingly stringent, synthetic substances with greater toxicological concern are gradually being phased out. Consequently, there is a growing need to identify plant protection solutions with a lower environmental impact. In recent years, PPPs of natural origin have emerged as safer alternatives compared to synthetic chemical compounds [1,2]. The term "Basic Substances" refers to certain "products or recipes" whose primary application is not in plant protection, yet they can provide supportive plant protection functions. These substances generally include common food-grade materials, such as sucrose, vegetable oils (e.g., sunflower or onion oil), and non-toxic plant extracts (*Salix* spp. and *Urtica* spp.). By definition, "Basic Substances" are authorized for plant protection, with the prerequisite that they cannot be commercialized. This ensures their cost-effectiveness since the farmer can readily prepare them independently and most importantly use them without a time limitation. These substances gained increasing research attention as potential alternatives to synthetic pesticides, exhibiting fungicidal, bactericidal and insecticidal activity [1,3]. The authorization of the "Basic Substances" is regulated under Article 23 of Regulation (EC) No 1107/2009 of the European Parliament, which governs the authorization of PPPs. Approval prerequisites require that they pose no risk to the environment or to human and animal health at the concentrations used, specifically avoiding endocrine, neurological, or immunological disruptions. As a result, they are not associated with residue-related concerns, and no maximum residue limits (MRLs) are established. Certain "Basic Substances" are also approved as active substances in low-risk PPPs—for example, sodium bicarbonate (baking soda), approved as a low-risk substance, and vinegar, whose primary component, acetic acid, is approved as an herbicide. A notable example is chitosan, which elicits plant defense mechanisms [1,3,4].

The first "Basic Substances" originated from food products, such as vinegar and whey. Subsequently, a hydroalcoholic extract of *Equisetum arvense* was investigated and registered in 2014 for its fungicidal activity. Substances of mineral origin (e.g., talc), animal origin (e.g., chitosan), and microorganisms (e.g., brewer's or baker's yeast) have also been registered as "Basic Substances"; as of the latest review (27 December 2025), 28 "Basic Substances" have been approved (Table 1) [5]. Due to their properties, these substances are suitable in organic farming, with 21 specifically authorized for this purpose. They can be applied directly or after dilution; water is the primary diluent used to date, although chitosan has also been approved for dissolution in vinegar. Moreover, the "Basic Substances" are compatible with other plant protection agents, including microorganisms and substances of natural, mineral, or animal origin [1,3,4]. Comprehensive guidelines for their use are provided in Table S1 of Supplementary Materials.

Table 1. "Basic Substances" approved for crop protection in the EU as of the latest review, 27 Dec 2025.

No.	Basic Substances	Target Pests	Source
1	*Allium cepa* L. bulb (extract)	Plant Pathogens	[6]
2	*Allium fistulosum* (processed)	Plant Pathogens	[7]
3	Beer	Pest Slugs and Snails	[8]
4	Calcium Hydroxide	Plant Pathogens	[9]
5	Chitosan	Plant Elicitor Against Plant Pathogens	[10]
6	Chitosan Hydrochloride	Plant Elicitor Against Plant Pathogens	[11]
7	Clayed Charcoal	Plant Pathogens	[12]
8	Cow Milk	Plant Pathogens	[13]
9	Diammonium Phosphate	Plant Pathogens	[14]
10	*Equisetum arvense* L.	Plant Pathogens and Insect Pests	[15]
11	Fructose	Plant Pathogens and Insect Pests	[16]
12	Hydrogen Peroxide	Plant Pathogens	[17]
13	L-cysteine	Leaf Cutting Ants	[18]
14	Lecithins	Plant Pathogens	[19]
15	Magnesium hydroxide E528	Plant Pathogens	[20]
16	Mustard Seeds Powder	Plant Pathogens	[21]
17	Sodium Cloride	Plant Pathogens	[22]
18	Onion Oil	Insect Pests	[23]
19	*Onobrychis viciifolia* (sainfoin) dried pellets	Phytoparasitic Nematodes	[24]

https://doi.org/10.3390/plants15030458

Table 1. *Cont.*

No.	Basic Substances	Target Pests	Source
20	*Salix* spp. Cortex	Plant Pathogens	[25]
21	Sodium Hydrogen Carbonate	Plant Pathogens	[26]
22	Sucrose	Plant Pathogens and Insect Pests	[27]
23	Sunflower Oil	Plant Pathogens	[28,29]
24	Talc	Plant Pathogens and Insect Pests	[30]
25	*Urtica* spp.	Insect Pests	[31]
26	Vinegar	Plant Pathogens	[32]
27	*Vitis vinifera* L. seed extract (grape seed extract)	Plant Pathogens	[33]
28	Whey	Plant Pathogens	[34]

Field efficacy trials of "Basic Substances" against pests and pathogens remain limited. However, the application of hydrochloric chitosan under field conditions has shown effects comparable to certain conventional PPPs [11]. Although the body of scientific literature on "Basic Substances" is still limited, it has been steadily growing in recent years. Their use is considered particularly important within the framework of the European Green Deal's Farm to Fork strategy [3,4]. While the nematicidal effects of salts and ionic stressors have been widely reported, the present study differs in its practical application focus. Beer, sodium bicarbonate, and sodium chloride are already registered for the control of some agricultural pests, yet their potential against plant-parasitic nematodes has not been evaluated. This regulatory status makes them particularly attractive candidates for timely repurposing in nematode management programs.

Rosa damascena, of the Rosaceae family, is commonly referred to as the Damask rose or the hundred-petaled rose and is recognized as one of the most significant aromatic plant species globally. While frequently cultivated as an ornamental species, its primary value lies in the diverse products obtained from its flowers. The species is widely grown across Asia and Europe, with Bulgaria, Turkey, and Iran being the principal producers.

Botanically, the Damask rose is a perennial, upright shrub that can reach 2.5 m in height, flowering from May to June. The blossoms typically contain around thirty petals, exhibiting vivid shades of pink. A fully mature shrub, aged four years or more, can yield approximately 500 to 600 flowers annually.

Fresh or dried petals, or entire flowers, are utilized to produce essential oil, rose water, extracts, or waxes. These derivatives are employed in the food, fragrance, cosmetic, and pharmaceutical industries. The rose exhibits a range of beneficial properties, including anticancer, antioxidant, and analgesic effects. Additionally, residues from the distillation process are repurposed as animal feed or compost [35]. Toxicological studies on Damask rose extracts have demonstrated a very low risk of toxicity to both animals and humans [36,37] supporting the potential for inclusion in the list of "Basic Substance".

The primary chemical constituents of the rose include monoterpenic alcohols such as citronellol, geraniol, and nerol; monoterpenes, including rose oxide; secondary metabolites of the shikimate pathway, such as methyl eugenol and phenylethanol; long-chain hydrocarbons; and metabolites derived from carotenoid degradation. Phenylethanol, present in high concentrations in the flowers, contributes to the characteristic aroma of the plant and is readily water-soluble [38].

In the context of exploring naturally derived substances with nematicidal activity, three "Basic Substances" and three rose extracts were selected and assessed for their potential to induce paralysis on *Meloidogyne incognita* second-stage juveniles, the infective stage of the

https://doi.org/10.3390/plants15030458

nematode. The "Basic Substances" tested included sodium bicarbonate, sodium chloride, and beer. On the other hand, dried petals from organically cultivated *Rosa damascena* were tested as sonicated water extract and hydrosol, the hydro-distillation byproduct. Chemical analysis of the ultrasound-assisted extract, proved to be the most effective, was conducted to characterize its constituent compounds. To the best of our knowledge these "Basic Substances" along with *Rosa domestica* extracts have not been previously investigated on root knot nematodes.

2. Results

2.1. Study of Inducing Paralysis of J2 Larvae with Basic Substances

2.1.1. Efficacy of Sodium Bicarbonate

Table 2 lists the EC_{50} values, as calculated for all immersion intervals of the J2 larvae in the sodium bicarbonate solutions, after statistical processing. At 24 h post-immersion, paralysis was induced with an EC_{50} value equal to 15.08 mg/mL. At 48 h, the EC_{50} value decreased to 6.45 mg/mL, and at 72 h, it stabilized at 7.88 mg/mL, at which point the paralyzed larvae were classified as dead, since the paralysis was irreversible.

Table 2. EC_{50} values and corresponding confidence limits, 24, 48 and 72 h after the immersion of *M. incognita* J2, in aqueous solutions of sodium bicarbonate.

Duration of Immersion ($NaHCO_3$)	R^2	EC_{50} (mg/mL)	95% Confidence Limits	Max Efficacy (%)	Min Efficacy (%)	Natural Paralysis (%)
24 h	0.93	15.08	13.83–16.33	84	0	7
48 h	0.79	6.45	5.05–7.84	90	21	12
72 h	0.80	7.88	6.39–9.36	94	8	12

2.1.2. Efficacy of Sodium Chloride

For all immersion intervals, the EC_{50} values obtained after statistical processing are reported in Table 3. At 24 h post-immersion of J2 larvae in the sodium chloride solutions paralysis was considerable, with an EC_{50} value equal to 7.04 mg/mL. Increasing the duration of immersion to 48 h, resulted in an EC_{50} value equal to 6.47 mg/mL, and 24 h later remained at a similar level (6.45 mg/mL). 72 h later, larvae classified as paralyzed were considered dead.

Table 3. EC_{50} values and corresponding confidence limits, 24, 48 and 72 h after the immersion of *M. incognita* J2, in aqueous solutions of sodium chloride.

Duration of Immersion (NaCl)	R^2	EC_{50} (mg/mL)	95% Confidence Limits	Max Efficacy (%)	Min Efficacy (%)	Natural Paralysis (%)
24 h	0.88	7.04	5.98–8.11	89	6	7
48 h	0.79	6.47	5.06–7.87	94	11	12
72 h	0.79	6.45	5.05–7.84	95	11	12

2.1.3. Efficacy of Beer

Table 4 presents the EC_{50} values calculated for alcohol-free beer solution paralysis on J2 larvae, as determined through statistical analysis. Alcohol-free beer was used to rule out potential toxicity of ethanol. At 24 h post-immersion of J2 larvae in the test solutions, the paralysis activity of the beer was evident, with the EC_{50} value calculated at 3.78% (v/v). At 48 and 72 h post-experiment establishment, the EC_{50} value was determined at similar levels, 3.35% and 3.07% (v/v), respectively, and the paralyzed larvae were classified as dead.

https://doi.org/10.3390/plants15030458

Table 4. EC_{50} values and corresponding confidence limits, 24, 48 and 72 h after the immersion of *M. incognita* J2, in aqueous solutions of alcohol-free beer concentrations.

Duration of Immersion (Alcohol-Free Beer)	R^2	EC_{50} % (v/v)	95% Confidence Limits	Max Efficacy (%)	Min Efficacy (%)	Natural Paralysis (%)
24 h	0.97	3.78	3.62–3.94	94	6	19
48 h	0.70	3.35	2.47–4.23	100	7	25
72 h	0.85	3.07	2.60–3.53	82	0	30

2.2. *Study of Inducing Paralysis of J2 Larvae with Rose Extracts*

2.2.1. Efficacy of Rose Petal Extract Produced in Sonicator

Table 5 reports the EC_{50} values calculated after J2 immersion in water extracts of rose petals, produced in the sonicator. The extract induced paralysis and the EC_{50} value was 25.82% (v/v), 24 h post-J2 immersion in test solutions. Forty-eight hours later, the EC_{50} value decreased to 11.06% v/v, and seventy-two hours later, the value was set to 2.66% (v/v), and paralyzed larvae were considered dead. The results presented a positive dose–effectiveness correlation over time.

Table 5. EC_{50} values and corresponding confidence limits, 24, 48 and 72 h after the immersion of *M. incognita* J2, in test solutions of dried rose petals water extract produced in sonicator.

Duration of Immersion (Sonicated Rose Petals in Water)	R^2	EC_{50} % (v/v)	95% Confidence Limits	Max Efficacy (%)	Min Efficacy (%)	Natural Paralysis (%)
24 h	0.72	25.82	20.4–31.2	100	1	16
48 h	0.85	11.06	9.09–13.02	100	0	25
72 h	0.74	2.66	1.18–4.13	100	62	33

2.2.2. Efficacy of Rose Petal Hydrosol as Clevenger by Product

Table 6 shows the EC_{50} values calculated after J2 immersion in rose petals' hydrosol solutions. A clear dose–response relationship was established between induced paralysis over time. At 24, 48 and 72 h post-larvae immersion in test solutions, the EC_{50} value was calculated at 7.73, 5.95 and 4.95% (v/v), respectively. Motility was not recovered after immersion in water; thus, the larvae were considered dead.

Table 6. EC_{50} values and corresponding confidence limits, after 24, 48 and 72 h of *M. incognita* J2 immersion in hydrosol solutions of rose petals.

Duration of Immersion (Rose Petals Hydrosol)	R^2	EC_{50} % (v/v)	95% Confidence Limits	Max Efficacy (%)	Min Efficacy (%)	Natural Paralysis (%)
24 h	0.82	7.73	6.27–9.18	80	0	23
48 h	0.75	5.95	3.69–8.21	87	33	25
72 h	0.74	4.94	3.80–6.07	91	29	24

2.2.3. Chemical Analysis of Sonicated Extract of Rose Petals

Table 7 shows the results of the chemical analysis of the rose petals' extract, with the main components containing (>1%), which constituted 82.2% of the total components. A total of 13.6% of them remained unidentified, while those that were present at <1% in the extract reached a percentage equal to 4.2%.

https://doi.org/10.3390/plants15030458

Table 7. The major constituents (>1%) as determined by chemical analysis of dried rose petals extracted with water in sonicator.

Spectra Database NIST	Retention Time (Rt)	Substance Content (%) over Total Identified
4-methylhexan-2-one	4.358	2.8
2-hexanone, 3.4-dimethyl-	4.809	1.0
phenylethylalcohol/Rose Oil	12.485	20.7
rose oxide	13.281	1.4
citronellol/rhodinol	19.161	21.3
β-citral	19.704	3.9
cis-geraniol	20.545	9.3
pelargic acid	21.310	6.4
carvacrol	22.946	1.4
eugenol	25.523	3.5
geranic acid/neric acid	26.028	6.2
n-decanoic acid	26.510	1.4
methyl eugenol	27.840	2.9

According to the results, the substances having the highest percentages over total identified compounds were citronellol (21.3%), phenylethyl alcohol (20.7%), cis-geraniol (9.3%), pelargonic acid (6.4%), geranic acid (6.2%), β-citral (3.9%), eugenol (3.5%) and methyleugenol (2.9%).

3. Discussion

The risks posed by pesticides constitute a significant problem for society and are related to their residues in the environment and food, with their use being the third greatest concern regarding food safety. For this reason, integrated pest management is essential, which includes the use of natural products as alternatives to synthetically developed industrial formulations [39]. "Basic Substances" and Low-Risk Plant Protection Products containing microorganisms and plant extracts, in addition to direct toxicity on plant parasites, also function as biostimulants, promoting host plant growth and enhancing soil health [40]. They can suppress plant pest development, stimulate the host plant's defense mechanisms, or promote growth by improving nutrient utilization. Some may not be effective when applied alone against plant pests but are useful when combined with other methods as part of a broader plant protection strategy. Notably, they can be used in organic farming [40]. "Basic Substances" are, by definition, substances that do not raise concern because they do not cause endocrine, neurological, or immunological disorders. They are not classified as plant protection products, but they are useful when applied directly or following dilution. Regarding their approval, compared to low-risk substances, whose approval does not exceed 15 years, "Basic Substances" have no such time limitation and do not have an expiration date [41]. According to Table S1 of Supplementary Materials, the "Basic Substances" approved so far relate to their use against phytopathogenic insects, fungi, bacteria, and viruses or for the disinfection of agricultural tools and have not been approved for nematodes' control. De Long et al. (2021), for example, in chrysanthemum cultivation, applied hydrochloric chitosan combined with garlic extract, approved as a low-risk nematicidal product in the EU, and demonstrated significant nematicidal action [42]. In this study we investigate the effect, through paralysis experiments, of three "Basic Substances" namely sodium bicarbonate, sodium chloride, and beer.

Sodium bicarbonate is an approved "Basic Substance" for use in the form of a water-soluble powder on vegetables, soft fruits, and ornamental plants to control fungi such as *Sphaerotheca* spp., *Oidium* spp., on grapevines against *Uncinula necator*, and on apple

https://doi.org/10.3390/plants15030458

trees against *Venturia inaequalis*. Additionally, it can be used post-harvest on various fruits to combat *Penicillium italicum* and *P. digitatum*, and in dry form on ornamental plants to suppress the growth of thalli such as *Lunularia cruciate* [25].

Other research highlights biological activities, such as combating fungi like *Didymellabryoniae* and *Podosphaera xanthii* in watermelon crops, pests such as *Aphis gossypii*, *Tetranychus urticae*, *Spodoptera exigua*, *S. litura*, and thrips [43]. In a study by Yassen & Sulaiman (2021), the application of 8000 mg/L of sodium bicarbonate solution resulted in a high mortality rate in *Culex pipiens molestus* mosquito larvae, four days after application [44]. Sodium bicarbonate shows synergistic effects when combined with other substances, such as linseed oil, resulting in the rapid immobilization of gregarious desert locusts on the first day of treatment [45].

Regarding sodium chloride, it has been approved as a "Basic Substance" for use in powder form soluble in water in vineyards against the fungus *Erysiphe necator* and as a granular formulation in mushroom cultivation for controlling species of *Cladobotryum* (e.g., *Mycophilum*), *Lecanicillium* (*Verticillium*) *fungicola*, and *Mycogone perniciosa* [21].

Concerning the pesticidal effect of sodium chloride, an example of adding NaCl to the plant nutrient solution resulted in an increase in osmotic potential, which reduced water availability for *Tuta absoluta* larvae, while simultaneously, due to the increased concentration of Na^+ and Cl^- ions in the leaves, it diminished the nutritional capacity of the larvae [46]. In a study of sodium chloride application on soybeans to treat against sucking insects (*Nezaraviridula*, *Piezodorusguildinii*, *Euschistusheros*), it acted as a repellant, and if combined with imidacloprid, it exhibited a synergistic effect [47,48].

The exposure of phytoparasitic nematodes to various salts and, consequently, to the ions they consist of affects their behavior and can cause toxicity. Specifically, *M. javanica* and *M. incognita* are not attracted to ions such as Na^+ and Cl^-, while CO_2 caused irreversible paralysis to *Heterodera* spp. [49]. The presence of ions like Na^+ or HCO_3^- can influence the parasitic ability of nematodes. The activity of nematodes depends on the concentration of the ions. For example, low concentrations of sodium chloride (NaCl) (0.001 M or 35.5 mg/L) enhanced the activity of *Hirschmanniella oryzae*. Sodium bicarbonate ($NaHCO_3$) has been used in a solution of 1000 mg/L to induce paralysis of *M. incognita*. Concentrations of $NaHCO_3$ and NaCl above 15,000 mg/L caused significant mortality in *M. incognita* and *M. javanica* larvae [49–51].

Beer, as a "Basic Substance", is approved for use in soil traps as a molluscicide and offers a good alternative to synthetic products due to its high effectiveness and low cost, without burdening the environment [7]. Beer contains volatile compounds that attract snails, such as CO_2, which regulates their chemoreceptors in the olfactory nerves. In fact, CO_2 content may also explain beer's effect on nematodes, as demonstrated by our experimental results. However, the composition of beer can vary, and therefore, its effectiveness can change not only from brand to brand but also between different batches of the same brand [52].

Comparing the effectiveness of sodium chloride (NaCl) and sodium bicarbonate ($NaHCO_3$), NaCl appeared more effective with an EC_{50} value of 7 mg/mL at 24 h post-experiment. After two additional days of J2 immersion in test solutions, NaCl and $NaHCO_3$ exhibited similar paralysis activity, with EC_{50} values calculated at 6.45 and 7.88 mg/mL. Interestingly the alcohol-free beer caused paralysis at lower concentrations, exhibiting an EC_{50} of 3 mg/mL after 72 h of J2 immersion in test solutions.

Damask rose is one of the most important aromatic plants in the world. It is used as an ornamental plant, but mainly for the extraction of essential oil from its petals and flowers. During the extraction process, by-products are also utilized, with applications in the food, perfumery, and pharmaceutical industries [38]. In our study, rose extracts,

 https://doi.org/10.3390/plants15030458

both in the form of petals' hydrosol and petals' aqueous extract prepared in a sonicator, demonstrated activity in *M. incognita* paralysis experiments. In particular, the hydrosol was faster in activity since the first day after the establishment of the experiment (EC_{50} 7.7% v/v), while the rose extract prepared in the sonicator followed (EC_{50} 25.8% v/v). Clear dose and time relationships were established for the sonicated extract and the hydrosol, with the EC_{50} values stabilizing at 72 h post-experiment at 2.6 and 4.9% (v/v), respectively. Rose petals are rich in polyphenols, specifically flavonols, including kaempferol and quercetin glycosides [53]. These compounds can be obtained both through aqueous extracts and in hydrodistillation by-products [54]. According to the chemical analysis of the dried rose petal extract conducted in our study, the main compounds identified were citronellol, phenylethanol, cis-geraniol, pelargonic acid, geranic acid, β-citral, eugenol, and methyleugenol. Among these compounds, geraniol and eugenol, as well as β-citral as monoterpenes, have been reported for their nematicidal activity against J2 larvae of *M. incognita*, while geraniol and citral additionally inhibit egg hatching [55].

4. Materials and Methods

4.1. Development and Maintenance of the M. incognita Population

The initial population of *M. incognita* nematodes was collected in the form of 2nd stage larvae (J2) from infected tomato roots (*Solanum lycopersicum* L.). Approximately 2000 J2 per plant were applied for artificial infection, when the plants had reached the fourth to fifth leaf stage. The infected plants were kept in a chamber under controlled conditions, with a temperature of 25–28 °C, a relative humidity of 60% and a photoperiod of 16 h. Plants were regularly monitored, watered every two to three days, and inspected for other infestations.

Method for Collecting Second Stage (J2) Larvae

The roots of the infected plants were monitored for gall formation from the end of the first thirty days onward. When yellow-orange nematode egg sacs appeared on the galls, the plants were selected for root washing. The above-ground parts of the plants were cut and discarded, and the roots were placed in water to remove soil residues. The roots were then cut into 2 cm sections. To collect the J2 larvae, a 4.5% sodium hypochlorite (NaClO) solution in water was used. The roots were immersed in this solution and shaken in a closed container for four minutes. Afterwards, the roots were placed in a double sieve (250 and 38 mesh) and rinsed under running water. From the lower sieve (38 mesh), the eggs released from the disrupted egg sacs were collected using a water extractor and placed in a modified Baermann funnel, to facilitate hatching, which was maintained at approximately 25 °C.

The first collection of this suspension occurred two days after setting up the Baermann funnel. This initial collection was not used in experiments because the resulting larvae were of mixed ages, and there was also the possibility of detecting eggs. The collection process was then repeated every 48 hours, so that the larvae used in experiments were two days old [56].

4.2. "Basic Substances" and Rose Extracts

Among the "Basic Substances", sodium bicarbonate ($NaHCO_3$), sodium chloride (NaCl) and non-alcoholic beer were used in paralysis experiments against *M. incognita* J2. Additionally, rose petals (*Rosa damascena*) water extracts produced in a sonicator along with the rose petal hydrosol, a hydro-distillation by-product, were tested on J2 paralysis bioassays. All bioassays were performed twice, and the results are pooled replicates across multiple assays.

https://doi.org/10.3390/plants15030458

4.3. *Paralysis Bioassays on M. incognita Second Stage (J2) Larvae*

The experiments were conducted on 96-well polystyrene plates. The larvae contained in each well were counted using an inverted microscope with a 40× objective lens, and the average number of J2 per well was 15, as reported in our previous studies [2]. The plates were sealed with a lid to avoid evaporation and kept in the dark at room temperature.

A completely randomized block design was implemented, with five treatments and six replicates for each experiment. To prevent contamination of the treatments, due to possible volatility, separate plates were used for each test concentration. The experimental control consisted of larvae immersed in water. After the last motility assessment, at 72 h of J2 immersion in test solutions, the larvae were transferred to plain water after being washed in tap water through a 20 μm pore screen to remove excess test compounds. If paralysis persisted and remained irreversible after additional 24 h, it was recorded as mortality from that point onward and thus individuals were considered dead. This type of mortality assessment has been reported previously [2].

4.3.1. Paralysis Experiments with Basic Substances

Preparation of Sodium Bicarbonate and Sodium Chloride Solutions

The test substances were dissolved in water and tested at five concentrations each. The procedure for preparing solutions was the same for both sodium bicarbonate ($NaHCO_3$) and sodium chloride (NaCl). In 1 mL of tap water, 40 mg of each substance was dissolved, and the mixture was stirred using a Vortex device until fully homogenized. This was followed by six successive 1:1 (v/v) ratio dilutions using the same procedure. As a result, concentrations double the desired final levels were obtained, calculated in mg/mL: 40, 20, 10, 5, 2.5, 1.25 and 0.62 (mg/mL). Then, 70 μL of each solution was taken six times, corresponding to the number of repetitions for each treatment, and added to the wells along with the J2 larval suspension in a 1:1 (v/v) ratio. The final test concentrations of sodium bicarbonate ($NaHCO_3$) and sodium chloride (NaCl) in the immersion wells calculated were 20, 10, 5, 2.5, 1.25, 0.62 and 0.31 (mg/mL).

Preparation of Beer Solutions

Beer solutions were prepared by dissolving Άλφα® with water by successive dilutions at 1:1 (v/v). The resulting concentrations were double the intended final concentrations and calculated as a percentage of the original solution (% v/v): 50, 25, 12.5, 6.25, 3.12, 1.56 and 0.78. Accordingly, 70 μL of each solution was taken six consecutive times and added to wells containing the J2 larval suspension in a 1:1 (v/v) ratio. Therefore, within the wells, the final desired concentrations, expressed as a percentage of the original solution (% v/v), were 25, 12.5, 6.25, 3.12, 1.56, 0.78 and 0.39.

4.3.2. Mobility Experiments with Rose Petal Extracts

Preparation of Petals Extracts Using Sonicator

Rose petals were first ground using a mill and the plant material was mixed with water at a ratio of 1:20 (v/v), due to the high water-holding capacity of dried rose. The extraction was performed in an ultrasonic bath for ten minutes at room temperature. Upon completion of the extraction, the mixture was filtered through cotton. This water extract of petals was consecutively diluted six times with water at a 1:1 (v/v) ratio to prepare solutions of 100, 50, 25, 12.5, 6.25, 3.12, 1.56 and 0.78 (% v/v). From each solution, 70 μL was added to the treatment well along with 70 μL of J2 larval suspension. The final test concentrations in the treatment well were 50, 25, 12.5, 6.25, 3.12, 1.56, 0.78 and 0.39 (% v/v).

 https://doi.org/10.3390/plants15030458

Preparation of Rose Petal Hydrosol Using Hydrodistillation

For the hydrodistillation of essential oil, 22 g of dried petals were initially weighed. The plant samples were placed in spherical flasks with a volume of 1 L, to which 500 mL of tap water was added. The flasks were placed in heating mantles and connected to a Clevenger apparatus. The hydrodistillation process was carried out for three hours. Due to the low yield of rose petals, no essential oil was produced and the hydrosol was used in J2 paralysis bioassays. The hydrosol was filtered from the plant materials and was then tested on J2 at 50, 25, 12.5, 6.25, 3.125, 1.56, 0.78 and 0.39 (% *v*/*v*).

Chemical Analysis of Ultrasound Extract of Dried Petals

For the chemical analysis of the rose petal water extract as prepared in sonication, it proved most effective according to paralysis bioassays. 15 g of dried rose petals was used, extracted with 270 mL of tap water in an ultrasonic bath for thirty minutes, without the application of heating. Filtration through cotton was performed, and 150 mL of extract was collected. To obtain the organic components and separate them from the aqueous solution, extraction with 10 mL of hexane in a separatory funnel followed. Then, 5 mL of the hexane extract was collected, concentrated to a final volume of 1.5 mL, and finally analyzed using a gas chromatography–mass spectrometry system. Separation and determination were performed using an Agilent 6890N GC (Agilent Technologies, Santa Clara, CA, USA) equipped with an Agilent 7683B Injector (Agilent Technologies, Santa Clara, CA, USA), a 30 m, 0.25 m.m i.d, HP-5MS capillary column (Agilent Technologies, Santa Clara, CA, USA) coated with 5% Phenyl-methylpolysiloxane (film thickness) 0.25 μm and an Agilent 5975 mass selective detector (MSD). The samples were injected in the pulsed splitless mode at an injection temperature of 280 °C. Mass spectrometry was acquired using the electron ionization (EI). The chromatograms were processed using appropriate software (Agilent, ChemStation LC B.04.03), while the identification was carried out according to the Total Ion Current (TIC) scan spectrum and compared with standard solutions and the NIST spectral library.

4.4. J2 Paralysis Evaluation

The paralysis of J2 larvae was assessed on an inverted microscope at 40× magnification and they were divided into two categories, mobile and paralyzed. Assessments of paralysis were carried out at 24, 48, and 72 h post-J2 immersion in test solutions. Throughout the experiments, the plates were maintained at approximately 25 °C in the dark.

4.5. Statistical Analysis

For the processing of paralysis results, the percentages of paralysis were calculated for each treatment, and the natural mortality of the control was removed according to Schneider-Orelli's formula: Increase in paralysis % = {(paralysis % in the treatment − paralysis % in the control)/(100 − paralysis % in the control)} × 100 [57]. Data from experimental replicates in time were analyzed (ANOVA) after being combined over time. As ANOVA indicated no significant treatment by time interaction, means were averaged over experiments. Corrected percentages of paralyzed J2 treated with rose extracts or "Basic Substances" were subjected to non-linear regression analysis. Data were analyzed using the SPSS program v30 in an analysis of variance (ANOVA) and applied to the log-logistic equation [58] to calculate EC_{50} values.

$$y = C + (D - C)/[1 + (x/EC_{50})^{b}]$$

where C = the lower limit, D = the upper limit, b = the slope of the line at the EC_{50} value, and EC_{50} = the concentration of the test substance(s) required for a 50% increase in paralyzed J2

https://doi.org/10.3390/plants15030458

compared to the control. In this dose–response equation, the concentration of the extract (μg/mL) was the independent factor (x) and the immobilized J2 (percentage increase over the control) was the dependent factor (y) [58].

5. Conclusions

In summary, the substances that were applied induced paralysis in *M. incognita* nematodes. Among "Basic Substances", sodium bicarbonate and sodium chloride showed similar effects and were more effective at lower concentrations compared to non-alcoholic beer. The presence of Na^+, Cl^- or HCO_3^- ions in the control solutions likely contributed to paralysis in J2 larvae. Although "Basic Substances" have not been approved for use against nematodes, they could be further investigated for their potential nematicidal activity. On the other hand, the rose petal extracts could be of interest for development into "Basic Substances" since they are foodstuffs and thus of *no concern*. In particular, the hydrosol used as a plant protectant could play a role in circular economy too, since it is a hydrodistillation by-product. Nonetheless, although the tested substances have clear potential and practical value in sustainable and integrated field applications, their use under real field conditions might be constrained by variability, limited stability, and formulation challenges. Variability in beer and rose material may result in fluctuations in extract composition and potency, while environmental factors can rapidly reduce stability and efficacy. Limited standardization, formulation, and scalability of the "Basic Substances" may further impact reproducibility and shelf life, thus requiring frequent applications and higher costs, alongside complex regulatory requirements despite their natural origin. On the other hand, although irreversible paralysis is a practical and widely applied endpoint in in vitro assays, it does not necessarily reflect true mortality, as juveniles that appear irreversibly paralyzed under laboratory conditions may retain residual metabolic activity or regain functionality under more favorable environments, such as soil or in association with host roots. To address this limitation, paralysis-based mortality estimates are usually supplemented with additional endpoints like infectivity assays on host plants. We are currently performing pot bioassays with the tested substances presented herein, so as to evaluate effects on nematode life cycle completion in tomato root hosts and thereby link in vitro findings to biologically relevant outcomes under field-relevant conditions.

Supplementary Materials: The following supporting information can be downloaded at: https://www.mdpi.com/article/10.3390/plants15030458/s1, Table S1: "Basic Substances" authorized in the EU as at the date 25 December 2025.

Author Contributions: Conceptualization, N.N. and N.G.T.; methodology, R.A., K.D.G., N.N. and N.G.T.; software, N.N.; validation, N.N. and N.G.T.; formal analysis, N.N.; investigation, R.A. and N.N.; resources, K.D.G. and N.N.; data curation, R.A. and N.N.; writing—original draft preparation, R.A. and N.N.; writing—review and editing R.A. and N.N.; visualization, N.N.; supervision, N.N.; project administration, N.N. and N.G.T. All authors have read and agreed to the published version of the manuscript.

Funding: This research received no external funding.

Data Availability Statement: The original contributions presented in this study are included in the article. Further inquiries can be directed to the corresponding authors.

Acknowledgments: We are grateful to Aitheria Company, based in Velvento, Kozani, Greece, and Ethericon Greek Herbs Company, based in Larisa, Greece, for the kind offer of rose petals. We are also grateful to Bill Vogiatzis and Etherikon, Greek Herbs for the kind offer of aromatic material.

Conflicts of Interest: The authors declare no conflicts of interest.

https://doi.org/10.3390/plants15030458

References

1. Marchand, P.A. Basic Substances: An Opportunity for Approval of Low Concern Substances under EU Pesticide Regulation. *Pest Manag. Sci.* **2015**, *71*, 1197–1200. [CrossRef]
2. Ntalli, N.G.; Menkissoglu-Spiroudi, U.; Giannakou, I. Nematicidal activity of powder and extracts of *Melia azedarach* fruits against *Meloidogyne incognita*. *Ann. Appl. Biol.* **2010**, *156*, 309–317. [CrossRef]
3. Romanazzi, G.; Orçonneau, Y.; Moumni, M.; Davillerd, Y.; Marchand, P.A. Basic substances, a sustainable tool to complement and eventually replace synthetic pesticides in the management of pre- and postharvest diseases: Reviewed instructions for users. *Molecules* **2022**, *27*, 13484. [CrossRef] [PubMed]
4. Constantini, E.; La Torre, A. Regulatory framework in the European Union governing the use of basic substances in conventional and organic production. *J. Plant Dis. Prot.* **2022**, *129*, 715–743. [CrossRef]
5. European Commission. *EU Pesticides Database—Active Substances (Including Basic Substances)*; European Commission Website. Available online: https://ec.europa.eu/food/plant/pesticides/eu-pesticides-database/start/screen/active-substances (accessed on 27 December 2025).
6. European Commission. *Final Review Report for the Basic Substance Allium cepa L. Bulb Extract (Onion Bulbs), Finalised by the Standing Committee on Plants, Animals, Food and Feed on 22 October 2020, in View of the Approval of Allium cepa L. Bulb Extract (Onion Bulbs) as a Basic Substance in Accordance with Regulation (EC) No 1107/2009*; European Commission: Brussels, Belgium, 2020; pp. 1–8.
7. European Commission. *Final Review Report for the Basic Substance Allium fistulosum, Processed, Finalised by the Standing Committee on Plants, Animals, Food and Feed at Its Meeting on 3 October 2024, in View of the Approval of Allium fistulosum, Processed as a Basic Substance in Accordance with Regulation(EC) No 1107/2009*; European Commission: Brussels, Belgium, 2024; pp. 1–15.
8. European Commission. *Final Review Report for the Basic Substance Beer, Finalised in the Standing Committee on Plants, Animals, Food and Feed at Its Meeting on 6 October 2017, in View of the Approval of Beer as a Basic Substance in Accordance with Regulation (EC) No 1107/2009*; European Commission: Brussels, Belgium, 2017; pp. 1–6.
9. European Commission. *Final Review Report for the Basic Substance Calcium Hydroxide, Finalised in the Standing Committee on Plants, Animals, Food and Feed at Its Meeting on 20 March 2015 and Amended on 24 March 2021, in View of the Approval of Calcium Hydroxide as a Basic Substance in Accordance with Regulation (EC) No 1107/2009*; European Commission: Brussels, Belgium, 2021; pp. 1–10.
10. European Commission. *Final Review Report for the Basic Substance Chitosan, Finalised by the Standing Committee on Plants, Animals, Food and Feed on 28 January 2022, in View of the Approval of Chitosan as a Basic Substance in Accordance with Regulation (EC) No 1107/2009*; European Commission: Brussels, Belgium, 2022; pp. 1–14.
11. European Commission. *Final Review Report for the Basic Substance Chitosan Hydrochloride, Finalised in the Standing Committee on the Food Chain and Animal Health at Its Meeting on 20 March 2014, Amended in the Standing Committee on Plants, Animals, Food and Feed at Its Meeting on 25 January 2021, and Corrected on 5 July 2021 and 23 March 2023, in View of the Approval of Chitosan Hydrochloride as a Basic Substance in Accordance with Regulation (EC) No 1107/2009*; European Commission: Brussels, Belgium, 2023; pp. 1–10.
12. European Commission. *Final Review Report for the Basic Substance Clayed Charcoal, Finalised in the Standing Committee on Plants, Animals, Food and Feed at Its Meeting on 24 January 2017 and Amended on 25 January 2021, in View of the Approval of Clayed Charcoal as a Basic Substance in Accordance with Regulation (EC) No 1107/2009*; European Commission: Brussels, Belgium, 2021; pp. 1–8.
13. European Commission. *Final Review Report for the Basic Substance Cow Milk, Finalised by the Standing Committee on Plants, Animals, Food and Feed on 19 May 2020, in View of the Approval of Milk as a Basic Substance in Accordance with Regulation (EC) No 1107/2009*; European Commission: Brussels, Belgium, 2020; pp. 1–7.
14. European Commission. *Review Report for the Basic Substance Diammonium Phosphate, Finalised in the Standing Committee on Plants, Animals, Food and Feed at Its Meeting on 8 March 2016, in View of the Approval of Diammonium Phosphate as a Basic Substance in Accordance with Regulation (EC) No 1107/2009*; European Commission: Brussels, Belgium, 2016; pp. 1–6.
15. European Commission. *Review Report for the Basic Substance Equisetum arvense L., Finalised in the Standing Committee on the Food Chain and Animal Health at Its Meeting on 20 March 2014, in View of the Approval of Equisetum arvense L. as a Basic Substance in Accordance with Regulation (EC) No 1107/2009*; European Commission: Brussels, Belgium, 2017; pp. 1–16.
16. European Commission. *Final Review Report for the Basic Substance Fructose, Finalised in the Standing Committee on Plants, Animals, Food and Feed at Its Meeting on 14 July 2015 and Amended on 17 July 2020, in View of the Approval of Fructose as a Basic Substance in Accordance with Regulation (EC) No 1107/2009*; European Commission: Brussels, Belgium, 2020; pp. 1–8.
17. European Commission. *Final Review Report for the Basic Substance Hydrogen Peroxide, Finalised in the Standing Committee on Plants, Animals, Food and Feed at Its Meeting on 24 January 2017, in View of the Approval of Hydrogen Peroxide as a Basic Substance in Accordance with Regulation (EC) No 1107/2009*; European Commission: Brussels, Belgium, 2017; pp. 1–8.
18. European Commission. *Final Review Report for the Basic Substance L-Cysteine, Finalised in the Standing Committee on Plants, Animals, Food and Feed on 24 March 2020, in View of the Approval of L-Cysteine as a Basic Substance in Accordance with Regulation (EC) No 1107/2009*; European Commission: Brussels, Belgium, 2020; pp. 1–8.

https://doi.org/10.3390/plants15030458

19. European Commission. *Final Review Report for the Basic Substance Lecithins, Finalised in the Standing Committee on Plants, Animals, Food and Feed at Its Meeting on 25 May 2015, in View of the Approval of Lecithins as a Basic Substance in Accordance with Regulation (EC) No 1107/2009*; European Commission: Brussels, Belgium, 2018; pp. 1–8.
20. European Commission. *Final Review Report for the Basic Substance Magnesium Hydroxide E528, Finalised by the Standing Committee on Plants, Animals, Food and Feed at Its Meeting on 31 January 2024, in View of the Approval of Magnesium Hydroxide E528 as a Basic Substance in Accordance with Regulation (EC) No 1107/2009*; European Commission: Brussels, Belgium, 2024; pp. 1–12.
21. European Commission. *Final Review Report for the Basic Substance Mustard Seeds Powder, Finalised in the Standing Committee on Plants, Animals, Food and Feed at Its Meeting on 6 October 2017, in View of the Approval of Mustard Seeds Powder as a Basic Substance in Accordance with Regulation (EC) No 1107/2009*; European Commission: Brussels, Belgium, 2017; pp. 1–6.
22. European Commission. *Final Review Report for the Basic Substance Sodium Chloride, Finalised in the Standing Committee on Plants, Animals, Food and Feed at Its Meeting on 20 July 2017 and Amended on 25 January 2021, in View of the Approval of Sodium Chloride as a Basic Substance in Accordance with Regulation (EC) No 1107/2009*; European Commission: Brussels, Belgium, 2021; pp. 1–11.
23. European Commission. *Final Review Report for the Basic Substance Onion Oil, Finalised in the Standing Committee on Plants, Animals, Food and Feed at Its Meeting on 20 July 2018, in View of the Approval of Onion Oil as a Basic Substance in Accordance with Regulation (EC) No 1107/2009*; European Commission: Brussels, Belgium, 2018; pp. 1–7.
24. European Commission. *Final Review Report for the Basic Substance Onobrychisviciifolia (Sainfoin) Dried Pellets, Finalised in the Standing Committee on Plants, Animals, Food and Feed at Its Meeting on 23 May 2024, in View of the Approval of Onobrychisviciifolia (Sainfoin) Dried Pellets as a Basic Substance in Accordance with Regulation (EC) No 1107/2009*; European Commission: Brussels, Belgium, 2024; pp. 1–14.
25. European Commission. Review Report for the Basic Substance *Salix spp. Cortex; Finalised in the Standing Committee on Plants, Animals, Food and Feed at Its Meeting on 29 May 2015 in View of the Approval of Salix spp. Cortex as Basic Substance in Accordance with Regulation (EC) No 1107/2009*; European Commission: Brussels, Belgium, 2015; pp. 1–6.
26. European Commission. *Final Review Report for the Basic Substance Sodium Hydrogen Carbonate, Finalised in the Standing Committee on Plants, Animals, Food and Feed at Its Meeting on 9 October 2015, in View of the Approval of Sodium Hydrogen Carbonate as a Basic Substance in Accordance with Regulation (EC) No 1107/2009*; European Commission: Brussels, Belgium, 2018; pp. 1–7.
27. European Commission. *Final Review Report for the Basic Substance Sucrose, Finalised in the Standing Committee on Plants, Animals, Food and Feed at Its Meeting on 11 July 2014 and Amended on 17 July 2020, in View of the Approval of Sucrose as a Basic Substance in Accordance with Regulation (EC) No 1107/2009*; European Commission: Brussels, Belgium, 2020; pp. 1–9.
28. European Commission. *Review Report for the Basic Substance Sunflower Oil, Finalised in the Standing Committee on Plants, Animals, Food and Feed at Its Meeting on 7 October 2016, in View of the Approval of Sunflower Oil as a Basic Substance in Accordance with Regulation (EC) No 1107/2009*; European Commission: Brussels, Belgium, 2016; pp. 1–7.
29. EFSA. *Outcome of the Consultation with Member States and EFSA on the Basic Substance Application for Approval of Sunflower Oil for the Extension of Use in Plant Protection as a Fungicide on Vegetables (Common Bean, Cucumber), Rosaceae (like Prunus,* Fragaria, Rosa, Rubus *etc.), Apple, Pear, Grapevine, Wheat, Barley, Potato and Carrot*; EFSA: Parma, Italy, 2020; pp. 1–39.
30. European Commission. *Final Review Report for the Basic Substance talc E553b, Finalised in the Standing Committee on Plants, Animals, Food and Feed at Its Meeting on 22 March 2018, in View of the Approval of talc E553b as a Basic Substance in Accordance with Regulation (EC) No 1107/2009*; European Commission: Brussels, Belgium, 2018; pp. 1–9.
31. European Commission. *Final Review Report for the Basic Substance Urtica spp., Finalised in the Standing Committee on Plants, Animals, Food and Feed at Its Meeting on 24 January 2017, in View of the Approval of Urtica spp. as a Basic Substance in Accordance with Regulation (EC) No 1107/2009*; European Commission: Brussels, Belgium, 2017; pp. 1–13.
32. European Commission. *Final Review Report for the Basic Substance Vinegar, Finalised in the Standing Committee on Plants, Animals, Food and Feed at Its Meeting on 29 May 2015 and Amended on 13 December 2018, 26 January 2021 and 28 January 2022, in View of the Approval of Vinegar as a Basic Substance in Accordance with Regulation (EC) No 1107/2009*; European Commission: Brussels, Belgium, 2022; pp. 1–12.
33. European Commission. Final Review Report for the Basic Substance *Vitis vinifera L. Seed Extract (Grape Seed Extract), Finalised by the Standing Committee on Plants, Animals, Food and Feed at Its Meeting on 5 December 2024, in View of the Approval of Vitis vinifera L. Seed Extract (Grape Seed Extract) as a Basic Substance in Accordance with Regulation (EC) No 1107/2009*; European Commission: Brussels, Belgium, 2024; pp. 1–16.
34. European Commission. *Final Review Report for the Basic Substance Whey, Finalised in the Standing Committee on Plants, Animals, Food and Feed at Its Meeting on 8 March 2016 and Amended on 25 March 2021, in View of the Approval of Whey as a Basic Substance in Accordance with Regulation (EC) No 1107/2009*; European Commission: Brussels, Belgium, 2021; pp. 1–8.
35. Kolev, N.; Ivanova, M.; Balabanov, A.; Vlahova-Vangelova, D.; Kišová, A.; Vizzarri, F. Utilization of agro-industrial residues from the *Rosa damascena* Mill. oil industry: A literature review on biomass potential for food and feed ingredients. *Processes* **2025**, *13*, 1945. [CrossRef]

36. Akbari, M.; Kazerani, H.R.; Kamrani, A.; Mohri, M. A preliminary study on some potential toxic effects of *Rosa damascena* Mill. *Iran. J. Vet. Res.* **2013**, *14*, 232–236.
37. Peron, S.; Hadi, F.; Azarbani, F.; Murthy, H.C.A. Antimicrobial, antioxidant, anti-glycation and toxicity studies on silver nanoparticles synthesized using *Rosa damascena* flower extract. *Green Chem. Lett. Rev.* **2021**, *14*, 519–533. [CrossRef]
38. Nunes, H.S.; Miguel, M.G. *Rosa damascena* essential oils: A brief review about chemical composition and biological properties. *Trends Phytochem. Res.* **2017**, *1*, 111–128.
39. Robin, D.C.; Marchand, P.A. Evolution of the biocontrol active substances in the framework of the European pesticide regulation (EC) No. 1107/2009. *Pest Manag. Sci.* **2019**, *75*, 950–958.
40. Matyjaszczyk, E. "Biorationals" in integrated pest management strategies. *J. Plant Dis. Prot.* **2018**, *125*, 523–527. [CrossRef]
41. European Commission. *Study Supporting the REFIT Evaluation of the EU Legislation on Plant Protection Products and Pesticide Residues (Regulation (EC) No 1107/2009 and Regulation (EC) No 396/2005)*; European Commission: Brussels, Belgium, 2018; pp. 1–336.
42. De Long, J.R.; Streminska, M.A.; Persijn, A.; Huisman, H.M.I.; van der Salm, C. Biological control of *Meloidogyne* spp. in glasshouse-grown chrysanthemum. *J. Nematol.* **2021**, *52*, e2020-125. [CrossRef] [PubMed]
43. Kim, Y.S.; Han, J.W.; Jeong, T.G.; Noh, S.J.; Park, S.W.; Jeon, Y.M.; Hue, Y.S.; Kim, T.I. Effect of organic materials for the control of diseases and pests in organic watermelon cultivation. *Proc. Spring Int. Conf. KSAE* **2019**, *4*, 124.
44. Yassen, A.T.; Sulaiman, K.A. Insecticidal activity of some chemicals against mosquitoes *Culex pipiens molestus* Forskal. *Baghdad Sci. J.* **2021**, *18*, 716–721. [CrossRef]
45. Abdelatti, Z.A.S.; Hartbauer, M. Plant oil mixtures as a novel botanical pesticide to control gregarious locusts. *J. Pest Sci.* **2020**, *93*, 341–353. [CrossRef]
46. Han, P.; Desneux, N.; Becker, C.; Larbat, R.; Le Bot, J.; Adamowicz, S.; Zhang, J.; Lavoir, A.V. Bottom-up effects of irrigation, fertilization and plant resistance on *Tuta absoluta*: Implications for integrated pest management. *J. Pest Sci.* **2019**, *92*, 1359–1370. [CrossRef]
47. Corso, I.C.; Gazzoni, D.L. Sodium chloride: An insecticide enhancer for controlling pentatomids on soybeans. *Pesq. Agropec. Bras.* **1998**, *33*, 1563–1571.
48. Rodrigues, H.S.; Haddi, K.; Campos, M.O.; Ferreira-Filho, A.; Guedes, R.N.C.; Newland, P.L.; Oliveira, E.E. Synergism and unintended effects of the association between imidacloprid and sodium chloride (NaCl) on the management of *Euschistusheros*. *Pest Manag. Sci.* **2021**, *77*, 417–424.
49. Xiang, N.; Lawrence, K.S. Optimization of in vitro techniques for distinguishing between live and dead second-stage juveniles of *Heteroderaglycines* and *Meloidogyne incognita*. *PLoS ONE* **2016**, *11*, e0154818.
50. Castro, C.E.; Belser, N.O.; McKinney, H.E.; Thomason, I.J. Strong repellency of the root-knot nematode, *Meloidogyne incognita*, by specific inorganic ions. *J. Chem. Ecol.* **1990**, *16*, 1199–1205. [CrossRef]
51. Zaki, M.J.; Khan, D.; Abid, M. Nematodes in the saline environment: A mini overview. *Int. J. Biol. Biotechnol.* **2012**, *9*, 99–113.
52. Piechowicz, B.; Rębisz, E.; Grodzicki, P.; Zaręba, L.; Balawejder, M.; Pieniążek, M.; Stawarczyk, K.; Piechowicz, I. The smell of beer as a factor affecting the emission of carbon dioxide by *Arion lusitanicus* auct. non-Mabille. *Ann. Anim. Sci.* **2016**, *16*, 463–476. [CrossRef]
53. Shikov, V.; Kammerer, D.R.; Mihalev, K.; Mollov, P.; Carle, R. Antioxidant capacity and colour stability of texture-improved canned strawberries as affected by the addition of rose (*Rosa damascena* Mill.) petal extracts. *Food Res. Int.* **2012**, *46*, 552–556. [CrossRef]
54. Dina, E.; Sklirou, A.D.; Chatzigeorgiou, S.; Manola, M.S.; Cheilari, A.; Louka, X.P.; Argyropoulou, A.; Xynos, N.; Skaltsounis, A.L.; Aligiannis, N.; et al. An enriched polyphenolic extract obtained from the by-product of *Rosa damascene* hydrodistillation activates antioxidant and proteostatic modules. *Phytomedicine* **2021**, *93*, 153757. [CrossRef]
55. Liu, Z.; Li, Q.X.; Song, B. Pesticidal activity and mode of action of monoterpenes. *J. Agric. Food Chem.* **2022**, *70*, 4556–4571. [CrossRef]
56. Hussey, R.S.; Barker, K.R. A comparison of methods of collecting inocula of *Meloidogyne* species, including a new technique. *Plant Dis. Rep.* **1973**, *57*, 1025–1028.
57. Puntener, W.; Zahner, O. *Manual for Field Trials in Plant Protection*, 2nd ed.; Ciba-Geigy: Basle, Switzerland, 1981; p. 205.
58. Seefeldt, S.S.; Jensen, J.E.; Fuerst, E.P. Log-logistic analysis of herbicide rate response relationships. *Weed Technol.* **1995**, *9*, 218–227.

https://doi.org/10.3390/plants15030458

plants

MDPI

Article

Chemical Composition and Antifungal Activity of *Artemisia sieversiana* Essential Oil Growing in Jilin Against Black Spot on Yanbian Pingguoli Pear in China

Rong Zhang †, Ti-Yan Zheng † and Yu Fu *

Department of Chemistry, Yanbian University, Yanji 133002, China; 15034734254@163.com (R.Z.); 19315219926@163.com (T.-Y.Z.)

* Correspondence: fuyu@ybu.edu.cn

† These authors contributed equally to this work.

Abstract

Black spot disease substantially impairs both the aesthetic quality and commercial viability of affected Pingguoli pears. Previous studies have shown that *Alternaria alternata* and *A. tenuissima* are the pathogens that cause black spot disease. Essential oils represent novel alternatives to synthetic fungicides to control these pathogens. This study extracted *Artemisia sieversiana* essential oil (AsEO) by hydro-distillation using a crystal tower pure dew essential oil machine. The chemical compositions of AsEO were analyzed via gas chromatography–mass spectrometry (GC–MS). A total of 42 compounds were detected. 1,8-cineole, trans-caryophyllene, (1R,4S)-1,7,7-trimethylbicyclo [2.2.1] heptan-2-yl acetate, (±)-camphor, and β-myrcene were identified as the five main constituents. Moreover, the antifungal activity of AsEO was assessed against black spot on Yanbian Pingguoli pear in China. The minimum inhibitory concentration (MIC) and minimum fungicidal concentration (MFC) values were determined as 0.10% (*v*/*v*) and 0.12% (*v*/*v*), respectively. Scanning electron microscopy (SEM) analysis revealed that treatment with AsEO induced significant morphological aberrations in *A. alternata* and *A. tenuissima* mycelia, including surface roughening, hyphal collapse, and loss of structural integrity. Concurrently, a marked increase in alkaline phosphatase (AKP) enzyme activity and electrical conductivity was observed, a key indicator of cell wall and plasma membrane permeabilization and damage. When the concentration of AsEO was less than 120 μg/mL, there was no toxicity to keratinocytes (HaCaTs) and skin fibroblasts (NHSFs). In summary, this study provides a theoretical basis for the development of AsEO as a fungicide against black spot disease on Pingguoli pear in China.

Keywords: essential oil; antifungal activity; black spot; Pingguoli

1. Introduction

Members of the *Alternaria* genus, classified within the order Pleosporales (Ascomycota), exhibit a cosmopolitan distribution and function ecologically as both phytopathogens and saprophytes. *Alternaria* comprises over 300 described species that have caused serious economic losses in global horticulture through diseases like black spot and black rot [1]. Normally, *Alternaria* species are categorized into two obviously distinct groups: large-spored and small-spored *Alternaria* species [2]. The conidial bodies of large-spored species typically measure 60–100 μm in length, while the small-spored species are less than 60 μm. Several small-spored species can infect humans and animals, causing diseases such as tinea, onychomycosis, and osteomyelitis of the jawbone [3]. Phylogenetically diverse *Alternaria* species demonstrate remarkable biosynthetic versatility, producing over 150 secondary metabolites with taxonomic significance [4]. This chemical repertoire encompasses: alternariol (AOH), alternariol monomethyl ether (AME), altenuene (ALT), altertoxin I, II, and III (ATXI, II, and III), tenuazonic acid (TEA), iso-tenuazonic acid (iso-TEA), and tentoxin (TEN). AOH and AME exhibit dose-dependent cytotoxicity and genotoxicity at micromolar concentrations, primarily through mechanisms such as DNA intercalation or oxidative stress induction [5]. In contrast, ATXs display markedly enhanced acute toxicity (e.g., lower LD_{50} values in murine models) and mutagenic potential relative to AOH/AME [6]. TEA demonstrates mild cytotoxicity and mutagenicity in vitro, but in vivo studies have shown that it exhibits acute toxicity and synergistic effects in the presence of other toxins [7]. TEN, a phytotoxin known to disrupt chlorophyll synthesis and induce chlorosis in plant seedlings via photosystem II inhibition, remains understudied in mammalian systems [8]. These *Alternaria* toxins are primarily produced by various species, such as *A. alternata*, *A. arborescens*, *A. brassicae*, *A. capsici-anui*, *A. citri*, *A. cucumerina*, *A. dauci*, *A. japonica*, *A. longipes*, *A. mali*, *A. solani*, *A. tenuissima*, and *A. tomato* [9]. Moreover, the metabolic products have been detected in different substrates (tomato, wheat, apple, pear, walnuts, etc.) [9]. Therefore, it is crucial to find effective biological disinfectants.

The taxonomically diverse genus *Artemisia* L. (*Asteraceae*), comprising approximately 500 species globally, exhibits remarkable biogeographical patterns, with significant endemism in China, where 190 taxa have been documented across diverse ecosystems [10]. *Artemisia sieversiana* Ehrhart ex Willd., known as "Da-Zi-Hao" in Chinese, is an annual or biennial herbaceous rhizome plant of the *Artemisia* family that grows in sandy clay, moist habitats, and is distributed in Liaoning, Jilin, Hebei, Shanxi, Gansu, Xinjiang, and Xizang province in China [11]. *A. sieversiana* contains sesquiterpenes, dimeric guaianolides, lignans, flavones, and essential oils [10]. *AsEO* is widely known for its various active properties, such as its antitumor, anti-inflammatory, antiparasitic and antifungal effects [12–14]. The components of plant essential oils vary significantly depending on factors such as region, temperature, geographic origin, photoperiod, edaphic influences, season, and microbial diversity [15]. Zhang et al. [10] found that the major components of AsEO growing in Hebei province, China, were neryl propanoate, β-nerol and β-cubebene. Suleimenov et al. [13] found that the principal components of AsEO from plants growing in West Kazakhstan Oblast were myrcene, 1,8-cineol, linalool, p-cymene, nerylisovalerate, and β-caryophyllene. Liu et al. [14] reported that the major chemical components of AsEO from Beijing were eucalyptol, geranyl butyrate, borneol, and camphor. Zhigzhitzhapova et al. [16] found that the principal constituents of AsEO from Buryatia were chamazulene, germacrene D, a-bisabolol, neryl-3-methylbutanoate, and neryl-2-methylbutanoate. Liu et al. [17] reported that the major chemical components of AsEO from Liaoning province, China, were eucalyptol, (–)-borneol, (+)-camphor, β-caryophyllene, and (E)-β-farnesene. Jiang et al. [18] found that the principal constituents of AsEO from Xinjiang, China, were a-thujone, eucalyptol, and santolina triene. Li et al. [19] found the major components of AsEO from Tibet were chamazulene, α-bisabolol, and α-phellandrene. Until now, the major components of AsEO from plants grown in Jilin province, China, has not been studied. Although numerous studies have shown that AsEO has an antifungal effect, little research has been conducted on *Alternaria* species.

Previous research confirmed that the pathogens that cause black spot disease in Yanbian Pingguoli pear were *A. tenuissima* and *A. alternate*. In this study, we used GC-MS to analyze the main chemical components of AsEO. The effect of AsEO on *Alternaria* pathogens was also assessed. These findings are anticipated to generate further scientific interest in the sustainable utilization of *A. sieversiana*.

https://doi.org/10.3390/plants15020207

2. Results

2.1. Chemical Composition of the AsEO

The blue essential oil was obtained from *Artemisia sieversiana*. Spectral authentication (>99% confidence level) was implemented via multi-dimensional spectral deconvolution of raw GC-MS data with the National Institute of Standards and Technology (NIST) mass spectral library (version 20), complemented by chromatographic co-elution validation using in-house synthesized reference standards. A total of 42 components were found in AsEO, accounting for 98.2% of the total oil (Table 1). The structural formulas of the various components in AsEO are shown in Supplementary Table S1. The main components of the essential oil were 1,8-cineole (34.03%), trans-caryophyllene (22.38%), and (1R,4S)-1,7,7-trimethylbicyclo [2.2.1] heptan-2-yl acetate (8.39%), followed by (±)-camphor (7.89%), and β-myrcene (3.90%). The chemical composition of the essential oil was different from that reported in other studies [17–19]. The distinctive chemical profiles of AsEOs extracted from plants growing in different localities could be explained by geographic locality, climate, stress (such as drought), harvest time, etc.

Table 1. Chemical composition of essential oils extracted from *Artemisia sieversiana*.

No.	Compound	Retention Time (min)	Molecular Formula	CAS	Relative Content (>0.10%)
1	α-thujene	10.00	$C_{10}H_{16}$	2867-05-2	0.33
2	α-pinene	10.23	$C_{10}H_{16}$	80-56-8	1.98
3	(±)-camphor	10.94	$C_{10}H_{16}O$	464-48-2	7.89
4	sabinene	11.90	$C_{10}H_{16}$	3387-41-5	1.97
5	β-pinene	12.06	$C_{10}H_{16}$	127-91-3	0.52
6	Oct-1-en-3-ol	12.27	$C_8H_{16}O$	3391-86-4	0.19
7	β-myrcene	12.63	$C_{10}H_{16}$	123-35-3	3.90
8	2-carene	13.14	$C_{10}H_{16}$	554-61-0	1.67
9	δ-4-carene	13.66	$C_{10}H_{16}$	29050-33-7	0.79
10	o-cymene	13.99	$C_{10}H_{14}$	527-84-4	0.46
11	(+)-limonene	14.15	$C_{10}H_{16}$	5989-27-5	0.22
12	1,8-cineole	14.37	$C_{10}H_{18}O$	470-82-6	34.03
13	γ-terpinene	15.32	$C_{10}H_{16}$	99-85-4	0.55
14	isoterpinolene	16.38	$C_{10}H_{16}$	586-63-0	0.42
15	linalool	17.05	$C_{10}H_{18}O$	78-70-6	1.98
16	nonanal	17.21	$C_9H_{18}O$	124-19-6	0.13
17	neo-alloocimene	18.07	$C_{10}H_{16}$	7216-56-0	0.10
18	2,4-hexadienyl isobutyrate	18.56	$C_{10}H_{16}O_2$	16491-24-0	0.12
19	cis-chrysanthenol	19.46	$C_{10}H_{16}O$	55722-60-6	0.42
20	terpinine-4-ol	20.12	$C_{10}H_{18}O$	562-74-3	0.54
21	α-terpineol	20.66	$C_{10}H_{18}O$	10482-56-1	0.82
22	(1R,4S)-1,7,7-trimethylbicyclo [2.2.1]heptan-2-yl acetate	20.90	$C_{12}H_{20}O_2$	92618-89-8	8.39
23	decanal	21.04	$C_{10}H_{20}O$	112-31-2	0.11
24	nerol	21.69	$C_{10}H_{18}O$	106-25-2	0.10
25	cis-3-hexenyl 2-methylbutanoate	22.09	$C_{11}H_{20}O_2$	53398-85-9	0.11
26	benzyl isobutanoate	24.23	$C_{11}H_{14}O_2$	103-28-6	0.10
27	2,4-decadienal	25.02	$C_{10}H_{16}O$	2363-88-4	0.11
28	α-amorphene	26.71	$C_{15}H_{24}$	20085-19-2	0.11
29	α-copaene	26.94	$C_{15}H_{24}$	3856-25-5	0.14
30	β-bourbonene	27.21	$C_{15}H_{24}$	5208-59-3	0.47
31	β-elemene	27.40	$C_{15}H_{24}$	515-13-9	0.96
32	cis-jasmone	27.49	$C_{11}H_{16}O$	488-10-8	0.11
33	trans-caryophyllene	28.43	$C_{15}H_{24}$	87-44-5	22.38
34	(E)-β-famesene	29.42	$C_{15}H_{24}$	18794-84-8	0.15
35	α-caryophyllene	29.54	$C_{15}H_{24}$	6753-98-6	0.12
36	cedrene	29.69	$C_{15}H_{24}$	11028-42-5	0.10
37	β-selinene	30.61	$C_{15}H_{24}$	17066-67-0	0.53
38	(+)-7-epi-sesquithujene	30.77	$C_{15}H_{24}$	159407-35-9	0.68
39	farnesene	31.07	$C_{15}H_{24}$	502-61-4	1.92
40	δ-cadinene	31.48	$C_{15}H_{24}$	483-76-1	0.31
41	methyl isocostate	35.12	$C_{16}H_{24}O_2$	132342-55-3	1.97
42	cembrene	39.49	$C_{20}H_{32}$	1898-13-1	1.10

https://doi.org/10.3390/plants15020207

2.2. Antifungal Activity of AsEO

As shown in Figure 1A,B, the colony diameter decreased from 7.85 ± 0.04 cm to 2.0 cm as the concentration of AsEO increased. No growth inhibition was observed in control Petri plates. Different concentrations (0.02–0.10%, v/v) of AsEO had significantly inhibitory effects on lesion diameter. There was no statistically significant difference mycelium growth diameter between *A. tenuissima* and *A. alternate* when treated with AsEO. Mycelial growth was completely inhibited when exposed to an AsEO concentration of 0.10% (v/v), which was defined as the MIC (Figure 1A). The mycelial growth in transferred potato dextrose agar (PDA) plates was completely inhibited when exposed to a concentration of 0.12% (v/v), which was defined as the MFC (Figure 1C). SEM analysis revealed striking ultrastructural differences between control and AsEO-treated *A. tenuissima* and *A. alternate* (Figure 1D). The control group exhibited intact, smooth hyphal surfaces with uniform cellular outlines, whereas AsEO exposure (1/2 MIC, 48 h) induced severe morphological aberrations: hyphae appeared shrunk, wrinkled, and rough. These alterations are indicative of AsEO's disruptive effects on fungal cell wall integrity and cytoskeletal organization in *A. tenuissima* and *A. alternate*. These results confirm that AsEO has the potential to inhibit black spot disease in Yanbian Pingguoli pear.

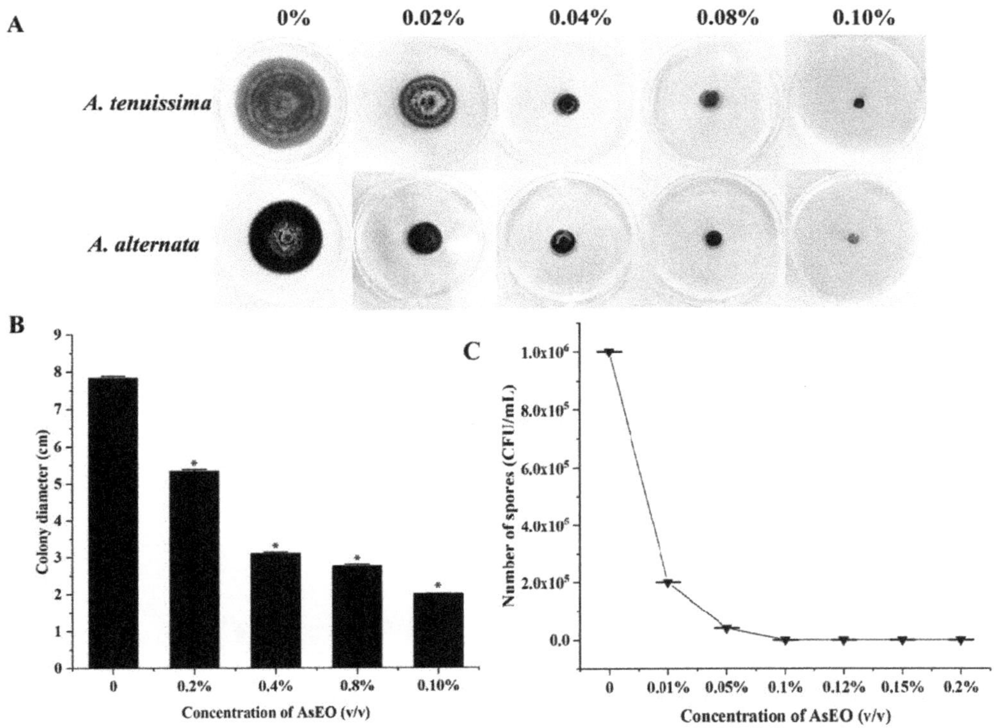

Figure 1. *Cont.*

https://doi.org/10.3390/plants15020207

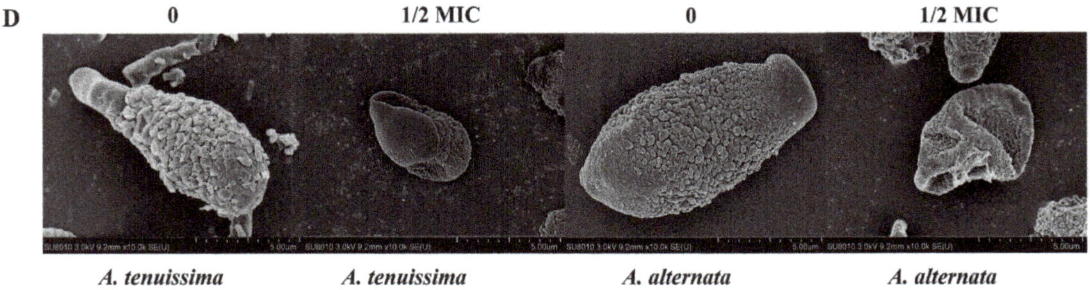

Figure 1. Antifungal activity of AsEO against *A. tenuissima* and *A. alternate*. (**A**) Morphology of *A. tenuissima* and *A. alternate*; (**B**) diameter of *A. tenuissima* colonies after 7 d with different concentrations of AsEO; (**C**) number of A. *tenuissima* spores treated with different concentrations of AsEO in potato dextrose broth (PDB) medium for 3 d; (**D**) SEM of control (0 *v*/*v*) and 1/2 MIC (0.10% *v*/*v*) carvacrol-treated *A. tenuissima* and *A. alternate* spores for 48 h. Values represent the mean ± SD (n = 3; * $p < 0.05$).

2.3. AsEO Disrupts Cell Wall Integrity

The structural integrity of microbial cell walls, which serve as protective barriers against environmental stressors, was systematically evaluated through AKP activity profiling. Positioned at the interface between the peptidoglycan layer and cytoplasmic membrane, AKP serves as a sensitive biomarker for cell wall permeabilization. As illustrated in Figure 2, exposure to AsEO induced a time-dependent increase in extracellular AKP activity ($R^2 = 0.96$), with a 2 MIC treatment demonstrating a 6.4 ± 0.5-fold elevation compared to controls after 6 h ($p < 0.001$).

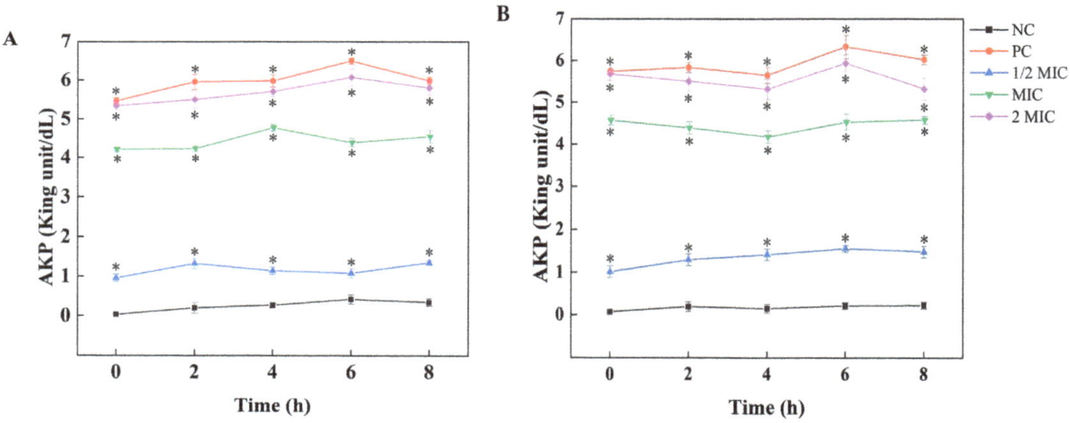

Figure 2. Effects of AsEO on AKP enzyme activity of the pathogenic fungi causing Yanbian Pingguoli pear black spot at 28 °C. (**A**) *A. tenuissima*; (**B**) *A. alternata*. Values represent the mean ± SD (n = 3; * $p < 0.05$).

2.4. AsEO Damages Cell Membrane Integrity

The cell membrane is a protective layer, and plays an extremely important role in organism metabolism. Cell membrane damaged, causes imbalances in osmotic pressure and pH, increases the permeability of the cell membrane, leads to electrolyte leakage, and can even cause cell death. Therefore, cell membrane permeability can be evaluated by measuring the conductivity of the cell culture medium. As shown in Figure 3, there was a significant increase in nucleic acid leakage (1.8-fold extracellular conductivity for 360 min

https://doi.org/10.3390/plants15020207

at 2 MIC). These results indicate that AsEO is capable of disrupting cell membrane integrity in *A. tenuissima* and *A. alternata*.

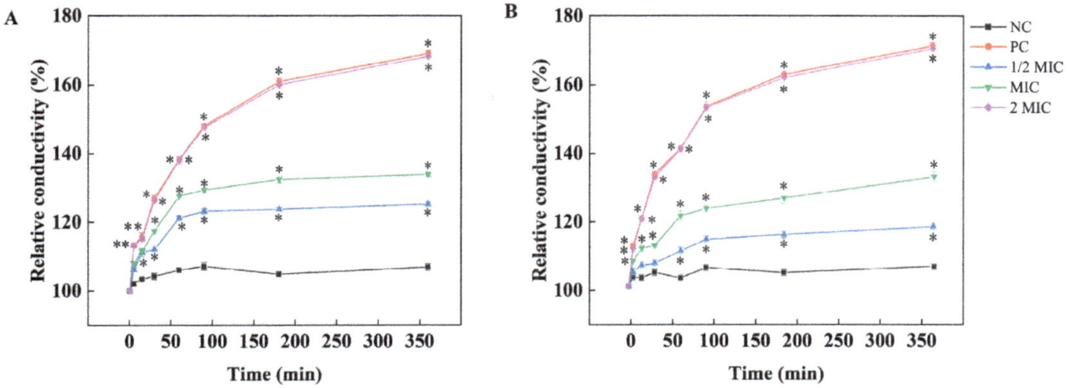

Figure 3. Impact of various concentrations of AsEO (0.5 MIC, MIC, and 2 MIC) on cell membrane integrity in *Alternaria* species. (**A**) *A. tenuissima*; (**B**) *A. alternata*. Values represent the mean ± SD (n = 3; * $p < 0.05$).

2.5. Safe Concentration of AsEO on Human Keratinocytes Cells (HaCaTs) and Skin Fibroblasts Cells (HSFs)

As shown in Table 2, cell viability under different concentrations of AsEO was determined using the CCK-8 assay. The analysis indicated that AsEO concentrations of less than 120 μg/mL had no effect on the activity of HaCaTs and NHSFs. As the concentration of the essential oil decreased, the safety to cells increased accordingly. At a concentration of 50 μg/mL, the essential oil promoted HaCaT proliferation; within the range of 10 to 120 μg/mL, it significantly promoted NHSF proliferation.

Table 2. The survival rates of different cells using CCK-8 assay.

Group	HaCaT	NHSF
1000 μg/mL	28.16 ± 2.15	12.82 ± 0.98
500 μg/mL	55.48 ± 1.18	20.44 ± 1.36
150 μg/mL	60.45 ± 3.28	54.16 ± 5.26
120 μg/mL	95.22 ± 4.58	102.24 ± 7.15
100 μg/mL	102.12 ± 2.36	106.18 ± 3.22
50 μg/mL	92.01 ± 1.98	118.36 ± 3.24
10 μg/mL	88.26 ± 1.08	98.04 ± 4.21
1 μg/mL	86.62 ± 1.26	99.87 ± 4.17
0.1 μg/mL	87.34 ± 3.17	109.38 ± 3.43
p value	0.03	0.04

3. Discussion

The Yanbian Pingguoli pear (*Pyrus pyrifolia* var. Yanbian Pingguoli pear), a cultivar with distinct apple-like morphology, was first introduced to Yanbian Korean Autonomous Prefecture (Jilin Province, China) through grafting experiments conducted in 1921 by Cui Changhao using scions imported from North Hamgyong Province, DPRK [20]. This historical introduction marked the origin of Asia's largest Pingguoli pear cultivation base, currently spanning 12,000 hectares with annual production exceeding 90,000 metric tons and generating nearly CNY ¥1 billion in economic value [20]. *Alternaria*-induced black spot disease represents a critical constraint in Yanbian Pingguoli pear cultivation, severely reducing the quality of the fruit's appearance and its economic value. This foliar pathogen manifests as initial circular, dark brown lesions (1–7 mm diameter) with chlorotic halos,

https://doi.org/10.3390/plants15020207

progressing into concentric rings of necrotic tissue (tan to brown-black pigmentation) and culminating in complete leaf abscission. When there are injuries on the trunk and roots of the tree, the pathogen will directly invade the tree tissues. The lesion will spread throughout the tree body, resulting in rotting and causing the tree to become weak or even die. When immature fruit is infected, there are occasionally several gray-brown spots on the surface of the fruit. Gradually, larger turbinate-shaped spots appear on the surface of the fruit, the fruit body begins to sink, and eventually, the fruit rots. Meanwhile, when immature fruits are infected, there are occasionally several gray-brown spots on the surface of the fruits. Subsequently, there is a significant increase in the number and size of black spots on the fruit surface. Large spots appear on the surface of the fruits and the fruit body begins to sink, eventually leading to fruit rotting. In the late stages of storage, the incidence of black spot disease is quite high, reaching 37% [21]. Chemical fungicides exhibit superior efficacy in limiting the dissemination of fungal phytopathogens and soil-borne infections due to their rapid biocidal action and environmental persistence. However, the excessive use of chemical disinfectants not only leads to the development of drug resistance in pathogenic fungi, but also causes environmental pollution [22]. Consequently, the development of novel, eco-friendly fungicides has become a critical priority in combating *Alternaria* species and their associated mycotoxin contamination.

Plant essential oils are aromatic secondary metabolites derived from plant materials. They exhibit a broad range of sources and chemical diversity, and demonstrate various biological activities, including antibacterial [23], anti-inflammatory [24], and antioxidant properties [25]. In recent years, extensive domestic and international research has demonstrated that plant essential oils possess significant antifungal efficacy against a wide spectrum of pathogenic fungi, positioning them as promising alternatives to conventional fungicides [26–28]. However, due to their high variability in composition and structural complexity, the underlying antibacterial mechanisms of these essential oils can differ substantially. Therefore, identifying the bioactive constituents responsible for antimicrobial activity and elucidating the modes of action of plant essential oils are crucial steps toward the development of effective natural antimicrobial agents. The extraction methods for plant essential oils include distillation, pressing, ultrasonication, microwave treatments, and ohmic heating [29]. Previous studies have all extracted AsEO via the steam distillation method [10,13,16–19]. *A. sieversiana* was subjected to hydro-distillation using a modified Clevenger-type apparatus for 2–6 h, with the extract quantity at 17–43. In this study, we first used a crystal tower pure dew essential oil machine (25L-DW, Hangzhou Hooloo Industrial Co., Hangzhou, China) for 4 h, and 42 compounds were obtained. Taking into account yield and distillation time, using a crystal tower pure dew essential oil machine could be an excellent approach for extracting essential oils.

Phytochemical analysis revealed that plant essential oils (EOs) exert antimicrobial effects through multiple mechanisms, including the cell walls and membrane integrity disruption, subcellular morphological alterations, nucleic acid impairment, enzymatic inhibition, and metabolic pathway perturbation [30]. Phylogenetic analyses reveal *A. tenuissima* and *A. alternata* as the primary etiological agents, implicated in black spot disease in Yanbian Pingguoli pear [20]. Many components of AsEOs play significant roles in the antifungal and antioxidant properties of the oils. 1,8-Cineole is a potential substitute for synthetic fungicides against *Alternaia tenuissima* [31]. Camphor can significantly inhibit fungal growth [32]. trans-Caryophyllene and β-myrcene (identified as predominant compounds in AsEO) are good indicators of antioxidant efficiency [33]. Terpenoids (such as limonene, linalool, etc.) and aromatic compounds (such as eugenol, cinnamaldehyde, etc.), identified as predominant compounds in AsEO, can disrupt membranes structure through hydrophobic interactions [34]. The antifungal properties of the main components of the

https://doi.org/10.3390/plants15020207

AsEO were verified in this study. Meanwhile, the integrity of fungal membrane systems could be quantitatively assessed through dual-parameter analysis involving extracellular AKP activity and ionic flux dynamics. In the study, exposure to AsEO increased AKP activity and electrical conductivity. The enhanced antifungal activity of AsEO was attributed to its dual-target mechanism, which simultaneously induces biochemical destabilization of fungal cell walls and structural disintegration of membranes.

4. Materials and Methods

4.1. Collection of Plant Materials

Artemisia sieversiana was grown in Yanji (42°54′32″ N, 129°28′47″ E), Jilin Province, China. The fresh aerial biomass (terminal branches with inflorescences) was harvested during the late vegetative phase (15 August 2023), ensuring optimal phenological representation. Specimen authentication was conducted by Professor Dong Weiwei (College of Agriculture, Yanbian University, Yanji, China) utilizing morphological characteristics and molecular barcoding.

4.2. Extraction and Isolation of AsEO

The plant materials (2.0 kg) were cut into 3–5 cm pieces, and transferred into crystal tower pure dew essential oil machine (25L-DW, Hangzhou Hooloo Industrial Co., Hangzhou, China). An amount of 4.0 L distilled water and defoaming agent (50 ppm of $CaCO_3$) was added, and the distillation rate was set at 25 mL/min. Then, the essential oil was obtained by hydrodistillation for 3 h. Moisture was removed from the final essential oil samples (anh. $MgSO_4$), which were stored in airtight containers at 4 °C.

4.3. GC-MS Analysis

The chemical compositions of the AsEO were evaluated via GC-MS on a Shimadzu QP 2010 Ultra system (Shimadzu, Tokyo, Japan) equipped with a DB-5MS column (30 m × 0.25 mm ID, 0.25 µm film thickness). The operating conditions were as follows [20]: the initial temperature was 45 °C for 4 min, and then the temperature increased 280 °C at a rate of 6 °C/min, and finalized with a 15 min isothermal hold. The GC-MS system was configured with injector/detector temperatures maintained at 280 °C. Ultra-high-purity helium (≥99%, 0.98 mL/min) served as the carrier gas, with splitless injection (2.0 µL) optimized for volatile compound analysis. Mass spectral acquisition utilized EI ionization (70 eV) with a scan range of 35–500 m/z. Compound identification integrated NIST 20 spectral library matching (similarity > 85%) against reference standards.

4.4. Fungal Strain and Culture Condition

The fungal pathogen species used in this study were *A. tenuissima* (NCBI GenBank Accession No: OQ001067) and *A. alternata* (NCBI GenBank Accession No: OQ001055), previously isolated from black spot disease on Yanbian Pingguoli pear [21]. The fungal species were subcultured on PDA medium, which was supplemented with 100 µg/L of streptomycin sulfate and incubated for 7 days at 28 °C. PDA medium: 200 g of potato, 15 g of agar powder, mixed with 1000 mL of deionized water, sterilized under high pressure at 121 °C for 25 min, placed upright, and cooled to 40 °C before pouring into disposable Petri dishes and left to stand for later use. The fungal pathogen species were cultured to logarithmic growth stage on potato dextrose Broth (PDB) medium. PDB medium: 200 g of potato, mixed with 1000 mL of deionized water, sterilized under high pressure at 121 °C for 25 min, cooled to room temperature and stored.

https://doi.org/10.3390/plants15020207

4.5. Antifungal Effect of the AsEO on Alternaria Species

The antifungal effect of AsEO was performed following the method described by Kou et al. [20]. The five concentrations of AsEO (v/v) used were 0%, 0.02%, 0.04%, 0.08%, and 0.10%. After incubation at 28 °C for 7 days, colony diameters were measured using the crossover approach to determine the percentage of inhibition for each concentration of AsEO treatment. The MIC was determined as the lowest concentration of the AsEO at which no visual fungal growth was observed. To determine the MFC of the AsEO, 10 μL of the AsEO (at the MIC and higher concentrations) was transferred to PDA plates. The plates were incubated for 48–72 h at 28 °C. The lowest concentration at which the fungus had no growth was considered as the MFC.

Spore samples were cultured on PDB medium supplemented with 0 and 1/2 MIC of AsEO, followed by sample collection for SEM analysis. The samples were fixed in 2.5% glutaraldehyde at 4 °C for 48 h and dehydrated using the method described by Kou et al. [20]. After dehydration, samples were mounted on the SEM stage, coated with gold for 60 s, and examined using SEM (SU-8010; Hitachi, Tokyo, Japan).

4.6. Effect of the AsEO on Cell Wall

The fungal pathogen species were cultured to the logarithmic growth stage on PDB medium, and different volume fractions of the AsEO were added to the spore suspension to achieve final AsEO volume fractions of 0.5 MIC, MIC, and 2 MIC. The fungal suspensions of *A. tenuissima* and *A. alternata* were 1×10^6 CFU/mL. The negative control (NC) group received an equivalent volume of ddH_2O, and was set at 100%, while the positive control (PC) group was treated with a lethal concentration of carbendazim solution. All experimental groups were incubated in a shaker at 37 °C and 120 rpm for 8 h. Samples were taken at 0 h, 2 h, 4 h, 6 h, and 8 h. After low-temperature centrifugation at 4 °C and 10,000 rpm for 10 min, the supernatants were collected and processed according to the instructions in the AKP kit (Beijing Bioss Technology Co., Beijing, China). Absorbance at 510 nm was measured using a UV-2550 ultraviolet–visible spectrophotometer (Shimadzu, Tokyo, Japan). The amount of phenol produced per gram of tissue per minute at 37 °C was taken as 1 μmol. The experiment was conducted in triplicate. Subsequently, a time-course kinetic profile was construct with reaction time (x-axis, 0–8 h) and AKP activity (y-axis, U/L) to analyze the effect of the AsEO on the cell walls of the two fungal pathogens.

The calculation formula for the AKP content in *Alternaria* species is as follows:

$$\mathrm{AKP} = \frac{\mathrm{A1} - \mathrm{A0}}{\mathrm{A2} - \mathrm{A0}} \times \mathrm{a} \times \mathrm{b} \tag{1}$$

where A1 is the absorbance of the sample, A0 is the absorbance of the blank, A2 is the absorbance of the standard sample, a is the concentration of phenol standard sample, and b is the dilution factor.

4.7. Effect of the AsEO on Cell Membrane

Electrical conductivity was assessed based on the foundational methodology of Min et al. [35]. A 200 μL aliquot of 1×10^6 CFU/mL spore suspensions of *A. tenuissima* and *A. alternata* were cultured in PDB at 28 °C for 4 days, after which grayish-black hyphae was collected. Subsequently, 0.5 g of hyphae sample was incubated in 5.0 mL distilled water with varying concentrations of AsEO (0.5 MIC, MIC, and 2 MIC) as the experimental groups; The negative and positive control groups were as previously described. Conductivity was measured at 0, 5, 15, 30, 60, 90, 180, and 360 min using a DDSJ-308F conductivity meter (Leizi, Shanghai, China). Next, the solution was boiled in water for the sterilization. After the solution cooled to room temperature, the conductivity was

https://doi.org/10.3390/plants15020207

measured again. The experiment was repeated three times. The relative conductivity at each time point was calculated using the conductivity formula. Variance analysis ($p < 0.05$) was completed using Graphpad Prism 7 software. To verify the effect of the AsEO on cell membrane permeability in *Alternaria* species, the trend graphs for each group were plotted. The culture time was on the x-axis and the conductivity was on the y-axis.

The formula for calculating relative conductivity is as follows:

$$C = \frac{C1 - C0}{C2} \times 100\% \quad (2)$$

where C is the relative conductivity (%), C1 is the electric conductivity at a certain time, C0 is the initial electric conductivity, and C2 is the final electric conductivity.

4.8. Determining the Safe Concentration of the AsEO

The safe concentration of the AsEO was determined using the cell counting kit-8 assay (Merck KGaA, Darmstadt, Germany). Human immortalized keratinocytes HaCaTs (BNCC101683) and skin fibroblasts HSFs (BNCC338008) were purchased from Beijing Beina Biotechnology Co., Ltd., Beijing, China. The HaCaT and HSF concentrations were adjusted to 5×10^3 CFU/mL, before inoculating cells into 96-well culture plates. A 100 μL aliquot of the cell suspension was added to each well. After 24 h of cultivation, a 50 μL aliquot of the culture medium containing different concentrations (0.1, 1, 10, 50, 100, 120, 150, 500, 1000 μg/mL) of the AsEO were added to each well. Control group (without adding AsEO) and blank groups (only culture medium, without cells) were also established. Each group had 4 duplicate wells. The absorbance (A value) at 450 nm was determined using the cell counting kit-8 [36].

The calculation formula for relative cell viability is as follows:

$$A = \frac{As - Ab}{Ac - Ab} \times 100\% \quad (3)$$

where A represents the relative cell viability (%), As represents the absorbance values of the cells containing different concentrations of the AsEO and the experimental wells containing the cell counting kit-8 medium. Ac represents the absorbance values of the control wells containing only cells and the cell counting kit-8 medium, and Ab represents the absorbance values of the blank wells containing only the cell counting kit-8 medium.

4.9. Statistical Analysis

All experimental procedures were performed in three independent replicates, with quantitative outcomes presented as mean values ± standard deviation (SD). Comparative analysis between experimental groups was conducted using Student's *t*-test ($\alpha = 0.05$), while parametric ANOVA analysis with Bonferroni post hoc correction was applied for multiple group comparisons. Statistical processing was performed using Statistical Package for the Social Sciences (SPSS) version 25.0 (IBM Corporation, Armonk, NY, USA), with significance thresholds set at $p < 0.05$.

5. Conclusions

In this study, the essential oil extracted from *Artemisia sieversiana* (Jilin biotype) was subjected to comprehensive GC-MS analysis with electron ionization (EI 70 eV) and a scan rate of 3 scans/s. A total of 42 bioactive volatiles were identified through NIST 20 spectral matching (similarity > 85%), with the key constituents being 1,8-cineole (34.03%), trans-caryophyllene (22.38%), (1R,4S)-1,7,7-trimethylbicyclo [2.2.1] heptan-2-yl acetate (8.39%), (±)-camphor (7.89%), and β-myrcene (3.90%). These bioactive constituents ex-

 https://doi.org/10.3390/plants15020207

hibited concentration-dependent antifungal activity against *A. tenuissima* and *A. alternate*, the causal agents of black spot disease in Yanbian Pingguoli pear. The MIC and MFC values were determined as 0.10% (v/v) and 0.12% (v/v), respectively. AsEO exhibited concentration-dependent cytoplasmic membrane permeabilization and cell wall degradation in *A. tenuissima* and *A. alternate*. This dual-target mechanism positions AsEO as a candidate biocontrol agent for managing *Alternaria*-induced black spot disease in Yanbian Pingguoli pear orchards. Further research is warranted to systematically assess the application of AsEO on fruits crops to optimize its biocontrol efficacy during both the growth and storage periods.

Supplementary Materials: The following supporting information can be downloaded at: https://www.mdpi.com/article/10.3390/plants15020207/s1, Table S1: Structural formulas of various components in the essential oil from *Artemisia sieversiana*.

Author Contributions: Conceptualization, T.-Y.Z.; methodology, Y.F.; resources writing—original draft preparation, R.Z.; writing—review and editing, Y.F.; supervision, Y.F. All authors have read and agreed to the published version of the manuscript.

Funding: This work was supported by the National Natural Science Foundation of China (No. 32360708).

Data Availability Statement: The original contributions presented in this study are included in the article/supplementary material. Further inquiries can be directed to the corresponding author.

Conflicts of Interest: The authors declare no conflicts of interest.

References

1. Nwe, Z.M.; Htut, K.N.; Aung, S.L.L.; Gou, Y.N.; Huang, C.X.; Deng, J.X. Two novel species and a new host record of *Alternaria* (*Pleosporales*, *Pleosporaceae*) from sunflower (*Compositae*) in Myanmar. *MycoKeys* **2024**, *105*, 337–354. [CrossRef]
2. Simmons, E.G. *Alternaria: An Identification Manual; Fully Illustrated and with Catalogue Raisonné*; CBS Fungal Biodiversity Centre: Utrercht, The Netherlands, 2007; pp. 1796–2007.
3. Pastor, F.J.; Guarro, J. *Alternaria* infections: Laboratory diagnosis and relevant clinical features. *Clin. Microbiol. Infec.* **2008**, *14*, 734–746. [CrossRef]
4. Louro, H.; Vettorazzi, A.; López de Cerain, A.; Spyropoulou, A.; Solhaug, A.; Straumfors, A.; Behr, A.C.; Mertens, B.; Žegura, B.; Fæste, C.K.; et al. Hazard characterization of *Alternaria* toxins to identify data gaps and improve risk assessment for human health. *Arch. Toxicol.* **2023**, *98*, 425–469. [CrossRef]
5. Pfeiffer, E.; Eschbach, S.; Metzler, M. *Alternaria* toxins: DNA strand-breaking activity in mammalian cells in vitro. *Mycotoxin Res.* **2007**, *23*, 152–157. [CrossRef]
6. Hickert, S.; Bergmann, M.; Ersen, S.; Cramer, B.; Humpf, H.U. Survey of *Alternaria* toxin contamination in food from the German market, using a rapid HPLC-MS/MS approach. *Mycotoxin Res.* **2016**, *32*, 7–18. [CrossRef]
7. Ismail, A.M.; Elshewy, E.S.; El-Ganainy, S.M.; Magistà, D.; Hamouda, A.F.; Alhudaib, K.A.; Ebrahim, W.; Almaghasla, M.I. Mycotoxins from tomato pathogenic *Alternaria alternata* and their combined cytotoxic effects on human cell lines and male albino rats. *J. Fungi* **2023**, *9*, 282. [CrossRef]
8. Escrivá, L.; Oueslati, S.; Font, G.; Manyes, L. *Alternaria* mycotoxins in food and feed: An overview. *J. Food Qual.* **2017**, *2017*, 1569748. [CrossRef]
9. Li, Y.P.; Cai, Y.P.; Yu, K.M.; Chao, Y.H.; Li, J.; Zhang, L.Y. Identification, mycotoxigenic ability and biosynthesis genes of *Alternaria* spp. from apples in China. *Int. J. Food Microbiol.* **2026**, *445*, 111491. [CrossRef]
10. Zhang, J.W.; Li, B.Y.; Lu, X.X.; Zheng, Y.; Wang, D.; Zhang, Z.; Zeng, D.; Du, S.S. Chemical diversity and anti-insect activity evaluation of essential oils extracted from five *Artemisia* species. *Plants* **2022**, *11*, 1627. [CrossRef]
11. Nuermaimaiti, M.; Turak, A.; Yang, Q.; Tang, B.X.; Zang, Y.; Li, J.; Aisa, H.A. Sesquiterpenes from *Artemisia Sieversiana* and their anti-inflammatory activities. *Fitoterapia* **2021**, *154*, 104996. [CrossRef]
12. Liu, S.J.; Liao, Z.X.; Tang, Z.S.; Cui, C.L.; Liu, H.B.; Liang, Y.N.; Zhang, Y.; Shi, H.X.; Liu, Y.R. Phytochemicals and biological activities of *Artemisia sieversiana*. *Phytochem. Rev.* **2017**, *16*, 441–460. [CrossRef]
13. Suleimenov, E.M.; Ozek, T.; Demirci, F.; Demirci, B.; Baser, K.H.C.; Adekenov, S.M. Component composition on essential oils of *Artemisia lercheana* and *A. sieversiana* of the flora of Kazakhstan. antimicrobial activity of *A. sieversiana* essential oil. *Chem. Nat. Compd.* **2009**, *45*, 120–123. [CrossRef]

14. Liu, Z.L.; Liu, Q.R.; Chu, S.S.; Jiang, G.H. Insecticidal activity and chemical composition of the essential oils of *Artemisia lavandulaefolia* and *Artemisia sieversiana* from China. *Chem. Biodivers.* **2010**, *7*, 2040–2045. [CrossRef]
15. Viteri, L.O.; Gonza'lez, M.J.; Silva, P.B.; Gomes, J.M.; Svacina, T.; Costa, L.T.M.; Valarezo, E.; Mantilla-Afanador, J.G.; Herrera, O.M.; Aguiar, R.W.S.; et al. GABA and octopamine receptors as potential targets for fumigant actions of Bursera graveolens essential oil against *Callosobruchus maculatus* and *Callosobruchus chinensis*. *J. Xenobiotics* **2025**, *15*, 91. [CrossRef] [PubMed]
16. Zhigzhitzhapova, S.V.; Popov, D.V.; Pintaeva, E.T.; Radnaeva, L.D.; Chimittsyrenova, L.I.; Randalova, T.É. Essential oil from *Artemisia sieversiana* Willd. and development of related oil-in-water emulsions. *Pharm. Chem. J.* **2017**, *51*, 388–390. [CrossRef]
17. Liu, J.Y.; Deng, Y.Q.; Zhao, Z.; Zhang, J.M.; Hua, J.; Luo, S.H. The chemical diversity of essential oils from galls of two *Artemisia* spp., and their insecticidal activity against the storage pest *Callosobruchus chinensis*. *Ind. Crop. Prod.* **2024**, *214*, 118505. [CrossRef]
18. Jiang, C.Y.; Zhou, S.X.; Toshmatov, Z.; Mei, Y.; Jin, G.Z.; Han, C.X.; Zhang, C.; Shao, H. Chemical composition and phytotoxic activity of the essential oil of *Artemisia sieversiana* growing in Xinjiang, China. *Nat. Prod. Res.* **2022**, *36*, 2434–2439. [CrossRef] [PubMed]
19. Li, H.L.; Pu, J.Y.; Zeng, L.Y.; Zhong, Y.; Xu, F.L.; Nan, P. Chemical composition of essential oil of *Artemisia sieversiana* from Tibet. *J. Essent. Oil Bear. Plants* **2017**, *20*, 1407–1412.
20. Kou, Z.X.; Dang, Y.; Liu, L.; Wu, X.H.; Fu, Y. Antifungal activity of *Artemisia capillaris* essential oil against *Alternaria* species causing black spot on Yanbian Pingguoli pear in China. *Plants* **2025**, *14*, 3146. [CrossRef]
21. Li, Y.C.; Bi, Y.; An, L.Z. Occurrence and latent infection of *Alternaria* rot of pingguoli pear (*Pyrus bretschneideri* Rehd. cv. Pingguoli) fruits in Gansu, China. *J. Phytopathol.* **2007**, *155*, 56–60. [CrossRef]
22. Miller, T.; Richmond, M.; Vann, M.; Hansen, Z. Sensitivity of *Phytophthora nicotianae* in Tennessee and North Carolina to Mefenoxam, Oxathiapiprolin, Mandipropamid, and Fluopicolide. *Plant Dis.* **2024**, *108*, 1612–1620. [CrossRef]
23. Cebi, N.; Erarslan, A. Determination of the antifungal, antibacterial activity and volatile compound composition of *Citrus bergamia* peel essential oil. *Foods* **2023**, *12*, 203. [CrossRef] [PubMed]
24. Subudhi, L.; Thatoi, H.; Banerjee, A. Anti-inflammatory activity of essential oil from medicinal plants: An insight into molecular mechanism, in-silico studies and signaling pathways. *Phytomedicine* **2025**, *138*, 156364. [CrossRef] [PubMed]
25. Cai, G.H.; Yu, M.; Liu, P.A.; Li, Z.Z.; Huang, M.Y.; Tang, X.D.; Huang, Q.Z.; Ye, B.N.; Guo, Z.X.; Sun, Y.Z. Compound plant essential oils improved antioxidative ability and immunity in juvenile largemouth bass *(Micropterus salmoides)*. *Aquacult. Rep.* **2025**, *45*, 103154. [CrossRef]
26. Yang, S.Z.; He, M.; Li, D.M.; Shi, J.; Peng, L.T.; Liu, J.J. Antifungal activity of 40 plant essential oil components against Diaporthe fusicola from postharvest kiwifruits and their possible action mode. *Ind. Crop. Prod.* **2023**, *194*, 116102. [CrossRef]
27. Liang, H.Y.; Yang, M.H.; Li, Q.; Zhang, L.; Zhao, X.S. A comprehensive review of the main components of plant essential oils and the mechanisms responsible for the inhibitory effects on fungal growth and aflatoxin synthesis. *Innov. Food Sci. Emerg.* **2024**, *96*, 103747. [CrossRef]
28. Yang, Y.; Liu, S.Y.; Wang, Y.; Yang, Q.; Wang, B.A.; Zhang, Y.T.; Tong, Z.K.; Zhang, J.H. Antifungal mechanisms of *Phoebe bournei* wood essential oil against *Botryosphaeria dothidea* and its application in apples. *Postharvest Biol. Technol.* **2025**, *229*, 113703. [CrossRef]
29. Zhou, W.; Li, J.; Wang, X.; Liu, L.; Li, Y.; Song, R.; Li, X. Research progress on extraction, separation, and purification methods of plant essential oils. *Separations* **2023**, *10*, 596. [CrossRef]
30. Yun, X.T.; Zhou, Y.N.; Shi, H.H.; Li, Z.J.; Yun, Y.H.; Xie, M.Z.; Chen, L.P. Combination of plant essential oils and ice: Extraction, encapsulation and applications in food preservation. *Food Chem.* **2025**, *493*, 145793. [CrossRef]
31. Singh, K.; Deepa, N.; Chauhan, S.; Tandon, S.; Verma, R.S.; Singh, A. Antifungal action of 1,8 cineole, a major component of Eucalyptus globulus essential oil against *Alternaria tenuissima* via overproduction of reactive oxygen species and downregulation of virulence and ergosterol biosynthetic genes. *Ind. Crop. Prod.* **2024**, *214*, 118580. [CrossRef]
32. Kong, W.B.; Huo, H.R.; Gu, Y.; Cao, Y.Q.; Wang, J.L.; Liang, J.Y.; Niu, S.Q. Antifungal activity of camphor against four phytopathogens of *Fusarium*. *S. Afr. J. Bot.* **2022**, *148*, 437–445. [CrossRef]
33. López, P.L.; Guerberoff, G.K.; Grosso, N.R.; Olmedo, R.H. Antioxidant-efficient indicator determinate by the relationship between β-myrcene/caryophyllene (α, β) on Hop (*Humulus lupulus*) essential oils under an accelerated oxidation test. *Ind. Crop. Prod.* **2023**, *205*, 117399. [CrossRef]
34. Hyldgaard, M.; Mygind, T.; Meyer, R.L. Essential oils in food preservation: Mode of action, synergies, and interactions with food matrix components. *Front. Microbiol.* **2012**, *3*, 12. [CrossRef] [PubMed]

https://doi.org/10.3390/plants15020207

35. Min, D.D.; Wu, H.Y.; Xu, M.; Leng, P.; Sun, J.; Liu, Y.G. Antifungal and mechanism of rose essential oil against *Monilinia fructicola* caused brown rot of peach fruit. *Postharvest Biol. Technol.* **2025**, *222*, 113398. [CrossRef]
36. Zhang, C.; Wang, C.W.; Bai, L.; Shi, S.L.; Yao, L. Chemical composition and antimicrobial activity of the essential oil of *Artemisia sieversiana* in Ningxia. *Chin. Agric. Sci. Bull.* **2020**, *36*, 62–68.

 https://doi.org/10.3390/plants15020207

plants

MDPI

Article

Organic Compounds as a Natural Alternative for Pest Control: How Will Climate Change Affect Their Effectiveness?

Virginia L. Usseglio [1,2,3], María P. Zunino [1,2,3,*], Vanessa D. Brito [1,3], Magalí Beato [1,3], Martin G. Theumer [4] and José S. Dambolena [1,2,3,*]

1 Instituto Multidisciplinario de Biología Vegetal (IMBiV-CONICET-UNC), Av. Velez Sarsfield 1611 (X5000OPB), Ciudad Universitaria, Córdoba X5016GCN, Argentina; vusseglio@imbiv.unc.edu.ar (V.L.U.); vbrito@imbiv.unc.edu.ar (V.D.B.); mbeato@imbiv.unc.edu.ar (M.B.)
2 Cátedra de Química Orgánica y Productos Naturales (FCEFyN-UNC), Av. Velez Sarsfield 1611 (X5000OPB), Ciudad Universitaria, Córdoba X5016GCN, Argentina
3 Instituto de Ciencia y Tecnología de los Alimentos (ICTA-FCEFyN-UNC), Av. Velez Sarsfield 1611 (X5000OPB), Ciudad Universitaria, Córdoba X5016GCN, Argentina
4 Centro de Investigación en Bioquímica Clínica e Inmunología (CIBIC-FCQ-UNC), Av. Medina Allende (X5000HUA), Ciudad Universitaria, Córdoba X5016GCN, Argentina; mgtheumer@unc.edu.ar
* Correspondence: paula.zunino.254@unc.edu.ar (M.P.Z.); jdambolena@imbiv.unc.edu.ar (J.S.D.)

Abstract

Climate change scenarios predict increased temperatures, potentially impacting the development of phytopathogenic fungi and the efficacy of their control. This study evaluated the effects of four natural organic compounds—carvacrol, eugenol, *trans*-cinnamaldehyde, and 1-heptyn-3-ol—on the growth of *Fusarium verticillioides* and the survival of *Sitophilus zeamais* under two temperature regimes (28 °C and 32 °C). Fungal growth was assessed through the lag phase duration and mycelial expansion, while insecticidal activity was determined by mortality of *S. zeamais*. Carvacrol (1 ppm) produced the most pronounced inhibitory effect on fungal growth, significantly extending the lag phase and reducing mycelial area, with eugenol showing similar effects at selected concentrations. Both compounds maintained or enhanced their antifungal activity at elevated temperatures. Trans-cinnamaldehyde and 1-heptyn-3-ol exhibited moderate or low effects, depending on concentration and temperature. Regarding *S. zeamais*, 1-heptyn-3-ol achieved complete mortality at all concentrations under both temperature scenarios, whereas carvacrol, eugenol, and *trans*-cinnamaldehyde showed dose-dependent effects at 28 °C and enhanced efficacy at 32 °C. Overall, these findings highlight the potential of these compounds as sustainable, climate-resilient alternatives for managing fungal pathogens and stored-product pests.

Keywords: biopesticides; *Fusarium verticillioides*; *Sitophilus zeamais*; global warming

1. Introduction

Since the Industrial Revolution, increased consumption of fossil fuels has led to significant imbalances in atmospheric composition. This disruption has contributed to the rise in global temperatures, a phenomenon recognized as a critical emergency by the United Nations Climate Change Conference (COP25) [1]. Climate change is associated with a range of environmental and biodiversity challenges, which, in turn, have profound impacts on human societies [2–4]. Agroecosystems are particularly vulnerable, facing reduced food security, declining crop yields, and shifts in productivity patterns [5,6]. Moreover, changing climatic conditions influence the development and adaptation of pests [7–10]. Under the Representative Concentration Pathways (RCP) 8.5 scenario, which predicts a global temperature increase of 2.5–3.5 °C above pre-industrial levels [11], crop yield losses due to insect pests are expected to double, as will losses associated with increased fungal infections and mycotoxin contamination [12,13].

The phytopathogenic fungus *Fusarium verticillioides* Sacc. (Nirenberg) is one of the major pests affecting maize production in temperate regions of the world. The presence of this fungus is associated with significant crop yield losses, ranging from 17% to 40% annually, either due to ear rot or the production of elevated levels of mycotoxins [14,15]. The presence of toxins produced by *F. verticillioides*, known as fumonisins, is correlated with the incidence of diseases in both farm animals and humans [15,16].

Sitophilus zeamais Motschulsky (Coleoptera: Curculionidae), the maize weevil, is one of the most important insect pests of stored grains worldwide, with a particularly severe impact in tropical and subtropical regions [17]. This insect infests maize both in the field and during storage, causing significant quantitative and qualitative losses. Its life cycle, characterized by larval development inside the kernel, hinders early detection and facilitates its spread [18,19]. In addition to direct endosperm consumption, its activity increases substrate moisture and temperature, favoring the growth of toxigenic fungi [20,21].

Traditionally, synthetic insecticides, systemic fungicides, or resistant cultivars have been used for pest control; however, these approaches are often neither fully effective nor eco-friendly [22–25]. For this reason, there is a growing search for natural alternatives for pest control. Natural organic compounds, identified from sources such as essential oils or intraspecific interactions, are being widely studied as semiochemicals of great agricultural importance [26–28]. However, there are few studies addressing the effectiveness of these compounds under predictive climate change scenarios.

With the increase in global temperature, mycotoxigenic fungi such as *F. verticillioides* and insect pests such *S. zeamais* have demonstrated high plasticity in adapting to climatic variations, altering their growth parameters and toxin production [29–33]. Nevertheless, the influence of global warming on the effectiveness of natural compounds for the control of these pests has yet to be explored. Therefore, the aim of this study was to evaluate the effectiveness of four selected organic compounds in controlling *S. zeamais* and *F. verticillioides* under global change scenarios, aiming to contribute to the design of ecologically sustainable strategies for the control of these phytopathogenic fungi.

2. Results

2.1. *Carvacrol Retains Antifungal Efficacy Under Global Warming Conditions*

The antifungal activity of carvacrol, eugenol, *trans*-cinnamaldehyde, and 1-heptyn-3-ol was assessed under two global warming scenarios: 28 °C (control condition) and 32 °C (RCP 8.5 scenario).

Variations in the mycelial growth rate of *F. verticillioides* under these conditions are shown in Figure 1. Control treatments (free compound) revealed that growth rate decreased on average by 1.5-fold with increasing temperature, suggesting that mycelial development may be delayed in a global warming context. This trend was consistent across most tested concentrations of the organic compounds, where growth rates at 28 °C (Figure 1A) were higher than those at 32 °C (Figure 1B).

Carvacrol exhibited the strongest antifungal activity. At higher concentrations, it almost completely inhibited the growth of *F. verticillioides* at 28 °C. Although strong antifungal activity was also observed at 32 °C, *F. verticillioides* growth rates were 5- to 14-fold lower than at 28 °C. These results indicate that the pronounced antifungal effect of carvacrol is not significantly affected by the temperature increase evaluated. Eugenol also showed significant antifungal activity at 1 ppm. The antifungal effect of eugenol was more

 https://doi.org/10.3390/plants15010048

pronounced at 28 °C, indicating that its efficacy may be compromised with increasing global temperatures.

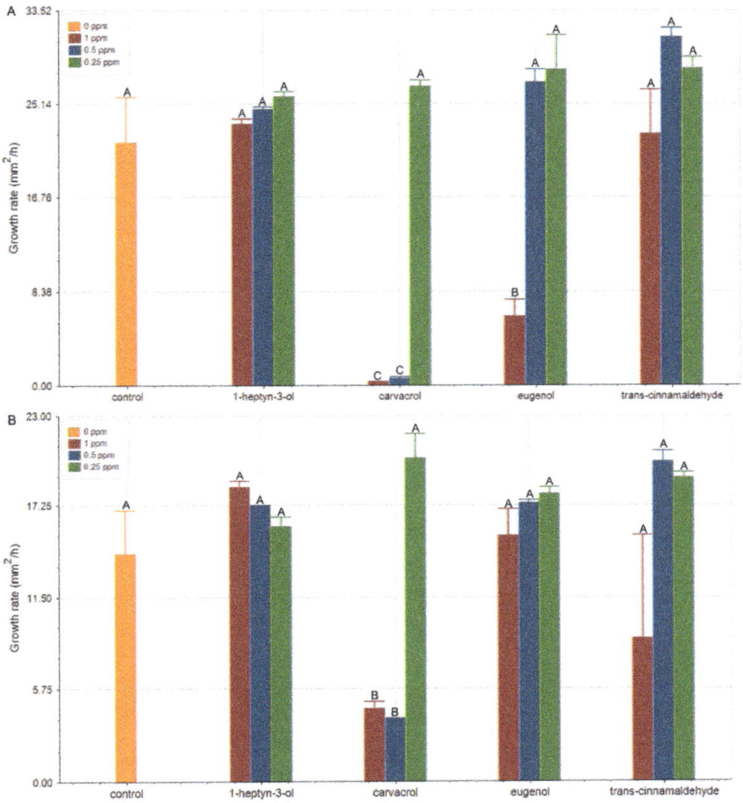

Figure 1. Growth rate (mm^2/h) of *F. verticillioides* exposed to increasing concentrations of carvacrol, eugenol, *trans*-cinnamaldehyde, and 1-heptyn-3-ol under two global warming scenarios: (**A**) 28 °C ($p < 0.0001$, df = 12; F = 55.70) and (**B**) 32 °C ($p < 0.0001$, df = 12; F = 8.34). Green bars: 0.25 ppm; Blue bars: 0.5 ppm; Red bars: 1 ppm; Orange bars: 0 ppm (controls). Different letters indicate statistically significant differences according to ANOVA followed by a post hoc DGC test.

In contrast, *trans*-cinnamaldehyde and 1-heptyn-3-ol did not show significant antifungal activity under the conditions evaluated.

Overall, these results demonstrate that the antifungal efficacy of the most bioactive organic compound would not be compromised by the temperature increases associated with climate change. In this context, compounds such as carvacrol emerge as promising natural alternatives for the control of *F. verticillioides*, even under future scenarios involving thermal stress.

2.2. Increased Temperature Does Not Compromise, and Enhances, the Toxicity of Volatile Compounds Toward S. zeamais

The acute effects of carvacrol, eugenol, *trans*-cinnamaldehyde, and 1-heptyn-3-ol on the survival of *S. zeamais* were evaluated under two global warming scenarios: 28 °C (control condition) and 32 °C (RCP 8.5 scenario). Mortality rates, calculated after 24 h of exposure and expressed as percentages, are presented in Figure 2. Control treatments under both scenarios showed mortality rates below 5%, indicating that the increase in temperature alone does not affect the survival of *S. zeamais*.

https://doi.org/10.3390/plants15010048

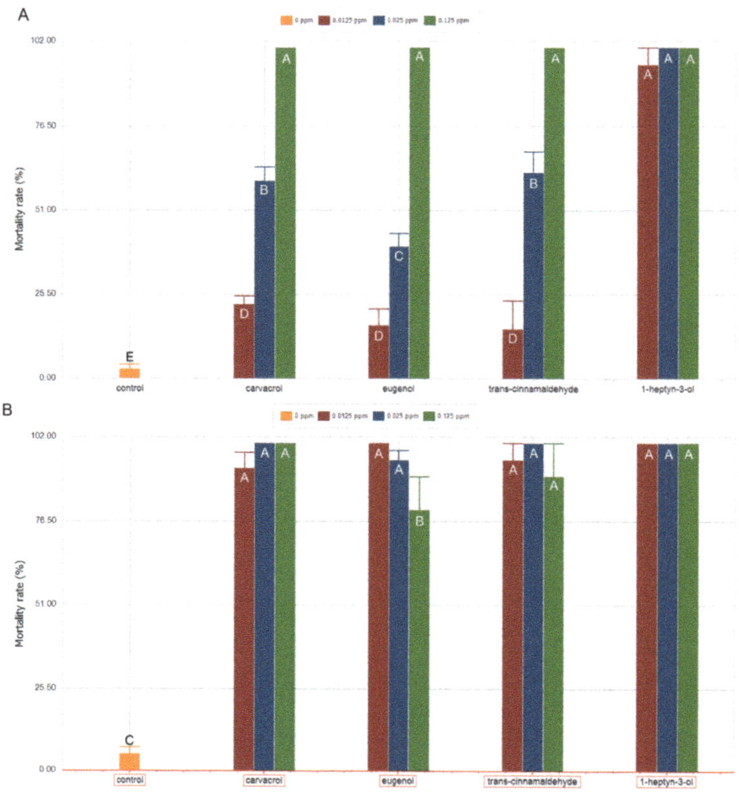

Figure 2. Mortality rate (%) of *S. zeamais* after 24 h of exposure to increasing concentrations of carvacrol, eugenol, *trans*-cinnamaldehyde, and 1-heptyn-3-ol under two global warming scenarios: (**A**) 28 °C ($p < 0.0001$, df = 12; F = 98.4) and (**B**) 32 °C ($p < 0.0001$, df = 12; F = 154.6). Green bars: 0.125 ppm; Blue bars: 0.025 ppm; Red bars: 0.0125 ppm; Orange bars: 0 ppm (controls). Different letters indicate statistically significant differences according to ANOVA followed by a post hoc DGC test.

At 28 °C (Figure 2A), 1-heptyn-3-ol was the most effective compound for controlling *S. zeamais*, whereas carvacrol, *trans*-cinnamaldehyde, and eugenol produced lower effects. At 0.125 ppm, the highest concentration tested, all compounds achieved 100% mortality; however, this effect persisted only for 1-heptyn-3-ol when the concentration was reduced five-fold. At this reduced concentration (0.025 ppm), carvacrol and *trans*-cinnamaldehyde exhibited similar effects, reaching mortality rates of 60.00 ± 4.08% and 62.50 ± 6.29%, respectively. At the same concentration, eugenol showed a mortality rate approximately 20% lower, a difference that was statistically significant compared with the other compounds. At 0.0125 ppm, the high mortality rate induced by 1-heptyn-3-ol was maintained, whereas mortality rates for the other three compounds decreased significantly, ranging between 15% and 22%, with carvacrol showing a slightly higher value.

Under the RCP 8.5 scenario at 32 °C (Figure 2B), 1-heptyn-3-ol maintained its effectiveness, reaching 100% mortality at all three concentrations tested (0.125, 0.025, and 0.0125 ppm), whereas the other compounds exhibited increased efficacy compared with the results observed at 28 °C. Carvacrol achieved 100% mortality at both 0.125 and 0.025 ppm, and at 0.0125 ppm the mortality rate reached 92.5%. In the case of *trans*-cinnamaldehyde and eugenol, no clear dose-dependent response was observed. *Trans*-cinnamaldehyde showed the lowest mortality percentage at the highest concentration tested (0.125 ppm), al-

https://doi.org/10.3390/plants15010048

though this value was not statistically different from those obtained at 0.025 and 0.0125 ppm. A similar pattern was observed for eugenol; however, in this case, the mortality rate at the highest concentration, 0.125 ppm, was 15–20% lower than those recorded for the other concentrations, and this difference was statistically significant. Despite these specific response patterns, mortality rates obtained under this global warming scenario were markedly higher than those recorded at 28 °C, indicating that, under projected increases in global temperature, the effectiveness of these natural compounds against *S. zeamais* would not be compromised and may even be enhanced.

3. Discussion

3.1. Sustained Efficacy of Carvacrol and Eugenol Against F. verticillioides Growth at Elevated Temperatures: Implications for Biocontrol on Climate Change Scenarios

Our findings reveal that the increase in global temperature resulted in a slight, non-significant reduction in fungal growth rate, suggesting that *F. verticillioides* may maintain its pathogenic potential under future climatic conditions. All tested organic compounds under simulated climate change conditions (32 °C) extended the lag phase, indicating an inhibitory effect on early adaptation and metabolic activation stages. These findings suggest that future atmospheric scenarios may influence the ecological fitness and growth rates of this phytopathogen.

Among the organic compounds tested, carvacrol exerted the most pronounced antifungal activity under simulated climate change conditions. The antifungal effect of carvacrol can be attributed to its high lipophilicity, which allows it to readily reach the enzymatic active sites, causing functional alterations and inducing oxidative stress [34–36]. The doubling of the lag phase by carvacrol at 32 °C suggests that this compound maintains, or even enhances, its efficacy under climatic change scenarios, making it a promising candidate for sustainable control strategies.

Eugenol ranked as the second most effective compound. Although its overall effect was lower than that of carvacrol, eugenol also caused a substantial delay in the onset of exponential growth. This effect may also be attributed to the presence of an aromatic ring supplemented with an OH group capable of forming hydrogen bonds with the active sites of various enzymes [37]. Notably, its antifungal activity remained stable across both temperatures, suggesting a temperature-independent mechanism of action. This stability suggests that formulations based on such bioactive compounds could retain their efficacy even under climate change-driven scenarios involving elevated temperatures, where pest incidence is expected to increase and food safety may be compromised [38].

In contrast, 1-heptyn-3-ol and *trans*-cinnamaldehyde exhibited more modest effects. The inhibitory activity of 1-heptyn-3-ol remained unchanged across concentrations and temperatures, while *trans*-cinnamaldehyde displayed a temperature-dependent behavior. Although the inhibitory effects observed for *trans*-cinnamaldehyde in this study were low, mainly at 28 °C, the antifungal properties of this compound are widely reported [39–43]. Conversely, there are no previous reports on the effects of 1-heptyn-3-ol on the growth of *F. verticillioides*. However, based on previous work on 1-octyn-3-ol [44], a structurally related compound, the compound was expected to exhibit strong antifungal activity due to the presence of a triple bond, which confers nucleophilic properties and makes it susceptible to addition reactions [45–47]. Although these reactions may occur inside fungal cells, the inhibitory effect of 1-heptyn-3-ol observed here is low, indicating that this compound is not a suitable candidate for controlling *F. verticillioides*.

Taken together, these findings provide important insights into how projected climate change conditions—namely elevated temperatures—may influence the growth dynamics of *F. verticillioides*, as well as the efficacy of natural antifungal compounds. The stable or

https://doi.org/10.3390/plants15010048

enhanced activity of carvacrol and eugenol under simulated climate conditions supports their potential as climate-resilient biocontrol agents. Previous studies have demonstrated the high efficacy of these compounds in inhibiting fumonisin B1 production [34,48]. Accordingly, the results obtained in the present study underscore the need to conduct further experiments to evaluate their performance under more realistic and dynamic environmental conditions.

Moreover, based on the background literature and the effects observed in this work, the design of a mixture combining trans-cinnamaldehyde with more bioactive compounds, such as carvacrol or eugenol, could be proposed to enhance antifungal efficacy and to assess the behavior of this mixture under climate change scenarios.

Collectively, these results reinforce the need for further research to explore the molecular mechanisms underlying these responses and to evaluate the field applicability of compound mixtures, particularly those combining carvacrol or eugenol with less active agents such as *trans*-cinnamaldehyde. Ultimately, the development of natural, climate-resilient formulations may offer a sustainable path forward for integrated strategies aimed at managing phytopathogens and reducing mycotoxin contamination in agricultural systems.

3.2. Enhanced Efficacy of 1-Heptyn-3-ol, Carvacrol, Eugenol, and Trans-Cinnamaldehyde Against S. zeamais Under Global Warming Scenarios

In the present study, we demonstrate that temperature plays a critical role in modulating the acute insecticidal activity of organic volatile compounds against *S. zeamais*. In both scenarios, mortality did not exceed 5%, indicating that the increase in temperature alone does not affect insect survival. Under control conditions (28 °C), 1-heptyn-3-ol exhibited the highest efficacy among the four compounds tested, producing complete mortality even at intermediate and low concentrations. In contrast, carvacrol, *trans*-cinnamaldehyde, and eugenol showed a clear dose-dependent pattern, with marked reductions in mortality at lower concentrations.

Exposure to a projected global warming scenario (32 °C, RCP 8.5) substantially altered these efficacy patterns. Notably, 1-heptyn-3-ol maintained maximum efficacy at all concentrations, indicating that its insecticidal action is robust to temperature increases. Moreover, carvacrol, *trans*-cinnamaldehyde, and eugenol all displayed enhanced performance at 32 °C compared with 28 °C. In particular, carvacrol reached nearly complete mortality even at reduced concentrations. Although these compounds did not display clear dose–response relationships at 32 °C, the overall increase in mortality suggests that moderate warming may amplify the toxic effects of these natural compounds.

In previous studies, we demonstrated the detrimental effect of 1-heptyn-3-ol against this insect [45]. The reactivity of this compound has been attributed not only to the length of its carbon chain and the presence of the hydroxyl group, but also to the triple bond in its structure [45,46,49–51]. The consistently high mortality observed under both temperature scenarios suggests that the elevated reactivity of this compound may be largely independent of temperature.

Similarly, carvacrol, eugenol, and *trans*-cinnamaldehyde are plant-derived bioactive compounds with demonstrated insecticidal activity against *S. zeamais* [52–55]. Their toxicity is primarily associated with disruptions in feeding and reproductive behavior and inhibition of acetylcholinesterase activity [52–59]. Interestingly, in the present study, an increase in mortality was observed at higher temperatures, a pattern not detected in control treatments, indicating that these compounds may exhibit enhanced bioactivity under projected global warming scenarios. This result is consistent with previous studies reporting that higher temperatures may increase volatilization rates, thereby enhancing insect exposure. Additionally, elevated temperatures may accelerate or modify respiratory rates or increase cuticle permeability in insects [60–62]. In this context, future studies incorporating

 https://doi.org/10.3390/plants15010048

direct measurements of vapor pressure or headspace concentration would help clarify the physicochemical effects associated with volatilization under elevated temperatures.

Overall, these results reinforce previous reports emphasizing the high potency of alkynyl alcohols against stored-product pests, while also confirming the moderate yet variable activity of phenylpropanoids and terpenoids under standard environmental conditions. Also, these findings suggest that the bioactivity of all tested compounds is mediated by multiple biochemical mechanisms, highlighting their potential as natural insecticides under diverse environmental conditions.

From an applied perspective, the finding that natural compounds retain or improve their insecticidal performance at elevated temperatures is particularly relevant, as climate change is expected to intensify pest pressures in stored-grain systems [31,33]. Identifying compounds whose activity is not diminished under warmer conditions is therefore crucial for future pest management strategies. The temperature resilience of 1-heptyn-3-ol and the temperature-enhanced efficacy of carvacrol, *trans*-cinnamaldehyde, and eugenol suggest that these compounds could be integrated into effective and sustainable control tools under warming climate conditions. Further research should investigate the biochemical mechanisms underlying temperature-mediated enhancements in toxicity, assess long-term stability and volatilization kinetics, and evaluate these compounds under realistic storage conditions to validate their suitability for climate-resilient pest management programs. Additionally, it will be essential to examine how the metabolism of successive insect generations exposed to projected global warming temperatures may shift, as potential physiological or metabolic adaptations could alter susceptibility patterns and ultimately influence the long-term effectiveness of these natural compounds.

4. Materials and Methods

4.1. Fungal Strain, Insects, and Natural Organic Compounds

The toxicogenic strain of *Fusarium verticillioides* (Sacc.) Nirenberg (M3125) [63] was obtained from the Department of Agriculture, Agricultural Research Service, National Center for Agricultural Utilization Research (Peoria, IL, USA), provided by Dr. Robert Proctor. For bioassays, conidial suspensions (1×10^6 CFU/mL) were prepared according to Dambolena et al., (2008) [64].

Adults of *Sitophilus zeamais* (Motschulsky) were reared under storage conditions on insecticide-free maize at 28 ± 2 °C and $70 \pm 5\%$ relative humidity [65].

The organic compounds used in this study were 5-Isopropyl-2-methylphenol (carvacrol, 98% purity), 1-heptyn-3-ol (97%), 2-methoxy-4-(2-propenyl)-phenol (eugenol, ≥98%), and trans-3-phenyl-2-propenal (*trans*-cinnamaldehyde, ≥99%) (Figure 3). Given that the pesticidal effectiveness of the natural organic compounds carvacrol, eugenol, and *trans*-cinnamaldehyde has been widely documented [42,48,52,54,66], these compounds were selected to evaluate their activity under simulated global warming scenarios. Although 1-heptyn-3-ol is a synthetic compound, it was included in this study due to the lack of reports on its activity against *F. verticillioides*. Previous findings have shown its insecticidal activity against *Sitophilus zeamais* and no phytotoxic effects on *Zea mays* seeds, supporting its potential as a candidate for integrated pest management strategies [45]. All compounds were purchased from Merck® (Buenos Aires, Argentina).

https://doi.org/10.3390/plants15010048

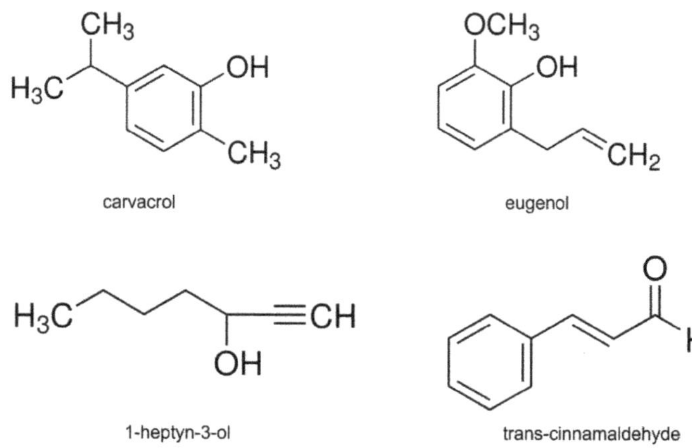

Figure 3. Molecular structure of selected organic compounds.

4.2. Climate Change Scenarios

To assess the impact of climate change on the development of *F. verticillioides* and the survival of *S. zeamais*, two temperature scenarios were considered. The temperatures to be tested were selected considering the RCP 8.5 scenario (3.5 °C for the period 2081–2100) in the central region of Argentina [11]. The control temperature corresponds to the average temperature during the maize sowing period (2010–2023) [67]. Accordingly, the experimental temperatures were set at 28 ± 1 °C and 32 ± 1 °C, control and treatment, respectively.

4.3. Effect of Organic Compounds on Fungal Growth

To determine the effects of natural organic compounds and 1-heptyn-3-ol on the vegetative growth of *F. verticillioides* the Minimum Inhibitory Concentration (MIC) technique was applied [68]. Briefly, the selected compounds were incorporated into Potato Dextrose Agar (PDA) at increasing concentrations (0, 0.25, 0.5, and 1 ppm) and poured into sterile Petri dishes (Ø = 9.0 cm). Each Petri dish was inoculated with 5 μL of conidial suspension (1×10^6 CFU/mL) and incubated as follows: three replicates of each concentration/compound were maintained at either 28 ± 1 °C or 32 ± 1 °C (temperature treatment).

For all replicates, mycelial diameter was measured daily until the control plates (0 ppm) were completely colonized. Lag phase and Inhibition percentage were subsequently calculated.

4.4. Insecticidal Effect of Organic Compounds Against S. zeamais

The insecticidal effect of the organic compounds under climate change conditions was evaluated following the fumigant toxicity protocol described by Herrera et al. (2015) [69]. Three concentrations of each compound (0.125, 0.025, and 0.0125 ppm) were applied to filter paper disks (Ø = 2 cm) and placed on the inner side of the lids of 30 mL amber glass flasks. Ten *S. zeamais* adults were then introduced into each flask and incubated under the corresponding environmental conditions. A nylon gauze was used to prevent direct contact between the insects and the compounds. Mortality was recorded after 24 h.

4.5. Statistical Analysis

For growth data analysis, linear regression was performed for each concentration and atmospheric condition to calculate growth rates (mm/h). Only the linear portion of each growth curve was used, with adjusted R^2 values ranging from 0.90 to 1 ($n_{temperature} = 33$). Subsequently, lag phase values were calculated and compared using

https://doi.org/10.3390/plants15010048

analysis of variance (ANOVA). When statistically significant differences were obtained (p-value ANOVA ≤ 0.05), a post hoc DGC test was applied.

Mortality data for the tested compounds were analyzed using ANOVA followed by a DGC post hoc test (ANOVA, $p \leq 0.05$). For each concentration, at least five replicates were performed.

For all statistical analyses, assumptions of normality and homogeneity of variance were verified ($p \geq 0.05$). All statistical analyses and graphical representations were performed using Navure software (Professional+ v 3.0) (Córdoba, Argentina) [70].

5. Conclusions

The present study demonstrates that both fungal and insect targets respond differently to temperature increases, yet natural bioactive compounds maintain or enhance their efficacy under elevated temperatures. Carvacrol and eugenol exhibited sustained antifungal activity against *F. verticillioides*, significantly delaying growth even under simulated climate change conditions. In contrast, 1-heptyn-3-ol, carvacrol, eugenol, and *trans*-cinnamaldehyde retained or improved their insecticidal activity against *S. zeamais* at higher temperatures.

Importantly, the identification of multifunctional compounds capable of controlling both fungal pathogens and insect pests is particularly relevant in stored-grain systems, where infestation by *S. zeamais* not only causes direct losses but also promotes fungal development by increasing grain damage and local humidity [20]. In this context, the dual efficacy of carvacrol highlights its potential as a broad-spectrum agent capable of mitigating two interacting threats.

Moreover, the development of formulations or compound mixtures with strong antifungal and insecticidal activity represents a promising avenue for future research, particularly under global warming scenarios. In addition, elucidating the biochemical mechanisms and physiological processes underlying the effectiveness of these natural compounds under increased temperature conditions constitutes an important next step.

The current agricultural production model is based on the construction of deeply simplified agroecosystems, in which ecosystem services are often overlooked and which depend on the constant input of external resources. Such systems are prone to rapid degradation due to biodiversity loss and deterioration of environmental services [71,72]. The use of natural bioactive compounds also offers significant environmental advantages. Their low persistence in the environment and reduced risk to non-target organisms make them safer alternatives to conventional synthetic pesticides [73–77].

Therefore, the results presented in this study provide valuable information for the development of rational and environmentally friendly pest control strategies that account for potential changes in pest aggressiveness in a context of global warming. Future studies should evaluate the effect of these strategies on non-target organisms under projected climate conditions.

Taken together, these findings underscore the importance of advancing both the study and application of natural bioactive compounds as sustainable and climate-resilient tools for protecting crop health and food safety in the face of climate change. Future research should also investigate how rising atmospheric CO_2 concentrations—another major driver of global change—may influence pest physiology, population dynamics, and the performance of natural compounds, as such interactions could further shape the efficacy of integrated control strategies under projected environmental conditions.

Author Contributions: Conceptualization: V.L.U., J.S.D. and M.P.Z. Methodology and formal analysis: V.L.U., V.D.B., M.B. and M.G.T. Investigation: V.L.U. Resources: V.L.U., J.S.D. and M.P.Z. Writing—original draft preparation: V.L.U., J.S.D., V.D.B., M.B., M.G.T. and M.P.Z. Supervision: J.S.D.

https://doi.org/10.3390/plants15010048

and M.P.Z. Funding acquisition: V.L.U., J.S.D. and M.P.Z. All authors have read and agreed to the published version of the manuscript.

Funding: This research was funded by the National Agency for Scientific and Technological Promotion, Argentina (ANPCyT), PICT-2021-I-INVI-00084, and SECyT CONSOLIDAR from the Secretary of Science and Technique (SECyT-UNC), Grant 33620230100243CB (SECyT-UNC), Argentina.

Data Availability Statement: The raw data supporting the conclusions of this article will be made available by the authors upon request (vusseglio@imbiv.unc.edu.ar).

Acknowledgments: The authors wish to express their gratitude to Luis Damián Barrionuevo (IMBiV-CONICET-UNC) for technical support to this research and to Silvina Colla for her professional assistance with the linguistic revision of the manuscript. The authors also thank the Fund for Scientific and Technological Research (FONCyT) and the Secretary of Science and Technology (SECyT-UNC).

Conflicts of Interest: The authors declare no conflicts of interest.

References

1. Naciones Unidas C.C. COP25. Available online: https://unfccc.int/event/cop-25#eq-12 (accessed on 23 November 2025).
2. Bezeng, B.S.; Yessoufou, K.; Taylor, P.J.; Tesfamichael, S.G. Expected Spatial Patterns of Alien Woody Plants in South Africa's Protected Areas Under Current Scenario of Climate Change. *Sci. Rep.* **2020**, *10*, 7038. [CrossRef]
3. Cheung, W.W.L.; Frölicher, T.L. Marine Heatwaves Exacerbate Climate Change Impacts for Fisheries in the Northeast Pacific. *Sci. Rep.* **2020**, *10*, 6678. [CrossRef]
4. Navarro-Racines, C.; Tarapues, J.; Thornton, P.; Jarvis, A.; Ramirez-Villegas, J. High-Resolution and Bias-Corrected CMIP5 Projections for Climate Change Impact Assessments. *Sci. Data* **2020**, *7*, 7. [CrossRef]
5. Schauberger, B.; Archontoulis, S.; Arneth, A.; Balkovic, J.; Ciais, P.; Deryng, D.; Elliott, J.; Folberth, C.; Khabarov, N.; Müller, C.; et al. Consistent Negative Response of US Crops to High Temperatures in Observations and Crop Models. *Nat. Commun.* **2017**, *8*, 13931. [CrossRef]
6. Bernacchi, C.J.; Ruiz-Vera, U.M.; Siebers, M.H.; DeLucia, N.J.; Ort, D.R. Short- and Long-Term Warming Events on Photosynthetic Physiology, Growth, and Yields of Field Grown Crops. *Biochem. J.* **2023**, *480*, 999–1014. [CrossRef]
7. Singh, B.K.; Delgado-Baquerizo, M.; Egidi, E.; Guirado, E.; Leach, J.E.; Liu, H.; Trivedi, P. Climate Change Impacts on Plant Pathogens, Food Security and Paths Forward. *Nat. Rev. Microbiol.* **2023**, *21*, 640–656. [CrossRef]
8. Subedi, B.; Poudel, A.; Aryal, S. The Impact of Climate Change on Insect Pest Biology and Ecology: Implications for Pest Management Strategies, Crop Production, and Food Security. *J. Agric. Food Res.* **2023**, *14*, 100733. [CrossRef]
9. Waheed, A.; Haxim, Y.; Islam, W.; Ahmad, M.; Muhammad, M.; Alqahtani, F.M.; Hashem, M.; Salih, H.; Zhang, D. Climate Change Reshaping Plant-Fungal Interaction. *Environ. Res.* **2023**, *238*, 117282. [CrossRef] [PubMed]
10. Casu, A.; Camardo Leggieri, M.; Toscano, P.; Battilani, P. Changing Climate, Shifting Mycotoxins: A Comprehensive Review of Climate Change Impact on Mycotoxin Contamination. *Compr. Rev. Food Sci. Food Saf.* **2024**, *23*, e13323. [CrossRef] [PubMed]
11. Secretaría de Ambiente y Desarrollo Sustentable de la Nación. *Tercera Comunicación Nacional Sobre Cambio Climático. "Cambio Climático En Argentina; Tendencias y Proyecciones"*; Secretaría de Ambiente y Desarrollo Sustentable de la Nación: Buenos Aires, Argentina, 2015.
12. Vila-Donat, P.; Marín, S.; Sanchis, V.; Ramos, A.J. A Review of the Mycotoxin Adsorbing Agents, with an Emphasis on Their Multi-Binding Capacity, for Animal Feed Decontamination. *Food Chem. Toxicol.* **2018**, *114*, 246–259. [CrossRef]
13. Deutsch, C.A.; Tewksbury, J.J.; Tigchelaar, M.; Battisti, D.S.; Merrill, S.C.; Huey, R.B.; Naylor, R.L. Increase in Crop Losses to Insect Pests in a Warming Climate. *Science* **2018**, *361*, 916–919. [CrossRef] [PubMed]
14. Xu, Y.; Zhang, Z.; Lu, P.; Li, R.; Ma, P.; Wu, J.; Li, T.; Zhang, H. Increasing Fusarium Verticillioides Resistance in Maize by Genomics-Assisted Breeding: Methods, Progress, and Prospects. *Crop J.* **2023**, *11*, 1626–1641. [CrossRef]
15. Omotayo, O.P.; Babalola, O.O. Fusarium Verticillioides of Maize Plant: Potentials of Propitious Phytomicrobiome as Biocontrol Agents. *Front. Fungal Biol.* **2023**, *4*, 1095765. [CrossRef] [PubMed]
16. Chen, J.; Wen, J.; Tang, Y.; Shi, J.; Mu, G.; Yan, R.; Cai, J.; Long, M. Research Progress on Fumonisin B1 Contamination and Toxicity: A Review. *Molecules* **2021**, *26*, 5238. [CrossRef] [PubMed]
17. Corrêa, A.S.; Vinson, C.C.; Braga, L.S.; Guedes, R.N.C.; de Oliveira, L.O. Ancient Origin and Recent Range Expansion of the Maize Weevil *Sitophilus zeamais*, and Its Genealogical Relationship to the Rice Weevil *S. Oryzae*. *Bull. Entomol. Res.* **2016**, *107*, 9–20. [CrossRef]
18. Rita Devi, S.; Thomas, A.; Rebijith, K.B.; Ramamurthy, V.V. Biology, Morphology and Molecular Characterization of *Sitophilus oryzae* and *S. zeamais* (Coleoptera: Curculionidae). *J. Stored Prod. Res.* **2017**, *73*, 135–141. [CrossRef]

 https://doi.org/10.3390/plants15010048

19. Demis, E.; Yenewa, W. Review on Major Storage Insect Pests of Cereals and Pulses. *Asian J. Adv. Res.* **2022**, *12*, 41–56.
20. Usseglio, V.L.; Dambolena, J.S.; Martinez, M.J.; Zunino, M.P. The Role of Fumonisins in the Biological Interaction Between *Fusarium verticillioides* and *Sitophilus zeamais*. *J. Chem. Ecol.* **2020**, *46*, 1059–1068. [CrossRef]
21. Ponce, M.A.; Lizarraga, S.; Bruce, A.; Kim, T.N.; Morrison, W.R. Grain Inoculated with Different Growth Stages of the Fungus, *Aspergillus flavus*, Affect the Close-Range Foraging Behavior by a Primary Stored Product Pest, *Sitophilus oryzae* (Coleoptera: Curculionidae). *Environ. Entomol.* **2022**, *51*, 927–939. [CrossRef]
22. Fai, P.B.A.; Ncheuveu, N.T.; Tchamba, M.N.; Ngealekeloeh, F. Ecological Risk Assessment of Agricultural Pesticides in the Highly Productive Ndop Flood Plain in Cameroon Using the PRIMET Model. *Environ. Sci. Pollut. Res.* **2019**, *26*, 24885–24899. [CrossRef]
23. Tao, H.; Fang, C.; Xiao, Y.; Jin, Y. The Toxicity and Health Risk of Chlorothalonil to Non-Target Animals and Humans: A Systematic Review. *Chemosphere* **2024**, *358*, 142241. [CrossRef]
24. Drosdoski, S.D.; Sinópolis Gigliolli, A.A.; Cabral, L.C.; Julio, A.H.F.; Bespalhok, D.D.N.; Santini, B.L.; Lapenta, A.S. Characterization of Esterases in the Involvement of Insecticide Resistance in *Sitophilus oryzae* and *Sitophilus zeamais* (Coleoptera: Curculionidae). *Int. J. Trop. Insect Sci.* **2024**, *44*, 1103–1115. [CrossRef]
25. Cortese, D.; de Oliveira, G.S.; Fernandes, M.G. Resistance Evaluation of Maize Varieties to *Sitophilus zeamais* Infestation Across Two Generations: Insights for Integrated Pest Management. *J. Stored Prod. Res.* **2024**, *109*, 102473. [CrossRef]
26. Razo-Belman, R.; Ozuna, C. Volatile Organic Compounds: A Review of Their Current Applications as Pest Biocontrol and Disease Management. *Horticulturae* **2023**, *9*, 441. [CrossRef]
27. Tonks, A.J.; Roberts, J.M.; Midthassel, A.; Pope, T. Exploiting Volatile Organic Compounds in Crop Protection: A Systematic Review of 1-Octen-3-Ol and 3-Octanone. *Ann. Appl. Biol.* **2023**, *183*, 121–134. [CrossRef]
28. Ayilara, M.S.; Adeleke, B.S.; Akinola, S.A.; Fayose, C.A.; Adeyemi, U.T.; Gbadegesin, L.A.; Omole, R.K.; Johnson, R.M.; Uthman, Q.O.; Babalola, O.O. Biopesticides as a Promising Alternative to Synthetic Pesticides: A Case for Microbial Pesticides, Phytopesticides, and Nanobiopesticides. *Front. Microbiol.* **2023**, *14*, 1040901. [CrossRef]
29. Medina, Á.; Rodríguez, A.; Magan, N. Climate Change and Mycotoxigenic Fungi: Impacts on Mycotoxin Production. *Curr. Opin. Food Sci.* **2015**, *5*, 99–104. [CrossRef]
30. Mshelia, P.L.; Selamat, J.; Iskandar Putra Samsudin, N.; Rafii, M.Y.; Abdul Mutalib, N.A.; Nordin, N.; Berthiller, F. Effect of Temperature, Water Activity and Carbon Dioxide on Fungal Growth and Mycotoxin Production of Acclimatised Isolates of *Fusarium verticillioides* and *F. Graminearum*. *Toxins* **2020**, *12*, 478. [CrossRef] [PubMed]
31. Sgrò, C.M.; Terblanche, J.S.; Hoffmann, A.A. What Can Plasticity Contribute to Insect Responses to Climate Change? *Annu. Rev. Entomol.* **2016**, *61*, 433–451. [CrossRef] [PubMed]
32. Rodrigues, Y.K.; Beldade, P. Thermal Plasticity in Insects' Response to Climate Change and to Multifactorial Environments. *Front. Ecol. Evol.* **2020**, *8*, 271. [CrossRef]
33. Skendžić, S.; Zovko, M.; Živković, I.P.; Lešić, V.; Lemić, D. The Impact of Climate Change on Agricultural Insect Pests. *Insects* **2021**, *12*, 440. [CrossRef]
34. Dambolena, J.S.; López, A.G.; Meriles, J.M.; Rubinstein, H.R.; Zygadlo, J.A. Inhibitory Effect of 10 Natural Phenolic Compounds on *Fusarium verticillioides*. A Structure–Property–Activity Relationship Study. *Food Control* **2012**, *28*, 163–170. [CrossRef]
35. Gao, Q.; Feng, Z.; Wang, Z.; Zhao, F.; Ju, J. Carvacrol Induces Apoptosis in *Aspergillus niger* Through ROS Burst. *World J. Microbiol. Biotechnol.* **2025**, *41*, 225. [CrossRef]
36. Divband, K.; Shokri, H.; Khosravi, A.R. Down-Regulatory Effect of *Thymus Vulgaris* L. on Growth and *Tri4* Gene Expression in *Fusarium oxysporum* Strains. *Microb. Pathog.* **2017**, *104*, 1–5. [CrossRef]
37. Campaniello, D.; Corbo, M.R.; Sinigaglia, M. Antifungal Activity of Eugenol Against *Penicillium*, *Aspergillus*, and *Fusarium* Species. *J. Food Prot.* **2010**, *73*, 1124–1128. [CrossRef] [PubMed]
38. van der Spiegel, M.; van der Fels-Klerx, H.J.; Marvin, H.J.P. Effects of Climate Change on Food Safety Hazards in the Dairy Production Chain. *Food Res. Int.* **2012**, *46*, 201–208. [CrossRef]
39. Wei, J.; Bi, Y.; Xue, H.; Wang, Y.; Zong, Y.; Prusky, D. Antifungal Activity of Cinnamaldehyde Against *Fusarium sambucinum* Involves Inhibition of Ergosterol Biosynthesis. *J. Appl. Microbiol.* **2020**, *129*, 256–265. [CrossRef] [PubMed]
40. Jaramillo Jimenez, B.A.; Awwad, F.; Desgagné-Penix, I. Cinnamaldehyde in Focus: Antimicrobial Properties, Biosynthetic Pathway, and Industrial Applications. *Antibiotics* **2024**, *13*, 1095. [CrossRef] [PubMed]
41. Zhou, L.R.; Hu, H.J.; Wang, J.; Zhu, Y.X.; Zhu, X.D.; Ma, J.W.; Liu, Y.Q. Cinnamaldehyde Acts as a Fungistat by Disrupting the Integrity of *Fusarium oxysporum* Fox-1 Cell Membranes. *Horticulturae* **2024**, *10*, 48. [CrossRef]
42. Xie, Y.; Huang, Q.; Wang, Z.; Cao, H.; Zhang, D. Structure-Activity Relationships of Cinnamaldehyde and Eugenol Derivatives Against Plant Pathogenic Fungi. *Ind. Crops Prod.* **2017**, *97*, 388–394. [CrossRef]
43. Wang, H.; Peng, Z.; Sun, H. Antifungal Activities and Mechanisms of Trans-cinnamaldehyde and Thymol Against Food-spoilage Yeast *Zygosaccharomyces rouxii*. *J. Food Sci.* **2022**, *87*, 1197–1210. [CrossRef]
44. Brito, V.D.; Achimón, F.; Zunino, M.P.; Pizzolitto, R.P. Control of *Fusarium verticillioides* in Maize Stored in Silo Bags with 1-Octyn-3-Ol. *J. Stored Prod. Res.* **2024**, *106*, 102279. [CrossRef]

 https://doi.org/10.3390/plants15010048

45. Cano, M.C.; Beato, M.; Usseglio, V.L.; Merlo, C.; Zunino, M.P. Bioactive Paints with Volatile Organic Alcohols for the Control of *Sitophilus zeamais*. *J. Stored Prod. Res.* **2024**, *109*, 102423. [CrossRef]
46. Schultz, T.W.; Yarbrough, J.W. Trends in Structure–Toxicity Relationships for Carbonyl-Containing α,β-Unsaturated Compounds. *SAR QSAR Environ. Res.* **2004**, *15*, 139–146. [CrossRef]
47. Chang, R. *Química*, 9th ed.; Mc Graw Hill Interamericana: Columbus, OH, USA, 2007.
48. Dambolena, J.S.; Zygadlo, J.A.; Rubinstein, H.R. Antifumonisin Activity of Natural Phenolic CompoundsA Structure–Property–Activity Relationship Study. *Int. J. Food Microbiol.* **2011**, *145*, 140–146. [CrossRef]
49. Bohman, B.; Nordlander, G.; Nordenhem, H.; Sunnerheim, K.; Borg-Karlson, A.-K.; Unelius, C.R. Structure–Activity Relationships of Phenylpropanoids as Antifeedants for the Pine *Weevil hylobius* Abietis. *J. Chem. Ecol.* **2008**, *34*, 339–352. [CrossRef]
50. Struck, S.; Schmidt, U.; Gruening, B.; Jaeger, I.S.; Hossbach, J.; Preissner, R. Toxicity vs Potency: Elucidation of Toxicity Properties Discriminating Between Toxins, Drugs, and Natural Compounds. *Genome Inform.* **2008**, *20*, 231–242. [CrossRef]
51. Min, B.; Gu, L.; McClung, A.M.; Bergman, C.J.; Chen, M. Free and Bound Total Phenolic Concentrations, Antioxidant Capacities, and Profiles of Proanthocyanidins and Anthocyanins in Whole Grain Rice (*Oryza sativa* L.) of Different Bran Colours. *Food Chem.* **2012**, *133*, 715–722. [CrossRef]
52. Zaio, Y.P.; Gatti, G.; Ponce, A.; Saavedra Larralde, N.A.; Martinez, M.J.; Zunino, M.P.; Zygadlo, J.A. Cinnamaldehyde and Related Phenylpropanoids, Natural Repellents, and Insecticides Against *Sitophilus zeamais* (Motsch.). A Chemical Structure-Bioactivity Relationship. *J. Sci. Food Agric.* **2018**, *98*, 5822–5831. [CrossRef] [PubMed]
53. Chaubey, M.K.; Kumar, N. Role of Carvacrol and Menthone in Maize Weevil *Sitophilus zeamais* (Coleoptera: Curculionidae) Management. *Eur. J. Biol. Res.* **2023**, *13*, 181–190. [CrossRef]
54. Rodríguez, A.; Beato, M.; Usseglio, V.L.; Camina, J.; Zygadlo, J.A.; Dambolena, J.S.; Zunino, M.P. Phenolic Compounds as Controllers of *Sitophilus zeamais*: A Look at the Structure-Activity Relationship. *J. Stored Prod. Res.* **2022**, *99*, 102038. [CrossRef]
55. Saad, M.M.G.; Abou-Taleb, H.K.; Abdelgaleil, S.A.M. Insecticidal Activities of Monoterpenes and Phenylpropenes Against *Sitophilus oryzae* and Their Inhibitory Effects on Acetylcholinesterase and Adenosine Triphosphatases. *Appl. Entomol. Zool.* **2018**, *53*, 173–181. [CrossRef]
56. Huang, Y.; Ho, S.H.; Lee, H.C.; Yap, Y.L. Insecticidal Properties of Eugenol, Isoeugenol and Methyleugenol and Their Effects on Nutrition of *Sitophilus zeamais* Motsch. (Coleoptera: Curculionidae) and *Tribolium castaneum* (Herbst) (Coleoptera: Tenebrionidae). *J. Stored Prod. Res.* **2002**, *38*, 403–412. [CrossRef]
57. Zhang, C.; Liu, H.; Wang, X.; Long, X.; Huang, A.; Zhang, J.; Geng, J.; Yang, L.; Huang, Z.; Dong, P.; et al. Inhibitory Effects and Mechanisms of Cinnamaldehyde Against *Fusarium oxysporum*, a Serious Pathogen in Potatoes. *Pest Manag. Sci.* **2024**, *80*, 3540–3552. [CrossRef]
58. Saad, M.M.G.; Abdelgaleil, S.A.M. Effectiveness of Monoterpenes and Phenylpropenes on *Sitophilus oryzae* L. (Coleoptera: Curculionidae) in Stored Wheat. *J. Asia. Pac. Entomol.* **2018**, *21*, 1153–1158. [CrossRef]
59. Chaubey, M.K. Insecticidal Activities of Natural Volatile Compounds Against Pulse Beetle, Callosobruchus Chinensis (Bruchidae). *Acta Sci. Biol. Sci.* **2024**, *46*, e68787. [CrossRef]
60. Beament, J.W.L. The Waterproofing Mechanism of Arthropods. *J. Exp. Biol.* **1959**, *36*, 391–422. [CrossRef]
61. Neven, L.G. Physiological Responses of Insects to Heat. *Postharvest Biol. Technol.* **2000**, *21*, 103–111. [CrossRef]
62. Oladipupo, S.O.; Wilson, A.E.; Hu, X.P.; Appel, A.G. Why Do Insects Close Their Spiracles? A Meta-Analytic Evaluation of the Adaptive Hypothesis of Discontinuous Gas Exchange in Insects. *Insects* **2022**, *13*, 117. [CrossRef]
63. Leslie, J.; Summerel, B.A. Species Concepts in Fusarium. In *Fusarium Laboratory Manual*; John Wiley & Sons: Hoboken, NJ, USA, 2006; pp. 88–95. ISBN 9780470278376.
64. Dambolena, J.S.; López, A.G.; Cánepa, M.C.; Theumer, M.G.; Zygadlo, J.A.; Rubinstein, H.R. Inhibitory Effect of Cyclic Terpenes (Limonene, Menthol, Menthone and Thymol) on *Fusarium verticillioides* MRC 826 Growth and Fumonisin B1 Biosynthesis. *Toxicon* **2008**, *51*, 37–44. [CrossRef]
65. Usseglio, V.L.; Dambolena, J.S.; Merlo, C.; Peschiutta, M.L.; Zunino, M.P. Insect-Corn Kernel Interaction: Chemical Signaling of the Grain and Host Recognition by *Sitophilus zeamais*. *J. Stored Prod. Res.* **2018**, *79*, 66–72. [CrossRef]
66. Peng, S.; Zhang, H.; Xu, R.; Chang, X.; Zhou, Z.; Yang, Y.; Xiang, H.; Li, Y. Inhibitory Effect of Eugenol on *Fusarium oxysporum* and Transcriptome Analysis. *Tob. Sci. Technol.* **2025**, *58*, 38. [CrossRef]
67. Servicio Meteorológico Nacional Servicio Meteorológico Nacional. Available online: https://www.smn.gob.ar/ (accessed on 23 July 2020).
68. Pizzolitto, R.P.; Jacquat, A.G.; Usseglio, V.L.; Achimón, F.; Cuello, A.E.; Zygadlo, J.A.; Dambolena, J.S. Quantitative-Structure-Activity Relationship Study to Predict the Antifungal Activity of Essential Oils Against *Fusarium verticillioides*. *Food Control* **2020**, *108*, 106836. [CrossRef]
69. Herrera, J.M.; Zunino, M.P.; Dambolena, J.S.; Pizzolitto, R.P.; Gañan, N.A.; Lucini, E.I.; Zygadlo, J.A. Terpene Ketones as Natural Insecticides Against *Sitophilus zeamais*. *Ind. Crops Prod.* **2015**, *70*, 435–442. [CrossRef]

70. Di Rienzo, J.A. Navure Professional+ Version 3.0. 2024. Available online: https://www.navure.com/inicio/ (accessed on 23 November 2025).
71. FAO-ONU Las Interacciones En La Agroecología. 2016. Available online: http://www.fao.org/documents/card/en/c/8a99dfd7-e8e1-452f-b61b-abb1966887b4/ (accessed on 23 November 2025).
72. Rodríguez, D.A. Interacciones Bióticas de Los Agroecosistemas del Semiárido Bonaerense. Ph.D. Thesis, Universidad Nacional del Sur, Bahía Blanca, Argentina, 2018.
73. Mossa, A.T.H. Green Pesticides: Essential Oils as Biopesticides in Insect-Pest Management. *J. Environ. Sci. Technol.* **2016**, *9*, 354–378. [CrossRef]
74. Isman, M.B. Botanical Insecticides in the Twenty-First Century-Fulfilling Their Promise? *Annu. Rev. Entomol.* **2020**, *65*, 233–249. [CrossRef]
75. Isman, M.B. Plant Essential Oils for Pest and Disease Management. In *Crop Protection*; Elsevier: Amsterdam, The Netherlands, 2000; Volume 19, pp. 603–608, ISBN 1604822864.
76. Gupta, S.; Dikshit, A.K. Biopesticides: An Ecofriendly Approach for Pest Control. *J. Biopestic.* **2010**, *3*, 186–188.
77. Pavela, R.; Benelli, G. Essential Oils as Ecofriendly Biopesticides? Challenges and Constraints. *Trends Plant Sci.* **2016**, *21*, 1000–1007. [CrossRef] [PubMed]

https://doi.org/10.3390/plants15010048

plants

MDPI

Article

Antimicrobial Potential of Six Plant Essential Oils Against *Pseudomonas syringae* pv. *actinidiae*: *In Vitro* Activity and *In Planta* Efficacy Do Not Always Align

Marta Nunes da Silva [1,2], Miguel G. Santos [1], Marta W. Vasconcelos [2] and Susana M. P. Carvalho [1,*]

[1] GreenUPorto—Sustainable Agrifood Production Research Centre/Inov4Agro, Departamento de Geociências, Ambiente e Ordenamento do Território (DGAOT), Faculty of Sciences, University of Porto, Rua da Agrária 747, 4485-646 Vila do Conde, Portugal; mansilva@ucp.pt (M.N.d.S.); mgsantos@fc.up.pt (M.G.S.)

[2] CBQF—Centro de Biotecnologia e Química Fina—Laboratório Associado, Escola Superior de Biotecnologia, Universidade Católica Portuguesa, Rua Diogo Botelho 1327, 4169-005 Porto, Portugal; mvasconcelos@ucp.pt

* Correspondence: susana.carvalho@fc.up.pt

Abstract: Plant essential oils (EOs) are attracting interest as ecofriendly alternatives to antibiotics and copper-based control of kiwifruit bacterial canker (KBC), caused by *Pseudomonas syringae* pv. *actinidiae* (*Psa*). This study chemically profiled six EOs (anise, basil, cardamom, cumin, fennel, and laurel) and evaluated their antimicrobial activity both *in vitro* and *in planta*. The *in vitro* assay targeted four strains, two of *Psa* and two of the low-virulent *P. syringae* pv. *actinidifoliorum* (*Pfm*), whereas the *in planta* assay focused on the highly virulent *Psa*7286 strain, assessed under preventive and curative application regimes (i.e., 14 days pre- or post-inoculation, respectively). Cumin, with cuminaldehyde as its major component (48%), was the most effective EO in vitro, significantly inhibiting growth at 5–10% concentration, whereas anise, rich in anethole (89%), was consistently the least effective one. However, the *in planta* application of the EOs produced antimicrobial effects that differed markedly from *in vitro* results and showed strong dependence on the timing of application. Preventive treatment significantly reduced *Psa* endophytic populations in basil (70%), anise (54%), laurel (42%), and cumin (35%) compared to untreated plants. In contrast, when the EOs were applied post-inoculation (curative treatment), a significant decrease in *Psa* colonization was observed in laurel (81%), cardamon (70%), cumin (31%) and fennel (29%). Although plant EOs are gaining momentum in the control of *Psa* and other diseases, translation from *in vitro* to *in planta* efficacy is not direct and is strongly timing-dependent, which underscores the need to perform validation trials *in planta* and to fine-tune application schedules for the integrated management of KBC.

Keywords: antimicrobial activity; crop protection; kiwifruit bacterial canker; *Pseudomonas syringae* pv. *actinidifoliorum*

1. Introduction

Kiwifruit bacterial canker, caused by *Pseudomonas syringae* pv. *actinidiae* (*Psa*), is a major constraint to plant growth and productivity, driving substantial economic losses and demanding rigorous orchard management [1]. This pathogen is now reported from all major kiwifruit-producing regions and can infect virtually all *Actinidia* species, including cultivated *A. chinensis* and *A. arguta* [2]. By contrast, *P. syringae* pv. *actinidifoliorum* (*Pfm*) is less virulent; although it does not typically cause severe yield losses, it compromises plant fitness through necrotic leaf spots [3,4]. Control of both Gram-negative pathovars

https://doi.org/10.3390/plants14243825

is important for crop performance, yet research attention has focused predominantly on *Psa* due to its global impact on kiwifruit health [5]. These efforts have delivered several candidate control strategies, including plant defense inducers and microbial biological control agents, at varying levels of technological maturity, with few products currently commercialized [6]. In practice, management has relied largely on copper-based compounds and, where permitted, antibiotics such as streptomycin. However, their continued use is increasingly discouraged due to limited and inconsistent efficacy and to environmental and human health concerns [7–10]. Despite incremental advances, robust and reliable control of *Psa* remains elusive, and antibacterial options for *Pfm* have received comparatively little attention [11]. In fact, closely related nonpathogenic or mildly virulent populations, such as *Pfm*, can act as reservoirs of virulence factors, enabling the emergence of more aggressive variants [5]. There is, therefore, a clear need to broaden the portfolio of safe and effective tools for kiwifruit disease management. In particular, molecules that prevent infection and/or lower endophytic bacterial loads in already infected plants are a priority for integrated programs [6].

Plant essential oils (EO) are promising candidates because they combine direct antimicrobial activity with the capacity to elicit plant defense responses and may also enhance copper efficacy when used in nanoformulations [6,12–14]. These complex mixtures can include substances from tens of chemical groups, with one or a few bioactive constituents often accounting for most of the composition [15–17]. Reported antibacterial mechanisms include disruption of the cell envelope that increases membrane permeability and leads to leakage of cellular contents, as well as interference with motility, the type III secretion system, and quorum sensing [18–20]. Many essential oils and their constituents are active at low concentrations, a practical advantage since phytotoxicity and poor water miscibility often necessitate low-concentration emulsions stabilized with surfactants [21,22]. However, *in planta* evaluations of EOs have focused mainly on fungal and oomycete diseases, while bacterial targets are less studied. Seed and soil treatments with plant EOs have reduced disease caused by diverse fungi in several horticultural crops, although effective concentrations, carriers, and timings vary [23–28]. For example, seed priming with EOs from cumin, basil, and geranium at 4% reduced root rot in cumin caused by *Fusarium* spp. [23], while seed priming with oregano EO at 12% also lowered the severity of *Sclerotinia sclerotiorum* in lima bean [28]. In potted tomato seedlings, oregano and clove EOs at concentrations up to 0.01% significantly decreased symptoms caused by *Botrytis cinerea* and *Ralstonia solanacearum*, respectively [24,25]. Likewise, clove EO emulsions at up to 10% limited disease caused by *Fusarium oxysporum* f. sp. *lycopersici* in potted tomato seedlings [26]. In greenhouse trials, a soil drench with lemon EO at 50 mL per plant reduced symptom severity caused by *Phytophthora* sp. on pepper, cucumber, and melon relative to untreated controls [27]. More recently, garlic and cinnamon essential oils have been encapsulated in silver nanoparticles, which enhanced their fungicidal activity against Botrytis cinerea, a major pathogen of horticultural crops [29]. The nanoformulated oils inhibited mycelial growth and conidial germination and caused cell wall disruption and deformed hyphae. In addition, chitosan nanoparticles loaded with lemon or spinach seed essential oils showed strong antifungal activity against *Penicillium expansum* (blue mold of apple) and *Podosphaera fusca* (powdery mildew of cucumber), while inducing defense-related and antioxidant enzymes, including polyphenol oxidase, peroxidase and phenylalanine ammonia lyase, without compromising fruit quality [30,31].

Most work on *Psa* has evaluated the antibacterial activity of PEOs and plant extracts *in vitro* [15–17,19], whereas *in planta* efficacy against *Psa* remains limited [13,20,32]. In kiwifruit, foliar applications of EOs from savory, thyme, and pennyroyal have shown inconsistent outcomes across studies [13,32]. Under greenhouse conditions, applying savory and thyme EOs 24 h before *Psa* inoculation, followed by nine further applications over 15 weeks, reduced both the number of diseased leaves and the affected leaf area. In contrast, in micropropagated plants treated 3 days before inoculation, winter savory at 5 mg.mL^{-1} produced negligible inhibition relative to pennyroyal [13]. Also, cinnamon EO-based emulsion markedly decreased disease frequency and severity, although protection was weaker when cinnamon was combined with oregano and when treatment was applied 24 h after inoculation rather than 24 h before [20]. Under field conditions, cinnamon EO applied at bloom and then at three monthly intervals was more effective than cinnamon or a cinnamon plus oregano mixture in lowering the disease index for up to four months after treatment. However, when oregano was combined with cinnamon, the mixture performed worse than cinnamon alone, indicating that oregano can diminish the protective effect of cinnamon [20]. Together with the weaker protection observed when cinnamon was applied after inoculation rather than before, these results point to complex, non-additive interactions among EOs and emphasize that mixture composition and application timing critically shape outcomes. In addition, variability across studies likely arises from differences in EO chemistry, the genotypes of both bacteria and host plants, and the formulation, concentration, application method and timing, and growth conditions used. These sources of heterogeneity underscore the need for standardized comparative trials that explicitly contrast preventive and curative schedules.

In this context, chemical profiling is essential to interpret bioassays, identify candidate bioactive constituents, and improve comparability across studies; however, many reports do not include full composition data. In several cases, chemical analyses of PEOs with stronger anti-*Psa* activity identified piperitenone oxide in apple mint, pulegone and menthone in pennyroyal, borneol in rosemary, caryophyllene in sage, camphene and cinnamyl acetate in laurel, and terpinene-4-ol in tea tree as the most abundant constituents [13,16]. Individual compounds such as carvacrol and juglone also showed strong inhibition of *Psa* by disrupting cell membrane permeability and integrity [33,34], and cinnamaldehyde, eugenol, estragole, and methyl-eugenol also inhibited in vitro, with efficacy varying across *Psa* strains [15]. Yet, focusing on single constituents does not fully explain whole-oil performance because constituent interactions can be synergistic or antagonistic. For example, eugenol, estragole, and methyl-eugenol exhibited high inhibitory activity, but extracts of West Indian Bay tree and Jamaica pepper, both rich in these molecules, were only as effective as Chinese cinnamon, where these compounds were not detected [15]. Similarly, although bergamot oil contained more thymol than a closely related wild species (59–64% versus 38–42%), their anti-*Psa* activities were comparable, likely due to carvacrol present in wild bergamot (at about 3.9%) [17]. These patterns underline that overall antimicrobial potential reflects the full chemical matrix rather than any single dominant component. In addition, the potential of plant EOs against *Pfm* remains largely unexplored. Although this pathovar is less destructive than *Psa*, its management is relevant in orchards where multiple *Pseudomonas* pathovars may co-occur. Moreover, despite the recent withdrawal of the European Commission's proposal to halve pesticide use by 2030 under the "Farm to Fork Strategy" due to lack of political consensus, the need to develop sustainable alternatives remains urgent. In this regard, incorporating plant EOs as both preventive and curative control strategies against KBC will contribute to achieving this objective. As such, this study tests the hypothesis that EOs effective *in vitro* against *Psa* and *Pfm* strains will also exhibit activity *in planta* against *Psa*, with their performance being conditioned by the application timing. Six widely used spice and herb oils—anise, cumin, fennel, basil, laurel, and cardamom—were therefore evaluated to: (i) quantify *in vitro* antibacterial activity against both pathovars; (ii) compare the sensitivity of strains differing in virulence; (iii) validate preventive and post-infection efficacy *in planta* against *Psa*7286; and (iv) characterize chemical composition to relate constituents to antimicrobial performance and to identify candidates for future management tools.

2. Results

2.1. In Vitro and In Planta Antimicrobial Activity Against Psa and Pfm

Across the four strains and six essential oils, *in vitro* inhibition increased with oil concentration, with strong concentration–response relationships (R^2 = 0.67–0.99). The exception was anise against *Psa*7286 and Pfm18804, which showed weaker fits (R^2 = 0.50 and 0.42, respectively; Supplementary Materials Table S1). Overall, cumin was the most effective in vitro, producing large inhibition halos at comparatively low concentrations (Figure 1).

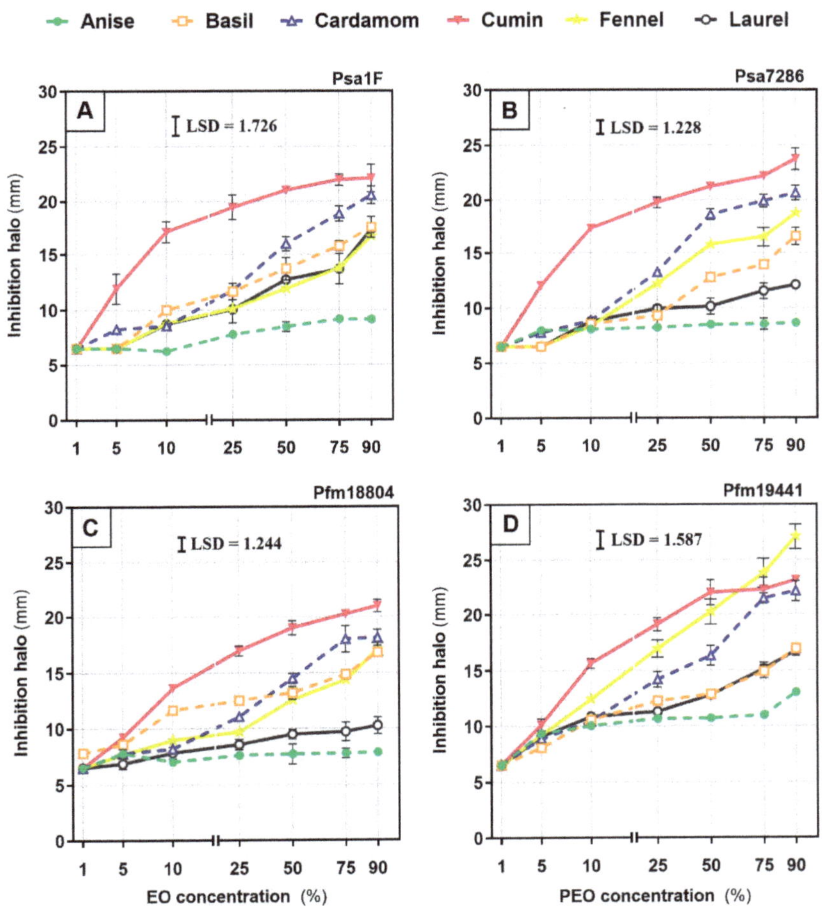

Figure 1. Inhibition zone diameter (mm) of *Pseudomonas syringae* pv. *actinidiae* (*Psa*; strains 1F—(**A**), and 7286—(**B**)) and *P. syringae* pv. *actinidifoliorum* (*Pfm*; strains 18804—(**C**), and 19441—(**D**)) measured after 48 h exposure to six essential oils: anise, basil, cardamom, cumin, fennel, and laurel. Oils were tested at 1, 5, 10, 25, 50, 75 and 90% v/v. Each symbol is the mean of three biological replicates, and error bars indicate the standard error of the mean. The least significant difference (LSD) at $p < 0.05$ is shown in each panel.

Cumin generally yielded inhibition halos from 9.2 ± 0.6 mm at 5% to 23.7 ± 1.0 mm at 90% across all strains, with fennel significantly outperforming cumin against *Pfm*19441 at 90% (reaching 27.1 ± 1.1 mm). In contrast, anise was consistently the least active EO, with the smallest halos even at 90%, ranging from 8.7 ± 0.4 mm for *Psa*7286 to 13.0 ± 0.2 mm for *Pfm*19441. Basil, cardamom, and laurel showed intermediate activity. At 5%, their halos reached up to 8.6 ± 0.2 mm, 8.8 ± 0.4 mm, and 9.1 ± 0.2 mm, respectively, and at 90% they reached 17.5 ± 1.0 mm, 22.1 ± 0.9 mm, and 17.3 ± 0.3 mm, respectively. However, some strain-specific differences were evident, including within pathovars. For example, at 90%, among *Psa* strains, *Psa*1F was more sensitive to laurel than *Psa*7286 ($p < 0.0001$) but less sensitive to fennel ($p < 0.01$), and among *Pfm* strains, *Pfm*19441 was more sensitive than *Pfm*18804 to anise ($p < 0.0001$), cardamom and cumin ($p < 0.05$), fennel and laurel ($p < 0.01$).

Concerning the *in planta* trials, in untreated inoculated plants (control), the endophytic population of *Psa* increased over time, reaching 15 ± 4.0 × 10^7 colony-forming units per gram (CFU·g^{-1}) of shoot tissue at 14 days post inoculation (dpi) and 36 ± 6.2 × 10^7 CFU·g^{-1} at 28 dpi (Figures 2B and 2C, respectively).

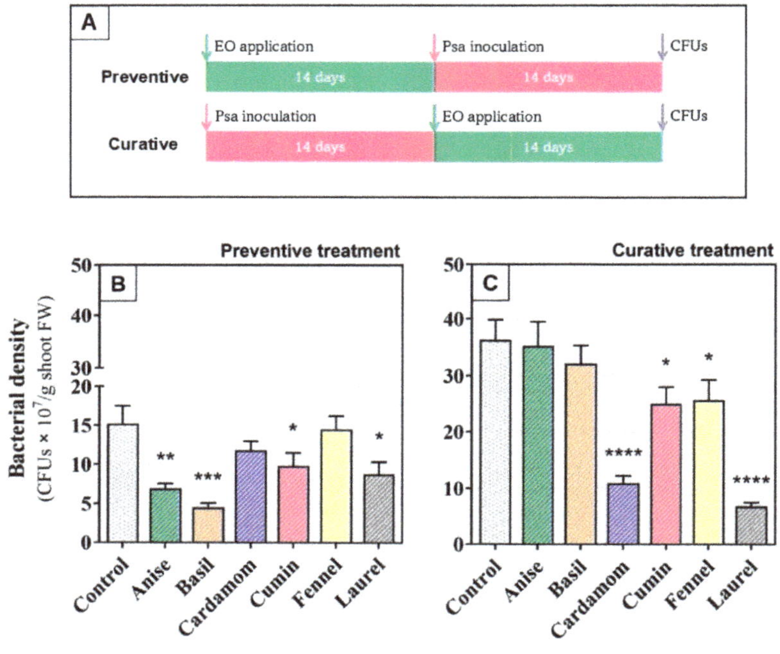

Figure 2. (**A**) In planta application of six plant essential oils at 0.1% (*v*/*v*), 14 days pre-inoculation (preventive treatment) or 14 days post-inoculation (curative treatment) and the respective (**B**,**C**) endophytic population of *Pseudomonas syringae* pv. *actinidiae* in shoots of *Actinidia chinensis* var. *deliciosa* 'Tomuri' expressed as CFU·g^{-1} fresh weight. Control plants were inoculated but received no essential oil treatment. Bars represent the means of three biological replicates, and error bars indicate the standard error of the means. Different letters denote significant differences among treatments within each panel (ANOVA with Fisher's LSD, $p < 0.05$). Significance levels are indicated as follows: ****, $p < 0.0001$; ***, $p < 0.001$; **, $p < 0.01$; *, $p < 0.05$.

Preventive application of EOs significantly reduced *Psa* colonization for basil (70%), anise (54%), laurel (42%), and cumin (35%), relative to the untreated control, whereas cardamom and fennel produced no significant change (Figure 2B). In contrast, when EOs were applied as curative treatment against *Psa* basil and anise had no significant effect, whereas laurel, cardamom, cumin, and fennel lowered *Psa* loads by 81, 70, 29, and 31%, respectively (Figure 2C). Overall, anise and basil only worked out as preventive treatments, whereas cardamom and fennel were only effective as curative treatments. Interestingly, cumin and laurel had broader action, with significant efficacy in both preventive and curative application regimes.

2.2. Chemical Characterization of the EOs

Chemical profiling revealed differing levels of complexity among the studied EOs (Table 1, Supplementary Materials). Anise, basil and cardamom contained the largest number of identified constituents, from 23 to 24 compounds, whereas fennel, cumin and laurel had 18 to 20. A closer look at relative abundances highlights distinct chemical signatures (Table 1).

Table 1. Most abundant constituents of anise, basil, cardamom, cumin, fennel, laurel essential oils shown as relative composition percent from GC-MS analysis. Remaining identified compounds are listed in Supplementary Material. Abbreviations: ND—not detected.

Constituents	Anise	Basil	Cardamom	Cumin	Fennel	Laurel
Anethole	88.89	1.74	0.18	0.14	ND	78.02
3-Carene	ND	0.03	2.23	13.77	1.46	0.70
1,8-Cineole *	ND	3.19	0.14	3.39	68.71	ND
Cuminaldehyde	ND	ND	ND	47.95	ND	0.08
β-Cymene	ND	ND	1.20	18.67	1.18	1.10
Estragole	2.67	93.28	0.46	0.53	0.45	6.66
Fenchone	ND	0.10	ND	ND	0.03	9.08
β-Himachalene	0.37	ND	77.72	ND	17.15	0.07
D-Limonene	ND	0.20	7.42	0.40	2.84	1.84
Linalyl acetate	0.12	0.32	3.77	0.34	1.83	0.02
Longifolene	3.76	ND	ND	ND	ND	ND

* Also known as Eucalyptol.

Anise showed the highest levels of anethole (89%), while basil was characterized by very high estragole (93%). Cardamom was rich in β-himachalene (78%), with lower amounts of D-limonene (7.4%), and linalyl acetate (3.8%). Cumin showed elevated cuminaldehyde (48%), β-cymene (19%), and 3-carene (14%). Fennel was dominated by 1,8-cineole (69%; also known as eucalyptol), with β-himachalene (17%), bornyl acetate (1.7%), terpinen-4-ol and γ-terpinene (both at 1.0%). Laurel contained substantial anethole (78%), fenchone (9.1%), and estragole (6.7%). Despite these differences, all EOs shared four compounds: estragole (0.44% to 93%), linalyl acetate (0.02% to 3.8%), and terpinen-4-ol (0.02% to 1.0%) and ϱ-cymene (0.01% to 0.89%). On the other hand, several constituents were exclusive to a single EO. For example, longifolene, thunbergol and γ-himalachene were only detected in anise (at 3.8, 1.5, and 0.55%, respectively), β-isocumene and nerol acetate in cardamom (at 1.2 and 1.1%, respectively), and chrysanthenol and phellandral were detected only in cumin (at 9.2 and 0.69%, respectively).

3. Discussion

Kiwifruit bacterial canker caused by *Psa* requires a diversified toolkit of sustainable control molecules that can limit endophytic colonization and help manage the emergence of chemical resistance [9,10]. Although *Pfm* is less virulent and typically poses limited direct economic risk, coinfection could weaken plant fitness and potentially exacerbate the impact of *Psa* where both occur [35]. Within this context, the present study contributes to identify sustainable options by expanding our knowledge on the list of EOs with activity against *Psa* and, for the first time, exploring potential inhibitory effects against *Pfm*.

In the In vitro assays, all tested essential oils significantly inhibited both pathovars in a predominantly concentration-dependent manner. Nevertheless, cumin was consistently the most active EO, while anise showed the weakest activity. Overall, several oils showed steep increases in activity at low concentrations (≤10%), which is encouraging for practical use since low doses facilitate emulsification and reduce phytotoxic risk [22]. In this context, microencapsulation and other nanotechnology-based delivery systems could improve stability and provide controlled release, enabling equivalent or greater protective effects at even lower application rates and thereby reducing the total oil required [36,37]. among the oils with intermediate *in vitro* effects, fennel was the most effective against *Pfm*19441 at 75% and 90% concentrations. Nonetheless, for *Pfm*18804 and at all other concentrations, cumin remained the most active EO. Such variation, within and between pathovars, aligns with previous reports and argues for tailoring oil selection to pathogen lineage or using rational combinations that target multiple mechanisms simultaneously [15]. Indeed, synergistic activity from subinhibitory combinations has been shown for rosemary, tea tree, and apple mint against *Psa* [16] and, therefore, systematic testing across strain panels will be needed to map predictable response patterns and anticipate limitations. Our results substantiate prior evidence that basil, cumin, fennel, and laurel can inhibit *Psa in vitro* [16,19] and newly document activity for anise and cardamom. Earlier reports of no cardamom effect against *Psa* may reflect differences in chemotype or methodology, as composition was not disclosed [19]. Here, cardamom shared some constituents with published profiles, such as D-limonene and linalyl acetate, while being rich in β-himachalene, a feature also linked to antibacterial and antioxidant activity in Atlas cedar oil [38,39]. Conversely, discrepant findings for rosemary across studies illustrate how strain choice, plant chemotype, extraction method, concentration, and growth conditions can shift outcomes, underscoring the value of meta-analyses to resolve patterns and optimize use conditions [40–42]. Crucially, this study extends evidence from plates to plants. We show that cumin, fennel, basil, cardamom, laurel, and anise can reduce endophytic *Psa* loads *in planta*, but efficacy depends on timing. Preventive applications favored basil and anise, curative applications favored laurel and cardamom, while cumin performed in both windows, albeit with different magnitudes.

The mismatch between *in vitro* activity and *in planta* protection in some cases can be partly explained by additional barriers and host responses that govern outcomes in living tissues. At plant level the EOs can trigger defense mechanisms (e.g., improved antioxidant response and phytohormone modulation) that go beyond a direct pathogen growth inhibition [13,43]. Composition activity relationships help interpret these patterns but are not strictly additive. For instance, oils active against *Psa* often contain piperitenone oxide, pulegone, menthone, borneol, caryophyllene, camphene, cinnamyl acetate, and terpinene-4-ol, and single compounds such as carvacrol and juglone can disrupt membranes and strongly inhibit growth [13,16,33,34]. Yet interactions among constituents can be synergistic or antagonistic. Extracts rich in eugenol, estragole, and methyl eugenol were as effective as Chinese cinnamon extracts that lacked these compounds, and similar activity has been observed in oils with differing thymol and carvacrol proportions [15,17]. In this study, estragole was one of the few constituents present across all six oils, alongside very low amounts of ϱ-cymene and terpinene-4-ol. The three oils with the highest estragole, basil then laurel then anise, were also the most effective in preventive applications, which suggests a role for elicitation. Estragole-rich matrices have primed resistance in several species, mostly against fungal pathogens, and multiple essential oils have enhanced host defenses *in planta*, although validation for *Actinidia-Psa* is still needed [12,44–48]. Under our experimental conditions, a purely surface protective effect seems less likely, given the high volatility of most constituents and the 14-day interval between application and inoculation. However, less volatile EO components, such as fatty acids, sterols and waxes, may persist for longer and could be retained by the surfactant used for emulsification [22,49].

Cardamom and laurel were more effective when applied after infection (as a curative approach), with cardamom showing limited preventive effects. Although their dominant constituents differ, β-himachalene in cardamom and 1,8-cineole in fennel, both have recognized antibacterial activities, and the overall activity could also reflect interactions among minor constituents [21,39,50]. In contrast, laurel and cumin reduced *in planta* bacterial density in both timing windows. For cumin, this dual action may derive from a diversified profile rather than a single driver. Cumin contained five constituents above 5%, with cuminaldehyde as the major component. Cuminaldehyde inhibits biofilm formation and proteolysis, reduces virulence, and increase membranes permeability in bacteria, with cumin oils rich in 3-caren-10-al and cuminaldehyde showing strong antioxidant capacity that could support plant defenses [51–54]. Similarly, cinnamaldehyde, the dominant component of *Cinnamomum cassia* oil, also targets bacterial membranes, and it has also been implicated in elicitation in apple leaves [15,55–57]; however, whether cumin also elicits defenses in *Actinidia* remains to be tested. Laurel and anise, both dominated by anethole, behaved similarly *in vitro* against *Psa*7286 and as preventive treatments, consistent with the antioxidant properties of anethole that may aid host defenses [58]. Nevertheless, their protective behavior diverged: laurel exhibited strong curative activity while anise did not, reflecting differences in their chemical profiles. For example, laurel contained higher levels of estragole plus fenchone and D-limonene, which were absent in anise. Although these compounds typically exhibit limited activity against Gram-negative bacteria, estragole has documented bactericidal effects on *Psa* and likely contributed to laurel's superior curative efficacy [15,59–63].

Taken together, these results highlight that: (1) essential oil efficacy is timing dependent *in planta*, so preventive and curative windows should be tested explicitly; (2) composition alone does not dictate performance because interactions among constituents are common; selecting a single molecule based only on relative abundance is risky, and (3) cumin and laurel emerge as robust candidates against both *Psa* e and *Pfm*, while basil and anise appear most useful preventively and cardamom and fennel curatively. Future work should map strain level response surfaces, quantify elicitation markers in *Actinidia* following oil application, assess formulation variables that improve persistence without phytotoxicity, and test rational combinations at subinhibitory doses to exploit synergy while meeting practical constraints in kiwifruit orchards.

4. Materials and Methods

4.1. Essential Oil Emulsions

Plant essential oils from anise (*Pimpinella anisum* L., Ref.: AT155), basil (*Ocimum basilicum* L., Ref.: AT324), cardamom (*Elettaria cardamomum* L. Maton, Ref.: AT056), cumin (*Cuminum cyminum* L., Ref.: AT057), fennel (*Foeniculum vulgare* Mill., Ref.: AT111) and laurel (*Laurus nobilis* L., Ref.: AT116) were purchased from H. Reynaud & Fils, Montbrun Les Bains, France. For each oil, a 90% (*v*/*v*) stock emulsion was prepared by mixing the pure, undiluted oil with sterile distilled water containing 2% (*v*/*v*) Tween 20 as emulsifier. To control for surfactant effects, matching Tween 20 solutions without essential oil were prepared at the corresponding final concentrations and used as non-treated controls. All emulsions were shaken for 30 s immediately before being used to ensure homogeneity.

4.2. Bacterial Strains and In Vitro Antibacterial Assay

Two *Psa* strains, 7286 (CFBP, Italy) and 1F (ANSES, France), and two *Pfm* strains, 18,804 and 19,441 (both ICMP, New Zealand and Australia, respectively), were used [2,64]. Along the text, strains are referred to as *Psa*7286, *Psa*1F, *Pfm*18804 and *Pfm*19441. Cultures were maintained on nutrient sucrose agar (NSA) at 27 °C in the dark. For inoculum, a single colony of each strain was grown overnight in liquid Luria–Bertani medium at 27 °C and 75 rpm.

Antibacterial activity was assessed on NSA plates by paper disk diffusion [65]. Bacterial suspensions were adjusted to OD_{600} = 1.0, approximately 1×10^9 cells.mL^{-1}, and spread onto separate plates. Sterile blank antimicrobial susceptibility disks (Thermo Fisher Scientific, MA, USA) were loaded with 10 µL of each essential oil emulsion or the corresponding Tween 20 control and placed at the center of inoculated plates. Plates were incubated at 27 °C for 48 h in the dark. Zones of growth inhibition were measured as the diameter of the inhibition halo, in millimeters, and compared with the negative controls. Each treatment was tested with three biological replicates, each comprising three technical replicates.

4.3. *In Planta Validation of Antibacterial Activity Against Psa*

Micropropagated *Actinidia chinensis* var. *deliciosa* 'Tomuri' plants, each with a single shoot 5 to 6 cm tall and 8 to 12 leaves, were obtained from QualityPlant (Castelo Branco, Portugal). Plants were maintained *in vitro* on modified full-strength Murashige and Skoog agar medium in a climate chamber as previously described [4].

Plants received essential oil treatments either 14 days before inoculation (preventive) or 14 days after inoculation (curative), as schematically depicted in Figure 2A. For each timing, nine plants per essential oil were treated. An additional set of nine plants per timing received 2% Tween 20 without essential oil as non-treated controls, resulting in a total of 126 plants. Treatments consisted of dipping shoots for 15 s in 0.1% *v*/*v* essential oil emulsions prepared from the stocks in Section 2.1. This concentration was selected based on preliminary tests that revealed phytotoxicity at higher levels.

For inoculation, a fresh suspension of *Psa*7286 at 1–2 $\times 10^7$ CFU·mL^{-1} in sterile Ringer's solution was prepared on the day of challenge, which involved immersing plant shots in the inoculum for 15 s. After treatment in the preventive schedule or after inoculation in the curative schedule, plants were maintained for 28 days in a climate chamber (Fitoclima 5000 EH, Aralab, Rio de Mouro, Portugal) with a 16 h light photoperiod and a light intensity of 200 µmol.s^{-1}.m^{-2}, 22 °C during the light period and 20 °C during the dark period. Sampling occurred at 14 days post inoculation in the preventive experiment and at 28 days post inoculation in the curative experiment. For this, plants were removed from the growing media, and shoots were homogenized in Ringer's solution for endophytic *Psa* quantification by serial dilution and plating [4]. For each condition, three biological replicates were analyzed, each obtained by random pooling of three shoots.

4.4. *Chemical Characterization by GC–MS*

Essential oils were diluted in 10% ethanol (GC grade, Merck Group, Darmstadt, Germany). Samples were analyzed in triplicate on a Varian CP 3800 gas chromatograph with autosampler coupled to a Varian Saturn 4000 ion trap mass spectrometer, controlled by Varian software version 6.9.1. Separation used a VF 5 ms capillary column, 30 m $\times$ 0.25 mm $\times$ 0.25 µm, with high-purity helium as carrier at 1.0 mL.min^{-1} in splitless mode. The oven program was 40 °C for 1 min, ramp 5 °C min^{-1} to 250 °C with a 5 min hold, then 5 °C min^{-1} to 300 °C with no hold. Additional specifications followed Barros et al. [66]. Compounds were identified in a non-targeted approach by comparison with pure and mixed standards and by spectral matching against the NIST EPA NIH Mass Spectral Library version 2.2.

4.5. *Statistical Analysis*

Differences among means were evaluated by analysis of variance (ANOVA) followed by Fisher's least significant difference test at $p < 0.05$. Analyses were performed in GraphPad Prism version 10.4.1, GraphPad Software, Boston, MA, USA.

5. Conclusions

This study extends prior *in vitro* observations for basil, cumin, fennel and laurel by demonstrating their activity against *Psa* and, for the first time, showing inhibitory effects against *Pfm*. It also documents previously unreported activity for anise and cardamom. Crucially, *in planta* experiments show that EOs can reduce endophytic *Psa* populations, but that *in vitro* efficacy does not translate directly to *in planta* effectiveness and is strongly dependent on application timing. The results identify distinct functional roles among the oils tested: basil and anise were effective only as preventive treatments, cardamom showed curative efficacy, and cumin and laurel performed well in both preventive and curative contexts. Chemical profiling linked bioactivity to candidate constituents such as anethole, estragole, himachalene, cuminaldehyde and 1,8-cineole, while emphasizing that overall composition alone does not reliably predict biological outcome because interactions among constituents are common. Future work should focus on standardizing EO formulations and doses for field use to reduce reliance on copper and antibiotics while minimizing phytotoxicity. Key priorities include benchmarking preventive and curative windows across diverse *Psa* and *Pfm* strains and *Actinidia* genotypes, testing rational EO combinations at subinhibitory concentrations to identify synergistic blends, and quantifying host elicitation markers and physiological responses (transient and cumulative) to clarify mechanisms of action. Together, these steps will help translate the present findings into robust, scalable tools for kiwifruit disease management.

Supplementary Materials: The following supporting information can be downloaded at: https://www.mdpi.com/article/10.3390/plants14243825/s1.

Author Contributions: S.M.P.C. and M.W.V. were responsible for the conception and design of the experimental work. S.M.P.C. was responsible for funding acquisition that financed the bulk of this research. M.N.d.S. conducted the chemical profiling of plant essential oils and the *in vitro* assays, and M.N.d.S. and M.G.S. performed *in planta* treatment, inoculation and sampling. All authors have read and agreed to the published version of the manuscript.

Funding: This work was supported by National Funds from FCT—Fundação para a Ciência e a Tecnologia—through the projects PTDC/AGR-PRO/6156/2014, UID/5748/2025, UIDB/50016/2025, MGS's PhD scholarships (https://doi.org/10.54499/2020.08874.BD) and MNS's PhD scholarship (SFRH/BD/99853/2014) and CEEC (https://doi.org/10.54499/2023.06124.CEECIND/CP2855/CT0008).

Data Availability Statement: The original contributions presented in this study are included in the article/Supplementary Materials. Further inquiries can be directed to the corresponding author.

Acknowledgments: The authors gratefully acknowledge Bárbara Nascimento and Catarina Correia for the first *in vitro* screenings.

Conflicts of Interest: The authors declare that the research was conducted in the absence of any commercial or financial relationships that could be construed as a potential conflict of interest.

References

1. Froud, K.; Beresford, R.; Cogger, N. Risk factors for kiwifruit bacterial canker disease development in 'Hayward' kiwifruit blocks. *Austral. Plant Pathol.* **2017**, *46*, 421–431. [CrossRef]
2. Scortichini, M.; Marcelletti, S.; Ferrante, P.; Petriccione, M.; Firrao, G. *Pseudomonas syringae* pv. *actinidiae*: A re-emerging, multi-faceted, pandemic pathogen. *Mol. Plant Pathol.* **2012**, *13*, 631–640. [CrossRef]
3. Cunty, A.; Poliakoff, F.; Rivoal, C.; Cesbron, S.; Fischer-Le Saux, M.; Lemaire, C.; Jacques, M.A.; Manceau, C.; Vanneste, J.L. Characterization of *Pseudomonas syringae* pv. *actinidiae* (Psa) isolated from France and assignment of Psa biovar 4 to a de novo pathovar: *Pseudomonas syringae* pv. *actinidifoliorum* pv. nov. *Plant Pathol.* **2015**, *64*, 582–596. [CrossRef]
4. Nunes da Silva, M.; Vasconcelos, M.W.; Gaspar, M.; Balestra, G.M.; Mazzaglia, A.; Carvalho, S.M.P. Early pathogen recognition and antioxidant system activation contributes to *Actinidia arguta* tolerance against *Pseudomonas syringae* pathovars *actinidiae* and *actinidifoliorum*. *Front. Plant Sci.* **2020**, *11*, 1022. [CrossRef]

5. González, A.J.; Díaz, D.; Ciordia, M.; Landeras, E. Occurrence of *Pseudomonas syringae* pvs. *actinidiae*, *actinidifoliorum* and other *P. syringae* strains on kiwifruit in Northern Spain. *Life* **2024**, *14*, 208. [CrossRef]
6. Santos, M.G.; Nunes da Silva, M.; Vasconcelos, M.W.; Carvalho, S.M. Scientific and technological advances in the development of sustainable disease management tools: A case study on kiwifruit bacterial canker. *Front. Plant Sci.* **2024**, *14*, 1306420. [CrossRef] [PubMed]
7. Han, H.S.; Koh, Y.J.; Hur, J.S.; Jung, J.S. Occurrence of the strA-strB streptomycin resistance genes in *Pseudomonas* species isolated from kiwifruit plants. *J. Microbiol.* **2004**, *42*, 365–368.
8. Pietrzak, U.; McPhail, D.C. Copper accumulation, distribution and fractionation in vineyard soils of Victoria, Australia. *Geoderma* **2004**, *122*, 151–166. [CrossRef]
9. Colombi, E.; Straub, C.; Künzel, S.; Templeton, M.D.; McCann, H.C.; Rainey, P.B. Evolution of copper resistance in the kiwifruit pathogen *Pseudomonas syringae* pv. *actinidiae* through acquisition of integrative conjugative elements and plasmids. *Environ. Microbiol.* **2017**, *19*, 819–832.
10. Lee, Y.S.; Kim, G.H.; Koh, Y.J.; Jung, J.S. Mutation of *rpsL* gene in streptomycin-resistant *Pseudomonas syringae* pv. *actinidiae* biovar 3 strains isolated from Korea. *Res. Plant Dis.* **2022**, *28*, 26–31. [CrossRef]
11. Nunes da Silva, M.; Carvalho, S.M.P.; Rodrigues, A.M.; Gómez-Cadenas, A.; António, C.; Vasconcelos, M.W. Defence-related pathways, phytohormones and primary metabolism are key players in kiwifruit plant tolerance to *Pseudomonas syringae* pv. *actinidiae*. *Plant Cell Environ.* **2022**, *45*, 528–541. [CrossRef]
12. Banani, H.; Olivieri, L.; Santoro, K.; Garibaldi, A.; Gullino, M.L.; Spadaro, D. Thyme and savory essential oil efficacy and induction of resistance against *Botrytis cinerea* through priming of defense responses in apple. *Foods* **2018**, *7*, 11. [CrossRef]
13. Oliveira-Fernandes, J.; Oliveira-Pinto, P.R.; Mariz-Ponte, N.; Sousa, R.M.; Santos, C. *Satureja montana* and *Mentha pulegium* essential oils' antimicrobial properties against *Pseudomonas syringae* pv. *actinidiae* and elicitor potential through the modulation of kiwifruit hormonal defenses. *Microbiol. Res.* **2023**, *277*, 127490. [CrossRef] [PubMed]
14. Azarakhsh, M.; Gerivani, M. Antibacterial activity of chitosan/thyme essential oil/Copper nano-complexes against *Pseudomonas syringae* pv. *syringae*. *Med. Plant* **2025**, *48*, 1–11. [CrossRef]
15. Song, Y.-R.; Choi, M.-S.; Choi, G.-W.; Park, I.-K.; Oh, C.-S. Antibacterial activity of cinnamaldehyde and estragole extracted from plant essential oils against *Pseudomonas syringae* pv. *actinidiae* causing bacterial canker disease in kiwifruit. *Plant Pathol. J.* **2016**, *32*, 363. [CrossRef]
16. Vavala, E.; Passariello, C.; Pepi, F.; Colone, M.; Garzoli, S.; Ragno, R.; Pirolli, A.; Stringaro, A.; Angiolella, L. Antibacterial activity of essential oils mixture against PSA. *Nat. Prod. Res.* **2016**, *30*, 412–418. [CrossRef]
17. Mattarelli, P.; Epifano, F.; Minardi, P.; Di Vito, M.; Modesto, M.; Barbanti, L.; Bellardi, M.G. Chemical composition and antimicrobial activity of essential oils from aerial parts of *Monarda didyma* and *Monarda fistulosa* cultivated in Italy. *J. Essent. Oil-Bear. Plants* **2017**, *20*, 76–86. [CrossRef]
18. Diao, W.-R.; Hu, Q.-P.; Zhang, H.; Xu, J.-G. Chemical composition, antibacterial activity and mechanism of action of essential oil from seeds of fennel (*Foeniculum vulgare* Mill.). *Food Control* **2014**, *35*, 109–116. [CrossRef]
19. Pucci, N.; Orzali, L.; Modesti, V.; Lumia, V.; Brunetti, A.; Pilotti, M.; Loreti, S. Essential oils with inhibitory capacities on *Pseudomonas syringae* pv. *actinidiae*, the causal agent of kiwifruit bacterial canker. *Asian J. Plant Pathol.* **2018**, *12*, 16–26.
20. Danzi, D.; Thomas, M.; Cremonesi, S.; Sadeghian, F.; Staniscia, G.; Andreolli, M.; Bovi, M.; Polverari, A.; Tosi, L.; Bonaconsa, M.; et al. Essential oil-based emulsions reduce bacterial canker on kiwifruit plants acting as antimicrobial and antivirulence agents against *Pseudomonas syringae* pv. *actinidiae*. *Chem. Biol. Technol. Agric.* **2025**, *12*, 23. [CrossRef]
21. Verdeguer, M.; Sánchez-Moreiras, A.M.; Araniti, F. Phytotoxic effects and mechanism of action of essential oils and terpenoids. *Plants* **2020**, *9*, 1571. [CrossRef] [PubMed]
22. Chang, Y.; Harmon, P.F.; Treadwell, D.D.; Carrillo, D.; Sarkhosh, A.; Brecht, J.K. Biocontrol potential of essential oils in organic horticulture systems: From farm to fork. *Front. Nutr.* **2022**, *8*, 805138. [CrossRef]
23. Hashem, M.; Moharam, A.; Zaied, A.; Saleh, F. Efficacy of essential oils in the control of cumin root rot disease caused by *Fusarium* spp. *Crop Protect.* **2010**, *29*, 1111–1117.
24. Soylu, E.M.; Kurt, Ş.; Soylu, S. In vitro and in vivo antifungal activities of the essential oils of various plants against tomato grey mould disease agent *Botrytis cinerea*. *Int. J. Food Microbiol.* **2010**, *143*, 183–189. [CrossRef]
25. Lee, Y.-H.; Choi, C.-W.; Kim, S.-H.; Yun, J.-G.; Chang, S.-W.; Kim, Y.-S.; Hong, J.-K. Chemical pesticides and plant essential oils for disease control of tomato bacterial wilt. *Plant Pathol. J.* **2012**, *28*, 32–39. [CrossRef]
26. Sharma, A.; Rajendran, S.; Srivastava, A.; Sharma, S.; Kundu, B. Antifungal activities of selected essential oils against *Fusarium oxysporum* f. sp. *lycopersici* 1322, with emphasis on *Syzygium aromaticum* essential oil. *J. Biosci. Bioeng.* **2017**, *123*, 308–313. [CrossRef]
27. Amini, J.; Farhang, V.; Javadi, T.; Nazemi, J. Antifungal effect of plant essential oils on controlling *Phytophthora* species. *Plant Pathol. J.* **2016**, *32*, 16. [CrossRef]

28. Cibanal, I.L.; Fernández, L.A.; Rodriguez, S.A.; Pellegrini, C.N.; Gallez, L.M. Propolis extract combined with oregano essential oil applied to lima bean seeds against *Sclerotinia sclerotiorum*. *Eur. J. Plant Pathol.* **2022**, *164*, 33–43.
29. Bouqellah, N.A.; Abdulmajeed, A.M.; Rashed Alharbi, F.K.; Mattar, E.; Al-Sarraj, F.; Abdulfattah, A.M.; Hassan, M.M.; Baazeem, A.; Al-Harthi, H.F.; Musa, A.; et al. Optimizing encapsulation of garlic and cinnamon essential oils in silver nanoparticles for enhanced antifungal activity against *Botrytis cinerea* pathogenic disease. *Physiol. Mol. Plant Pathol.* **2025**, *136*, 102522. [CrossRef]
30. Dhanasekaran, S.; Liang, L.; Gurusamy, S.; Godana, E.A.; Yang, Q.; Zhang, H. Efficacy and mechanism of chitosan nanoparticles containing lemon essential oil against blue mold decay of apples. *Int. J. Biol. Macromol.* **2025**, *308*, 142633. [CrossRef] [PubMed]
31. Soleimani, H.; Mostowfizadeh-Ghalamfarsa, R.; Ghanadian, M. Characterization, biochemical defense mechanisms, and antifungal activities of chitosan nanoparticle-encapsulated spinach seed essential oil. *J. Agric. Food Res.* **2025**, *22*, 102016. [CrossRef]
32. Monchiero, M.; Gullino, M.L.; Pugliese, M.; Spadaro, D.; Garibaldi, A. Efficacy of different chemical and biological products in the control of *Pseudomonas syringae* pv. *actinidiae* on kiwifruit. *Austral. Plant Pathol.* **2015**, *44*, 13–23. [CrossRef]
33. Ni, P.; Wang, L.; Deng, B.; Jiu, S.; Ma, C.; Zhang, C.; Almeida, A.; Wang, D.; Xu, W.; Wang, S. Combined application of bacteriophages and carvacrol in the control of *Pseudomonas syringae* pv. *actinidiae* planktonic and biofilm forms. *Microorganisms* **2020**, *8*, 837. [CrossRef] [PubMed]
34. Han, Q.; Feng, L.; Zhang, Y.; Zhang, R.; Wang, G.; Zhang, Y. Effect of juglone against *Pseudomonas syringae* pv. *actinidiae* planktonic growth and biofilm formation. *Molecules* **2021**, *26*, 7580. [CrossRef] [PubMed]
35. Figueira, D.; Garcia, E.; Ares, A.; Tiago, I.; Veríssimo, A.; Costa, J. Genetic diversity of *Pseudomonas syringae* pv. *actinidiae*: Seasonal and spatial population dynamics. *Microorganisms* **2020**, *8*, 931. [CrossRef]
36. Baazeem, A.; Hassan, M.M.; Alqurashi, A.S.; Al Seberi, H.; Al-Harthi, H.F.; Al-Said, H.M.; Alhazmi, K.A.; Alnusaire, T.S.; AlShammari, W.; Alshammari, A.A.; et al. Peppermint and tarragon essential oils encapsulated in copper nanoparticles: A novel approach to sustainable fungal pathogen management in industrial crops. *Ind. Crops Prod.* **2025**, *232*, 121310. [CrossRef]
37. Weisany, W.; Samadi, S.; Tahir, N.A.-r.; Amini, J.; Hossaini, S. Nano-encapsulated with mesoporous silica enhanced the antifungal activity of essential oil against *Botrytis cinerea* (Helotiales; Sclerotiniaceae) and *Colletotrichum nymphaeae* (Glomerellales; Glomerellaceae). *Physiol. Mol. Plant Pathol.* **2022**, *122*, 101902. [CrossRef]
38. Derwich, E.h.; Benziane, Z.; Boukir, A. Chemical composition and in vitro antibacterial activity of the essential oil of *Cedrus atlantica*. *Int. J. Agric. Biol.* **2010**, *12*, 381–385.
39. El Hachlafi, N.; Mrabti, H.N.; Al-Mijalli, S.H.; Jeddi, M.; Abdallah, E.M.; Benkhaira, N.; Hadni, H.; Assaggaf, H.; Qasem, A.; Goh, K.W. Antioxidant, volatile compounds; antimicrobial, anti-inflammatory, and dermatoprotective properties of *Cedrus atlantica* (endl.) manetti ex carriere essential oil: In vitro and in silico investigations. *Molecules* **2023**, *28*, 5913. [CrossRef]
40. Mori, Y.; Kuwano, Y.; Tomokiyo, S.; Kuroyanagi, N.; Odahara, K. Inhibitory effects of Moso bamboo (*Phyllostachys heterocycla* f. *pubescens*) extracts on phytopathogenic bacterial and fungal growth. *Wood Sci. Technol.* **2019**, *53*, 135–150. [CrossRef]
41. Bincy, K.; Remesh, A.V.; Reshma Prabhakar, P.; Vivek Babu, C.S. Differential fumigant and contact biotoxicities of biorational essential oil of Indian sweet basil and its active constituent against pulse beetle, *Callosobruchus chinensis*. *Food Biosci.* **2023**, *51*, 102283. [CrossRef]
42. Simonetti, G.; Pucci, N.; Brasili, E.; Valletta, A.; Sammarco, I.; Carnevale, E.; Pasqua, G.; Loreti, S. In vitro antimicrobial activity of plant extracts against *Pseudomonas syringae* pv. *actinidiae* causal agent of bacterial canker in kiwifruit. *Plant Biosyst.* **2020**, *154*, 100–106. [CrossRef]
43. Rashad, Y.M.; Abdel Razik, E.S.; Darwish, D.B. Essential oil from *Lavandula angustifolia* elicits expression of three SbWRKY transcription factors and defense-related genes against sorghum damping-off. *Sci. Rep.* **2022**, *12*, 857. [CrossRef]
44. Sefu, G.; Satheesh, N.; Berecha, G. Effect of essential oils treatment on anthracnose (*Colletotrichum gloeosporioides*) disease development, quality and shelf life of mango fruits (*Mangifera indica* L.). *Am.-Eurasian J. Agric. Environ. Sci.* **2015**, *15*, 2160–2169.
45. Eke, P.; Adamou, S.; Fokom, R.; Nya, V.D.; Fokou, P.V.T.; Wakam, L.N.; Nwaga, D.; Boyom, F.F. Arbuscular mycorrhizal fungi alter antifungal potential of lemongrass essential oil against *Fusarium solani*, causing root rot in common bean (*Phaseolus vulgaris* L.). *Heliyon* **2020**, *6*, e05737. [CrossRef] [PubMed]
46. Danh, L.T.; Giao, B.T.; Duong, C.T.; Nga, N.T.T.; Tien, D.T.K.; Tuan, N.T.; Huong, B.T.C.; Nhan, T.C.; Trang, D.T.X. Use of essential oils for the control of anthracnose disease caused by *Colletotrichum acutatum* on post-harvest mangoes of Cat Hoa Loc variety. *Membranes* **2021**, *11*, 719. [CrossRef]
47. Sukegawa, S.; Shiojiri, K.; Higami, T.; Suzuki, S.; Arimura, G.i. Pest management using mint volatiles to elicit resistance in soy: Mechanism and application potential. *Plant J.* **2018**, *96*, 910–920. [CrossRef]
48. Rienth, M.; Crovadore, J.; Ghaffari, S.; Lefort, F. Oregano essential oil vapour prevents *Plasmopara viticola* infection in grapevine (*Vitis Vinifera*) and primes plant immunity mechanisms. *PLoS ONE* **2019**, *14*, e0222854. [CrossRef] [PubMed]
49. Kumar, A.; Singh, P.P.; Gupta, V.; Prakash, B. Assessing the antifungal and aflatoxin B(1) inhibitory efficacy of nanoencapsulated antifungal formulation based on combination of *Ocimum* spp. essential oils. *Int. J. Food Microbiol.* **2020**, *330*, 108766. [CrossRef]

50. Mączka, W.; Duda-Madej, A.; Górny, A.; Grabarczyk, M.; Wińska, K. Can eucalyptol replace antibiotics? *Molecules* **2021**, *26*, 4933. [CrossRef]
51. Chatterjee, B.; Vittal, R.R. Quorum sensing modulatory and biofilm inhibitory activity of *Plectranthus barbatus* essential oil: A novel intervention strategy. *Arch. Microbiol.* **2021**, *203*, 1767–1778. [CrossRef] [PubMed]
52. Monteiro-Neto, V.; de Souza, C.D.; Gonzaga, L.F.; da Silveira, B.C.; Sousa, N.C.; Pontes, J.P.; Santos, D.M.; Martins, W.C.; Pessoa, J.F.; Carvalho Junior, A.R. Cuminaldehyde potentiates the antimicrobial actions of ciprofloxacin against *Staphylococcus aureus* and *Escherichia coli*. *PLoS ONE* **2020**, *15*, e0232987.
53. Li, H.; Addo, K.A.; Yu, Y.-g.; Xiao, X.-l. Cuminaldehyde inhibits biofilm formation by affecting the primary adhesion of *Staphylococcus aureus*. *S. Afr. J. Bot.* **2023**, *156*, 13–20.
54. Ghasemi, G.; Fattahi, M.; Alirezalu, A.; Ghosta, Y. Antioxidant and antifungal activities of a new chemovar of cumin (*Cuminum cyminum* L.). *Food Sci. Biotechnol.* **2019**, *28*, 669–677.
55. Gill, A.O.; Holley, R.A. Mechanisms of bactericidal action of cinnamaldehyde against *Listeria monocytogenes* and of eugenol against *L. monocytogenes* and *Lactobacillus sakei*. *Appl. Environ. Microbiol.* **2004**, *70*, 5750–5755. [CrossRef]
56. Di Pasqua, R.; Hoskins, N.; Betts, G.; Mauriello, G. Changes in membrane fatty acids composition of microbial cells induced by addiction of thymol, carvacrol, limonene, cinnamaldehyde, and eugenol in the growing media. *J. Agric. Food Chem.* **2006**, *54*, 2745–2749. [CrossRef] [PubMed]
57. Werrie, P.-Y.; Durenne, B.; Delaplace, P.; Fauconnier, M.-L. Phytotoxicity of essential oils: Opportunities and constraints for the development of biopesticides. A review. *Foods* **2020**, *9*, 1291. [CrossRef]
58. Senatore, F.; Oliviero, F.; Scandolera, E.; Taglialatela-Scafati, O.; Roscigno, G.; Zaccardelli, M.; De Falco, E. Chemical composition, antimicrobial and antioxidant activities of anethole-rich oil from leaves of selected varieties of fennel [*Foeniculum vulgare* Mill. ssp. *vulgare* var. *azoricum* (Mill.) Thell]. *Fitoterapia* **2013**, *90*, 214–219. [CrossRef]
59. Gupta, A.; Jeyakumar, E.; Lawrence, R. Strategic approach of multifaceted antibacterial mechanism of limonene traced in *Escherichia coli*. *Sci. Rep.* **2021**, *11*, 13816. [CrossRef]
60. Friedman, M.; Henika, P.R.; Mandrell, R.E. Bactericidal activities of plant essential oils and some of their isolated constituents against *Campylobacter jejuni*, *Escherichia coli*, *Listeria monocytogenes*, and *Salmonella enterica*. *J. Food Prot.* **2002**, *65*, 1545–1560. [CrossRef]
61. Coêlho, M.L.; Ferreira, J.H.L.; Júnior, J.P.d.S.; Kaatz, G.W.; Barreto, H.M.; Cavalcante, A.A.d.C.M. Inhibition of the NorA multi-drug transporter by oxygenated monoterpenes. *Microb. Pathog.* **2016**, *99*, 173–177. [CrossRef] [PubMed]
62. Elkiran, O.; Telhuner, O. Chemical Profiles and Antimicrobial Activities of Essential Oil From Different Plant Parts of Fennel (*Foeniculum vulgare* Mill.). *Food Sci. Nutr.* **2025**, *13*, e70307. [CrossRef] [PubMed]
63. El Omari, N.; Balahbib, A.; Bakrim, S.; Benali, T.; Ullah, R.; Alotaibi, A.; El Mrabti, H.N.; Goh, B.H.; Ong, S.K.; Ming, L.C.; et al. Fenchone and camphor: Main natural compounds from *Lavandula stoechas* L., expediting multiple in vitro biological activities. *Heliyon* **2023**, *9*, e21222. [CrossRef]
64. He, R.; Liu, P.; Jia, B.; Xue, S.; Wang, X.; Hu, J.; Al Shoffe, Y.; Gallipoli, L.; Mazzaglia, A.; Balestra, G.M.; et al. Genetic diversity of *Pseudomonas syringae* pv. *actinidiae* strains from different geographic regions in China. *Phytopathology* **2019**, *109*, 347–357. [CrossRef]
65. Vincent, J.G.; Vincent, H.W.; Morton, J. Filter paper disc modification of the oxford cup penicillin determination. *Proc. Soc. Exp. Biol. Med.* **1944**, *55*, 162–164. [CrossRef]
66. Barros, E.P.; Moreira, N.; Pereira, G.E.; Leite, S.G.F.; Rezende, C.M.; de Pinho, P.G. Development and validation of automatic HS-SPME with a gas chromatography-ion trap/mass spectrometry method for analysis of volatiles in wines. *Talanta* **2012**, *101*, 177–186. [CrossRef] [PubMed]

plants

MDPI

Article

In Vitro Evaluation of the Antifungal Activity of *Trigonella foenum-graecum* Seed Extract and Its Potential Application in Plant Protection

Stelica Cristea [1], Alina Perisoara [2,*], Bianca-Maria Tihauan [3], Manuela Diana Ene [2], Mariana Constantin [4,5], Alexandru-Mihai Florea [2], Elena Ştefania Ivan [1], Relu Cristinel Zala [1], Bogdan Purcăreanu [2,6], Dan Eduard Mihaiescu [7] and Lucia Pirvu [8]

1 Plant Pathology Department, University of Agronomical Sciences and Veterinary Medicine, 59 Mărăşti Blvd., District 1, 011464 Bucharest, Romania; stelica.cristea@usamv.ro (S.C.); elena.ivan@qlab.usamv.ro (E.Ş.I.); cristinel.zala@usamv.ro (R.C.Z.)
2 Biotehnos SA, 3-5 Gorunului Street, 075100 Otopeni, Romania; diana.ene@biotehnos.com (M.D.E.); alexandru.florea@biotehnos.com (A.-M.F.); bogdanpb89@gmail.com (B.P.)
3 Research Institute, University of Bucharest-ICUB, Splaiul Independenței, No. 95, District 5, 050095 Bucharest, Romania; ciubuca.b@gmail.com
4 National Institute for Research & Development in Chemistry and Petrochemistry-ICECHIM, 202 Independentei Spl., 060021 Bucharest, Romania; mariana.constantin@icechim.ro
5 Faculty of Pharmacy, Titu Maiorescu University, 16 Bd. Gh. Șincai, 040441 Bucharest, Romania
6 Department of Science and Engineering of Oxide Materials and Nanomaterials, National University of Science and Technology Politehnica Bucharest, 011061 Bucharest, Romania
7 Department of Organic Chemistry, National University of Science and Technology Politehnica Bucharest, 011061 Bucharest, Romania; dan.mihaiescu@upb.ro
8 National Institute for Chemical-Pharmaceutical Research and Development, District 3, 031299 Bucharest, Romania; lucia.pirvu@yahoo.com
* Correspondence: pave.alinaa@gmail.com

Abstract: In the context of promoting ecological alternatives to synthetic pesticides, this study investigates the antifungal activity of *Trigonella foenum-graecum* L. seed extract and its potential application in plant protection. The extract, obtained by maceration in 40% ethanol, was analysed using UV-Vis spectrophotometric methods to assess its phytochemical composition, including phenolic compounds, reducing sugars, and soluble proteins, as well as antioxidant activity in acellular system (ABTS, DPPH, TEAC, and CUPRAC) and CAT, SOD, peroxidase, and lipid peroxidation in planting material lysate. Additionally, the extract was qualitatively analysed using ATR-FT-IR and FT-ICR-MS methods. The antifungal activity was tested in vitro against three fungal strains, revealing significant inhibitory effects, especially on *Fusarium graminearum* and *Monilinia laxa*. Following the biogermination study on wheat seeds, it was highlighted that the extract obtained from fenugreek seeds manifested a strong inhibitory effect, especially at the highest concentration (1.50%) studied, probably due to the high content of phenols and presence of steroidal saponins (diosgenin and precursor diosgenin–protodiosgenin) and pyridine alkaloids (trigonelline). These findings suggest that *Trigonella foenum-graecum* seed extract possesses potent antifungal properties, making it a promising candidate for the development of biofungicides in sustainable agriculture.

Keywords: eco-friendly biofungicides; *Trigonella foenum-graecum* seed extract; secondary metabolites

1. Introduction

Commercial agriculture relies primarily on chemical fungicides to protect crop plants against fungal pathogens by destroying and inhibiting their cells and spores. However, their accessibility and low cost result in overuse or repeated applications [1]. This excessive or abusive use of fungicides has toxic effects on beneficial living systems, human health, animals, and the environment too. Furthermore, the emergence of resistant strains of fungal phytopathogens makes treating fungal plant diseases increasingly difficult. Consequently, the development of healthy, non-toxic and environmentally friendly alternative approaches (green fungal control strategies) to chemical and synthetic fungicides is very useful and necessary in the control of fungal infections of plants [2,3]. The vast majority of known fungal species are strictly saprophytes; only very few species (less than 10% of identified fungi) can colonise plants. Phytopathogenic fungi are associated with an even smaller number of these plant colonisers. However, phytopathogenic fungi are key causal agents among phytopathogens for devastating epidemics of crop plants, in addition to leading to persistent and substantial losses in crop yield annually. Thus, phytopathogenic fungi are sought to be combated by both scientists and growers equally due to these important economic factors [4].

Moreover, in horticultural ecosystems, the pernicious action of phytopathogenic fungi influences the physiological resistance and fecundity of plant varieties. Among the most commonly encountered pathogens are *Fusarium*, *Monilinia*, and *Aspergillus* species [5].

Fusarium graminearum (*F. graminearum*), a prevalent fungal pathogen, induces *Fusarium* ear blight or *Gibberella* rot in cereals, causing grain discolouration, mould proliferation, and reduced grain viability, thus favouring significant production losses and product quality degradation [6,7]. In addition, its tendency to produce mycotoxins, especially deoxynivalenol (DON), raises numerous concerns regarding food safety and associated economic repercussions. Effective management strategies require an integrative approach that includes new cultural practices, chemical interventions, and an understanding of genetic resistance mechanisms [8] characteristic of this species. *Monilinia laxa* (*M. laxa*) is one of the main fungal phytopathogens that causes brown rot in stone fruit crops, especially peaches, nectarines, cherries, and apricots [9]. As a necrotrophic pathogen, *Monilinia laxa* secretes cell wall-degrading enzymes such as pectin methyl esterases and a series of phytotoxins, thereby opportunistically colonising fruits both before and after harvest, leading to substantial yield losses and quality degradation. After harvest, its persistence generates spoilage, exacerbating the ramifications of economic losses for growers and distributors. Effective management of this pathogen requires a comprehensive approach that includes improved cultural practices, chemical interventions, and sanitation measures to mitigate disease incidence and maintain orchard productivity [10,11].

The pathogenic potential of the *Aspergillus niger* (*A. niger*) species is manifested especially after harvest, as it colonises harvested crops, exploiting nutrient substrates and environmental conditions favourable for proliferation, degrading enzymatic plant tissues, and causing spoilage. Having the potential to synthesise mycotoxins, it thus compromises both the quality and safety of agricultural stores and commodities, requiring complex management strategies to mitigate economic losses and ensure food security [12].

A viable alternative for plant protection may be the use of biologically active products obtained from plant extracts, due to the multitude of active substances they contain [13]. In particular, the use of primary and secondary metabolites, such as volatile compounds (low-molecular-weight phenols), proteins, carbohydrates, fats, sterols, alkaloids, anthocyanins, anthraquinones, carotenoids, phenolic compounds, compounds containing sulphur molecules (phytoalexins), and aromatic hydrocarbons [14–16], which exhibit biological activity against pests, presents notable potential. According to the literature, plant extracts obtained from different parts (roots, seeds, fruits, leaves, flowers, rhizomes, or the whole plant) have been shown to exhibit antimicrobial activity against potential plant pests. This effect is attributed to their composition, which includes a range of phytochemical compounds that can act synergistically to inhibit the development of phytophagous strains [17]. Although studies have demonstrated the inhibitory effects of various plant extracts [15] on pathogenic strains such as *Aspergillus flavus*, *Fusarium oxysporum*, *Aspergillus fumigatus*, *Fusarium verticillioides*, *Penicillium expansum*, and *Penicillium brevicompactum* [18], there is still a limited number of commercially available products for plant protection based on biologically active substances (e.g., cinnamon oil, thyme oil, and citrus seed extract) [19].

Given the rich phytochemical composition of *Trigonella foenum-graecum* L. (*T. foenum-graecum*) seed extract, this study aims to evaluate its potential antifungal effect against three fungal strains (*Aspergillus niger*, *Monilinia laxa*, and *Fusarium graminearum*) with significant economic impact on horticulture and agriculture, to develop a potential natural product used in plant protection. For the following reasons, it is proposed that classic extraction technology of maceration at ambient temperature and thermal assistance (rotary evaporator) in the concentration phase, respectively, should be used, alongside the use of solvents from the category of concentrated alcohols (ethanol 70%): (1) the need to preserve the main active compounds with fungicidal effect as volatile compounds and (2) the solubility of the active compounds being the maximum in concentrated alcohols. The present study includes qualitative determinations (ATR-FT-IR and FT-ICR-MS methods) of biological compounds from the *Trigonella foenum-graecum* extract conditioned in 40% ethanol (the solvent concentration was chosen to minimise any potential negative impact on plants), a calculation of the reducing sugar and soluble protein content, and also evaluates the antioxidant properties of the extract in an acellular system using four different assays—CUPRAC, TEAC, DPPH, and ABTS—targeting distinct antioxidant mechanisms and CAT, SOD, peroxidase, and lipid peroxidation in the seedling lysate. The study also assesses the effects of the bioassays on wheat seed germination and highlights the antifungal activities of the extract against *Fusarium graminearum*, *Monilinia laxa*, and *Aspergillus niger*.

2. Results and Discussion

2.1. Physico-Chemical Screening of Fenugreek Extract in 40% Ethanol

2.1.1. Spectrum Evaluation by ATR-FT-IR

The drying process of the fenugreek seed extract in 40% ethanol resulted in a density of 1.08 $\pm$ 0.001 g/mL and a dry matter content of 42%. The qualitative profile of both the extract and the plant material was performed by ATR-FT-IR and is represented in Figure 1. FT-IR assessments provide easy and rapid information on biochemical samples regarding the composition and macromolecular structure [20].

In case of both the extract and plant material, the bands in the range of 4000–500 cm^{-1} highlight the presence of biochemical compounds belonging to the category of carbohydrates, lipids, proteins, and fibres, as well as aromatic compounds. The regions that reveal information about the content of the studied material are the region between the 900–1350 cm^{-1} bands, specific for carbohydrates; the amide I/II region, between the 1450–1650 cm^{-1} bands specific for proteins; the amide III region, specific for the 1230 cm^{-1} band [21,22]; the region between the 2800–3500 bands, specific for the O-H, N-H stretching vibrations; and the C skeleton fingerprint vibrations are attributed to lipids [23]. The bands 3330.37 cm^{-1} and 328.52 cm^{-1} are attributed to the O-H vibrational stretching, which reveals the presence of starch fibres and the N-H vibrational stretching, specific to protein amide A [24]. The bands 2971.58, 2924.23, 2927.08, and 2853.83 cm^{-1} are attributed to the alkyl C-H vibrational stretching. In the case of plant material (fenugreek seeds), the unique bands of 1743.85 cm^{-1}, attributed to the presence of saturated fatty aldehyde (C=O vibrational stretching, specific to lipids) [25], and 1237.58 cm^{-1}, which signals the presence of soluble fibres (pectin), were removed following the extraction process. The bands 1636.97, 1641.87, 1611.18 (C=O vibrational stretching, amide I), 1577.04, and 1541.37 cm^{-1} (N-H vibrational bending, amide II) indicate the presence of carbohydrates both in the extract and in the plant material, as well as the presence of phenolic compounds. The bands 1376.44–1377.64 cm^{-1} correspond to symmetric and asymmetric CH_3 bending [26]. The band in the region 1237.58–1042 cm^{-1}, due to C-O and C-OH stretching, reveals the presence of carbohydrates, phenolic compounds, and terpene compounds, but also the presence of starch, with a maximum absorption at 1042.60 cm^{-1}. The unique band at 873.95 cm^{-1} present in the extract indicates the presence of aromatic compounds (carboxylic acids), such as gallic, chlorogenic, coumaric, ferulic, sinapic, vanillic, syringic, caffeic, and cinnamic acids [27].

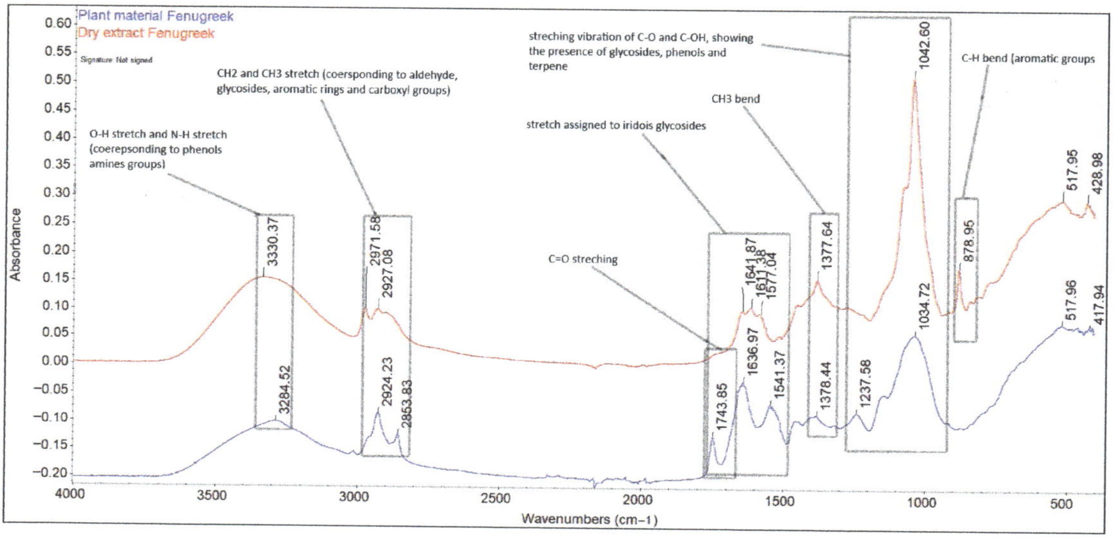

Figure 1. Identification of biological compounds in *Trigonella foenum-graecum* seed extract in 40% ethanol by ATR-FT-IR.

Several studies [28–30] reveal that ethyl alcohol is effective for the extraction of phenolic compounds, as well as their glycosides, but also phytochemical compounds belonging to the category of terpenoids, sterols, and alkaloids, while water contributes to the extraction of compounds belonging to the category of amino acids, organic acids, and carbohydrates, compounds also found in the *T. foenum-graecum* seed extract studied.

2.1.2. TPC and TFC Content Dosage

Phenolic acids are aromatic secondary plant metabolites associated with an array of functions, such as enzyme activity, structural integrity, nutrient uptake, and protein synthesis [31]. Flavonoids, too, are a type of aromatic secondary plant metabolites, whose role is to provide colour to the flower to attract pollinators, and in leaves, it is believed they confer protection against fungal pathogens and UV radiation and are involved in several growth regulation mechanisms [31]. For centuries, these two plant metabolites have proven themselves effective at combating different illnesses [31]. Following the study conducted on the *T. foenum-graecum* seed extract conditioned in 40% ethanol, it was highlighted that it has a high concentration of polyphenols and flavonoids (Table 1), with a content of 10.733% flavonoids from the total amount of quantified phenolic compounds. The total polyphenol content expressed as caffeic acid was determined by the Folin–Ciocalteu method and was found to be 36.817 ± 0.65 mg/mL, a significantly higher concentration than that found by Rahmani et al. [32] on four types of fenugreek extract (of different varieties) obtained in 96% methanol, which ranged from 1613 to 2083 GAE mg/g extract. It is likely that the type of solvent and the higher concentration of water used in the extraction system led to a higher content of phenolic compounds being obtained. At the same time, the extraction time can be an important factor in the extraction of phytochemical compounds, and this was highlighted in the study conducted by Turker et al. [33], obtaining a higher concentration of polyphenols (17.51 ± 0.28–18.30 ± 0.26 GAE mg/g dry extract) after 8–12 h of maceration of fenugreek seeds in absolute ethanol. Our results are in agreement with those obtained by Lohvina et al. [34], where the highest concentration of phenolic compounds was obtained in fenugreek seed extracts in 70% ethanol (from different varieties) compared to the extract variants in 30%, 50%, and 96% ethanol.

Table 1. Phenolic compounds content found in the *Trigonella foenum-graecum* seed extract in 40% ethanol (* mean ± SD, n = 3).

Parameter	*Trigonella foenum-graecum* Extract
TFC—Rutin mg/L	3.9517 ± 0.007 *
TPC—caffeic acid mg/L	36.817 ± 0.65 *

The flavonoid content expressed in rutin was determined by the spectrophotometric method, with a concentration of 3.9517 ± 0.007. Jokic et al. [35], in the study on the extract of some variants of soybean (*Glycine max*) in ethanol obtained by maceration at different concentrations (50, 60, 70, and 80%) and temperatures (25, 40, 50, 60, 70, and 80 °C), obtained a lower yield of flavonoid compounds (2.56 ± 0.04–0.60 ± 0.03 mgCE/gdb catechin equivalents on dry soybean basis). The highest value was found in the case of the variant obtained in 50% ethanol at a temperature of 80 °C. However, in the study conducted by Norziah et al. [36], it was highlighted that the aqueous extract of germinated fenugreek seeds recorded a better yield regarding the content of flavonoids (38.5 mg CE/g) compared to the extract variants in 75% ethanol and methanol. Ethanol and methanol are solvents with a lower polarity compared to water, and their use as the sole extraction solvent may make them less efficient in the extraction of phenolic compounds [25]. Additionally, the use of water (a polar solvent) in the extraction system may contribute to increasing their extraction yield. Polyphenols and flavonoids have different degrees of polarity [37]. Selecting an extraction solvent system with a polarity as close as possible to the phytochemical compounds to be extracted can ensure an efficient extraction [38]. At the same time, according to Rivas-García et al. [39], the extraction of polyphenols is more efficient when polar extraction solvents are used because the latter interact with polar functional groups. Fenugreek (*T. foenum-graecum*) seeds have a rich content of phytochemical compounds, and their most efficient extraction depends largely on the type of solvents used and the extraction technology applied [40]. Water–ethyl alcohol solvent mixtures exhibit higher polarity than solvents used in absolute concentration, leading to better extraction efficiency of phytochemical compounds [41,42].

In most cases, it is possible to identify organic compounds using only the obtained mass spectrum, due to the FT-ICR-MS ultra-high resolution and precision of the mass-to-charge ratio (m/z). Based on the molecular formula, ESI+ (1M + nH), the mass spectra of the bioactive compounds were generated using Bruker Compass Data Analysis software, which enabled their identification based on the isotopic pattern in the recorded spectra. According to the results obtained and shown in Table 2, compounds belonging to the categories of pyridine alkaloids (trigonelline), steroidal saponins (diosgenin Figure S5, proto-diosgenin—precursor of diosgenin Figure S2, tigogenin, and yamogenin), and phenolic compounds (vitexin, apigenin-7-O-glucoside, apigenin 6-C-galactoside 8-C-arabinoside, salicylic acid and luteolin-7-O-glucosides, rutin, abscisic acid, and t-resveratrol) were identified, compounds that are known for their antimicrobial, antioxidant [43], and anti-inflammatory properties. Contrary to the results of our study, Sakhira et al. [44] identified a larger number of phenolic compounds, in which the most important are catechin, epicatechin, vanillic acid, caffeic acid, coumaric acid, and cinnamic acid. Pasha et al. [45], regarding the study on two variants of fenugreek seed extract in different solvents (methanol, ethanol), highlights the fact that the methanol-rich extract had a higher composition in phenolic compounds (gallic acid, chlorogenic acid, p-coumaric acid, ferulic acid, sinapic acid, quercetin) compared to the ethanolic extract (gallic acid, vanillic acid, syringic acid, quercetin). Sigh et al. (2020) [46] conducted a study to identify the bioactive compounds present in Fenugreek seed extract using different LC-MS methods. Thus, five bioactive compounds (vitexin, isovitexin, trigonelline, isoorientin, and orientin) were identified using HPLC, and 25 compounds were identified using HPLC-ESI-QTOF-MS/MS based on retention time and molecular mass. Additionally, using fragment models of certain compounds, the following biological compounds were identified and quantified using UHPLC-ESI-MS/MS in multiple reaction monitoring (MRM) mode: pinitol, trigonelline, 4-hydroxyisoleucine, isoorientin, and isovitexin. As evidenced by numerous studies, saponins and phenolic compounds, especially flavonoids, exhibit a range of biological activities, including antiparasitic, anti-inflammatory, antimicrobial, allelopathic, and insecticidal properties [47]. Both benzoic acid and salicylic acid play an important role in the plant resistance process [48]. Hernandez-Guzman et al. (2024) [49], following their study of bioconversion by glycosylation and hydroxylation of diosgenin by the oleaginous yeast *Yarrowia lipolytica* P01a, aimed to improve the antioxidant, antifungal, and antiherbicidal performances of the obtained extracts. Thus, high amounts of protodiosgenin and soyagenin were obtained and they managed to highlight superior results regarding antioxidant activity (up to 97.02% of ABTS radicals and a 33.30% inhibition of DPPH radicals at 1000 mg L^{-1} of diosgenin) and, in subsequent antifungal studies, it had strong inhibitory effects on the strains *Botrytis cinerea* (67.34%), *Alternaria* sp. (35.63%), and *Aspergillus niger* (65.53%). At the same time, the bioconverted extracts also showed more herbicidal activity than to commercially available herbicide products. According to specialised studies [50], aqueous extracts of *T. foenum-graecum* obtained from different parts (seeds, stem cells, leaves, roots) have shown antifungal activity on strains of *Fusarium graminearum*, *Pythium aphanidermatum*, *Botrytis cinerea*, *Rhizoctonia solani*, and *Alternaria* sp. These results may make it a potential product for use in plant protection.

Table 2. The qualitative content of the biological compounds present in *Trigonella foenum-graecum* seed extract was determined by FT-ICR-MS.

Compound	Chemical Formula	(*m*/*z*) Measured	(*m*/*z*) Generated
Diosgenin	$C_{27}H_{42}O_3$	415.32089	415.32067
Trigonelline	$C_7H_7O_2$	138.05504	138.05495
Tigogenin	$C_{27}H_{44}O_3$	4.7.33657	417.33632
Yamogenin	$C_{27}H_{42}O_3$	415.32089	415.32067
Vitexin	$C_{21}H_{20}O_{10}$	433.11315	433.11292
Acid salicylic	$C_7H_6O_3$	139.03901	139.03897
Apigenin 6-C-galactoside 8-C—arabinoside	$C_{26}H_{28}O_{14}$	565.11543	565.11518
Luteoline-7-O-glucoside	$C_{21}H_{18}O_{12}$	463.08932	463.08710

Table 2. *Cont.*

Compound	Chemical Formula	(*m*/*z*) Measured	(*m*/*z*) Generated
Apigenin-7-O-glucoside	$C_{21}H_{20}O_{10}$	433.11315	433.11292
Protodiosgenin	$C_{51}H_{84}O_{22}$	1049.55408	1049.55270
Acid abscisic	$C_{15}H_{20}O_4$	265.14352	265.14344
Rutin	$C_{27}H_{30}O_{16}$	611.16110	611.16066
t-Resveratrol	$C_{14}H_{12}O_3$	228.07554	228.07810

2.1.3. Dosage of Soluble Carbohydrates and Proteins

Carbohydrates play an important role in plant growth and development, fulfilling nutrient functions, being at the same time a source of energy and carbon, but also as a central signalling or regulatory molecule that modulates the expression of genes related to metabolism (phytohormone), stress response, and resistance to phytophage agents [51–53]. Qualitative evidence of the presence of carbohydrate components in *T. foenum-graecum* seed extract was obtained by quantifying the content of reducing sugar. From the data in Table 3, it can be seen that using the DNS and Bradford method, a high concentration of reducing carbohydrates was obtained: 19.20 ± 0.35 D-glucose mg/mL and soluble proteins 1.96 ± 0.07 BSA mg/mL. This confirms the fact that the extract studied can function as a good nutrient and antioxidant for plant cells. The study by Dawood et al. [52], regarding the effect of foliar application of four types of aqueous fenugreek extracts of different concentrations (5%, 10%, and 15%) on two wheat varieties (Gimeza and Sakha), highlighted the fact that these had beneficial effects not only on the growth and development parameters of wheat seeds, but also on the yield of wheat grains and their biochemical constituents. The study conducted by Patel et al. [54] on 13 genotypes of fenugreek leaves indicates a soluble protein content ranging from 1.82 to 0.99% and a reducing sugar content ranging from 1.59 to 0.73%. Ghevariya et al. [55], following a comparative study of bioactive compounds found in leaves, stem cells, seeds, and microplants/shoots of *Trigonella foenum-graecum*, show that the highest content of reducing sugar was found in the seeds, and fenugreek leaves recorded the highest total protein content.

Table 3. Reducing sugar and soluble protein content found in *Trigonella foenum-graecum* seed extract in 40% ethanol (* mean ± SD, *n* = 3).

Parameter	*Trigonella foenum-graecum* Seed Extract
Total reducing sugar (mg D-glucose/mL)	19.20 ± 0.35 *
Total soluble proteins (mg BSA/mL)	1.96 ± 0.07 *

2.2. Determination of Antioxidant Activity

Free radicals, and more specifically ROS (reactive oxygen species) and RNS (Reactive Nitrogen species), are molecules highly dangerous to living cells, damaging DNA, proteins, and membrane lipids [56]. They include both free and non-free radical intermediates, such as hydroxyl radicals ($OH^{\bullet}$), hydrogen peroxide (H_2O_2), or peroxynitrite ($ONOO^-$) [57]. The three primary species of ROS are $O_2^{\bullet-}$, H_2O_2, and $HO^{\bullet}$. $O_2^{\bullet-}$ is produced enzymatically and is transformed by superoxide dismutase (SOD) into the less toxic H_2O_2. $O_2^{\bullet-}$ and H_2O_2 can react together to generate $HO^{\bullet}$ [58], the most reactive ROS. $O_2^{\bullet-}$ can also react with $^{\bullet}NO$ to form $ONOO^-$ [57]. As a matter of fact, mitochondria are both the main endogenous source of ROS [59] (bringing about up to 90% of the total ROS in the cell), which can be generated at the level of the electron transport chain, and a major target of these same molecules [60]. Up to 2% of the oxygen consumed during cellular respiration is turned into ROS [59]. Among the deleterious effects they have on mitochondria, or the cells in general, are mutations and deletions of the DNA, peroxidation of membrane lipids, and permeabilisation of the membrane due to the opening of the mitochondrial permeability pore [56]. For this reason, the cells need to regulate ROS and RNS concentrations tightly. An imbalance between these species and the cell's ability to detoxify them, via several specific enzymes such as SOD, Catalase, and Glutathione Peroxidase, or to repair the damage caused by them, is referred to as oxidative stress [61–65]. Phytochemicals present in plant extracts exhibit antimicrobial and antioxidant activity, and this is largely due to the presence of phenolic compounds [66]. According to our results, the fenugreek extract conditioned in 40% ethanol possessed a high concentration of phenolic compounds. In the present study, antioxidant activity was determined by four different methods, two of which, DPPH and ABTS, are methods based on the free radical scavenging mechanism, and the CUPRAC and FRAP methods are methods based on the reduction mechanism of ions (copper and iron). The CUPRAC method, unlike the FRAP method, is a less used and relatively new one, in which neocuproin is used as an oxidising agent [67]. In most studies, the antioxidant activity of plants is determined by the classical DPPH and ABTS methods. The protons of free radicals present in DPPH and ABTS reagents decrease rapidly in speed at the moment they come into contact with protons of radical scavengers [68]. Due to the high sensitivity of the ABTS method, the reaction kinetics are faster, and a large part of the phytochemical compounds show better antioxidant activity by this method, compared to the DPPH method [69]. At the same time, the ABTS radical cation shows reactivity with most antioxidants soluble in both organic and polar solvents; thus, lipophilic and hydrophilic antioxidants can be identified [56]. In the study conducted by Kaya and Akbaş [70] on the determination of the antioxidant activity of four variants of *Peganum harmala* seed extract, obtained in different solvents (water, ethanol, methanol and chloroform), it was highlighted that the best antioxidant activity was obtained in the case of the aqueous extract by applying the CUPRAC method (0.714), the methanolic extract showed better antioxidant activity by the DPPH method (74.06%), and the ethanolic extract by the ABTS method (72.06%). Wasim Bari et al. [67], by studying the antioxidant activity of the methanolic leaves extract of *Sphagneticola calendulacea* (L.) Pruski by five different methods (FRAP, CUPRAC, DPPH, ABTS, and nitric oxide free radical scavenging method) obtained better antioxidant activity by applying the ABTS (89.63%) and DPPH (79.43%) methods. Bukhari et al. (2025) [71] studied the antioxidant activity of four types of extracts obtained from fenugreek seeds in different solvents (methanol, ethanol, dichloromethane, acetone, hexane, and ethyl acetate) using the DPPH and ABTS methods, where they highlight the fact that all six types of extracts obtained showed a strong antioxidant effect, especially the extracts obtained in polar solvents, ethanol and methanol. The strongest antioxidant effect was recorded in the case of the extract in ethanol in both methods studied. This is due to the high concentration of phenolic compounds found in the composition of the extract.

In our experiment, the strongest antioxidant activity of the *T. foenum-graecum* seed extract studied was demonstrated by the DPPH method (Table 4), followed by ABTS, and the weakest antioxidant activity was recorded by the CUPRAC method. These results are in line with the trend of the majority of studies, indicating that ABTS is the method that best reveals the antioxidant activity of fenugreek extracts, followed by DPPH, while FRAP and CUPRAC show results with high variability but significance in some studies. Antioxidant activity tests (DPPH, ABTS, FRAP, CUPRAC) have an important biological significance in treatments with plant extracts to stimulate germination processes, as they highlight their ability to protect seeds against oxidative stress, which can inhibit germination of plant germplasm. The extract analysed in the present study, by its high antioxidant activity, as evidenced by the DPPH assay, suggests that the composition of the extract may protect the seeds from oxidative stress, which could inhibit root development or lead to slower germination. The high reducing potency demonstrated in the FRAP assay suggests that it also intervenes in the processes that maintain the cellular integrity of the seed germinating material during germination, protecting them from damage to membranes or other essential components that would hinder normal development. The results obtained from the CUPRAC test suggest a potential protective effect against potential damage caused by the presence of heavy metals or other contaminants in the soil, thus stimulating germination. All these experimental observations highlight the ability of the analysed extract with high antioxidant activity to help the planting material to better cope with adverse environmental conditions and may stimulate faster and healthier germination. According to our results, fenugreek seed extract can ensure the natural health state (homeostasis) of reactive oxygen species. These can act as signalling molecules with a role in plant growth and development [70].

Table 4. Antioxidant activity evaluation of *Trigonella foenum-graecum* extract.

Parameter	Volume of Extract Required to Reduce the Level of Reactive Species with 50% (μL)
DPPH	2.934 ± 0.023 *
ABTS	3.011 ± 0.013 *
FRAP	5.29 ± 0.34 *
CUPRAC	11.64 ± 0.64 *

* mean ± SD, $n = 3$.

2.3. The Effect of the *Trigonella foenum-graecum* Seeds Extract and the Solvent on Wheat Seeds (*Triticum Aestivum*)

A product/active substance with potential use in the plant protection and phytostimulation industry must be tested to determine possible toxic or phytostimulant effects that can manifest on plants [72]. The most convenient and widely used test for excess salinity or the presence of allelopathic substances (e.g., phenolic compounds) in plant extracts or compost is the seed germination bioassay [73–75].

According to the specialised literature [75–77], seeds can take up from the treatment medium substances that can fulfil a nutritional or allelopathic role, which enter into their composition, and this can be evidenced in the growth and development of their roots. According to Emino and Warman [78], the effect of the new substances/extracts studied on plants is reflected in the Gi value obtained. This, in turn, is obtained by quantifying the values of the germination index (RSG) and the relative root growth index (RRG) [76].

In the present study, the effect of *Trigonella foenum-graecum* seed extract at different concentrations (0.10%, 0.50%, 1.00%, and 1.50%) on wheat seeds was compared with its solvent (40% ethanol). This was performed to highlight the potential stimulant or phytotoxic effect of the extract on the treated seeds. Germination percentage (GP%) was determined by reporting the number of germinated seeds in the sample/control with the number of seeds studied [79]. GP indicates possible allelopathic effects on the studied seeds [80,81]. However, it is an indicator that cannot predict the possible delayed seed germination that could be caused by the presence of phytochemical compounds present in the plant extract. As can be seen from Table 5, the GP (%) values for both solvent and extract were in the range of 73.33–91.67%. It is noteworthy that the extract had a positive impact on wheat seeds compared to the solvent, except for the concentration of 1.50%, where the GP (%) value for the solvent (90 ± 2.47%) was higher compared to *Trigonella foenum-graecum* extract (85 ± 0.66%). Mominul Islam et al. [82] studied the phytotoxic activity of methanol obtained from *Ocimum tenuiflorum* (*Lamiaceae*) on six seed varieties (*Lepidium sativum, Medicago sativa, Lolium multiflorum, Echinochloa crusgalli, Phleum pratense, and Phleum pratense*) in different concentrations (3, 10, 30, and 100 mg extract equivalent mL^{-1}), highlighting that the germination percentage was significantly reduced with increasing extract concentration studied, for all seed varieties, except lettuce and yard grass seeds. Azizi et al. (2011) [83] conducted a study in which the effect of four aqueous extracts obtained from different parts of *Trigonella foenum-graecum* (leaf, stem, seed, pod, and whole plant) on four seed varieties (*Glycine max, Sesamus indicum, Amaranthus retroflexusI*, and *Abotilon theophrasti*) was monitored, where it was demonstrated that all the extract variants obtained presented inhibitory effects on the seeds studied. As our study reveals, there is a correlation between the germination percentage and the extract concentration studied. The most inhibitory effect was recorded in the case of the extracts obtained from the *Trigonella foenum-graecum* seeds and leaf.

Table 5. Results of germination percentage (GP%) of fenugreek seed extract and solvent (40% ethanol) (* mean ± SD, n = 3).

Sample Name	% Sample				Control
	0.10	0.50	1.00	1.50	
Fenugreek seed extract conditioned in 40% ethanol	91.67 ± 0.57 * RSD = 3.14	91.67 ± 2.08 * RSD = 11.35	88.33 ± 2.51 * RSD = 14.54	85 ± 0.66 * RSD = 5.88	96.67 ± 1.15 * RSD = 5.97
40% Ethanol (solvent)	91.66 ± 2.09 * RSD = 11.35	81.66 ± 2.51 * RSD =15.40	73.33 ± 1.52 * RSD = 10.41	90 ± 2.47 * RSD = 11.11	93.33 ± 1.15 * RSD = 6.18

Regarding the relative germination index RSG (%), it was interpreted by reporting the average of germinated seeds in the sample/solvent compared to the average of germinated seeds in the control sample (distilled water). The extract sample's RSG values (%) fell within the 94.80–87.93% range, and for the solvent, they fell within the range of 98.21–78.57% (Table 6). It is worth noting that the values of the relative germination index in the case of the extract decrease directly proportionally with the increase in the extract dose (non-significant statistically, p-value > 0.05). The same trend was also observed in the case of the solvent, except for the maximum tested concentration (96.42%), where the RSG value is higher than compared to the values obtained at the concentrations of 0.50% and 1.00%, but also than that recorded in the case of the extract (87.93%), at the maximum tested concentration (non-significant statistically, p-value > 0.05). It can be observed that the fenugreek seed extract conditioned in 40% ethanol at higher concentrations may have a slight inhibitory effect due to the higher content of phenolic compounds or residual solvents left after the extraction process. There are a series of studies [84] that have demonstrated the allelopathic effect manifested by the phenolic compounds present in plant extracts on the development and growth of plants, mainly intervening in physiological processes, such as the absorption of nutritional compounds, protein synthesis, plant cell permeability, respiration, and photosynthesis, as well as the water ratio. Our study can also be correlated with the results obtained by Xuan et al. [85], where it emerged that the aqueous extract of neem had inhibitory effects on the six seed varieties studied: alfalfa, beans, carrots, radishes, rice, and sesame.

The relative root growth index (%) is interpreted by comparing the average root length of germinated seeds from the extract/solvent sample to the average root length of the control sample. Data from the literature [86] reveal that the RRG index is a more sensitive indicator of the toxicity of biologically active compounds compared to RSG. One reason for this could be that roots can easily absorb the substances present in the treatment medium, and this can be observed in their development and growth [74]. As can be seen in Table 7, the RRG values (%) for the extract were within the range of 85–70–35.70%, and for the solvent between 109.10 and 70.48%. If we compare both the control and the solvent, we can see that the extract has strong allelopathic compounds that affect the growth of wheat seed roots; the greatest impact was seen at the maximum concentration tested (1.50%), where the lowest RRG value was recorded (35.79%, statistically significant $p < 0.05$, concerning both the solvent and the control sample, Figure 2). This result also correlates with the data obtained regarding RSG and GP from our study. Roy et al. [87] investigated the effect of aqueous extracts obtained from different parts of the banana plant (rhizome, root, leaf sheath, and leaf blade) at 15%, 25%, 50%, and 100% concentrations on six varieties of seeds (lettuce, red amaranth, amaranth, radishes, cucumbers, gourd, beans, and okra), where a strong inhibitory effect on root development, as well as on the germination capacity of seeds, was demonstrated for all extract variants, especially in the case of the extract obtained from rhizomes at the maximum concentration tested (100%).

Table 6. Results of the relative germination index (RSG%) of fenugreek seed extract and solvent (40% ethanol).

Sample Mean	% Sample				Control
	0.10	0.50	1.00	1.50	
Fenugreek seed extract conditioned in 40% ethanol	94.8 **	94.82 **	91.37 **	87.93 **	96.67 ± 1.15 RSD = 5.97
40% Ethanol (solvent)	98.21	87.5	78.57	96.42	93.33 ± 1.15 RSD = 6.18

+ statistically significant $p < 0.05$, compared to the control sample; ++ statistically significant $p < 0.05$ compared to the solvent. $n = 3$, ** ns = non-significant statistically.

Table 7. The relative root growth index results (RRG%) following the treatment of wheat seeds with fenugreek seed extract and solvent (40% ethanol); mean ± SD, ($n = 3$).

Sample Name	% Sample				Control
	0.10	0.50	1.00	1.50	
Fenugreek seed extract conditioned in 40% ethanol	+ 64.29% (1.19 ± 0.20 RSD = 17.24)	85.70% (1.58 ± 0.56 RSD = 35.43)	+ 50.00 (0.92 ± 0.14 RSD = 15.20)	+;++ 35.79% (0.66 ± 0.14 RSD = 22.11)	* 1.85 ± 0.17 RSD = 9.22
40% Ethanol (solvent)	70.48% (1.6 ± 0.47 RSD = 29.81)	91.18% (2.07 ± 0.23 RSD = 11.13)	84.87% (1.92 ± 0.71 RSD = 36.93)	109.10% (2.47 ± 0.47 RSD = 19.31)	2.27 ± 0.36 RSD = 16.08

* root length is measured in cm; + statistically significant $p < 0.05$, compared to the control sample; ++ statistically significant $p < 0.05$ compared to the solvent.

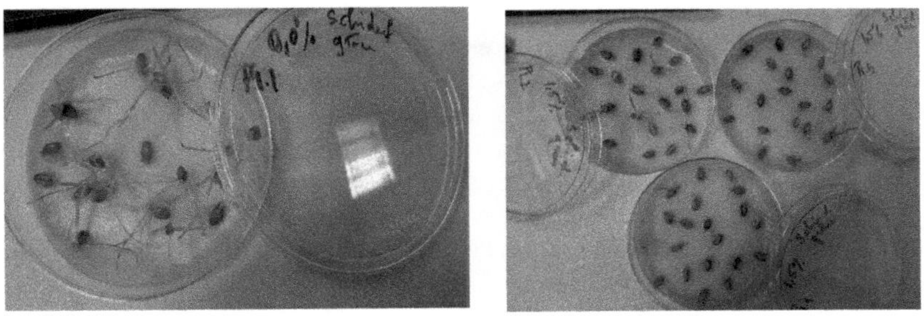

(a) (b)

Figure 2. Seedlings growth germination bioassay of wheat seeds using *Trigonellla foenum graecum* seed extract in 40% ethanol: (**a**) control and (**b**) 1.50% extract.

Interpretation of the results regarding the germination index (Gi% %) is based on the results obtained for the RSG index and the RRG index [80]. According to Cesaro et al. [88], the germination index is a parameter included in the list of quality assurance regulations in the Italian legislation on the marketing of composts with a role in plant protection and phytostimulation. Following the results obtained regarding the Gi index, we can interpret whether or not the fenugreek extract has a phytostimulating or phytotoxic effect on wheat seeds. The data obtained show that the *Trigonella foenum-graecum* seed extract exhibits a strong phytotoxic effect at the highest concentration tested (Gi 32.98%) (Figure 3), the Gi value recorded being below 50%. As for the solvent, it showed a slightly stimulatory activity at a concentration of 1.50% (Gi 105.73%), the Gi value being above 100%. Reigosa et al. [89], following their study on the impact of some small molecular size phenolic compounds (gallic acid, *p*-coumaric acid, *p*-vanillin, *p*-hydroxybenzoic acid, ferulic acid, and vanillic acid) on six weed varieties (*Chenopodium album* L., *Plantago lanceolata* L., *Amaranthus retroflexus* L., *Solanum nigrum* L., *Cirsium* sp., and *Rumex crispus* L.), demonstrated that they exhibited inhibitory effects only at the highest concentrations studied, but at low concentrations, they had no effect, and in some cases they exhibited stimulatory effects.

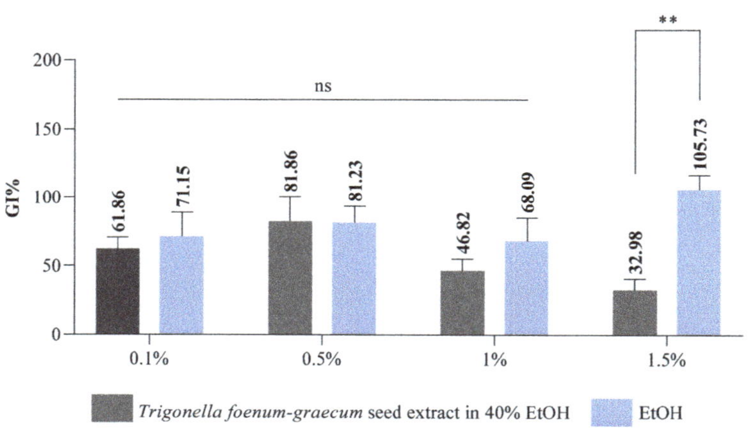

Figure 3. Germination index (Gi) results of *Trigonella foenum-graecum* seed extract variant on wheat seeds; ** p value = 0.005 (n = 3); ns = non-significant statistically.

Our results are also correlated with those obtained by Dai et al. [90], who investigated the effect of the aqueous extract of *Falveria bidentis* leaves on three plant varieties (Shanghai green, barnyard grass, and wheat), demonstrating that the inhibitory response of the extract was stronger at higher concentrations. It displayed inhibitory effects on both the germination rate and the development of plant roots. As we have mentioned before, phenolic compounds, used at certain concentrations, can exhibit undesirable effects on plants, and these can manifest themselves at molecular, physiological, and biochemical levels [91]. They can induce oxidative stress as a consequence of the high production of ROS (reactive oxygen species). It is known that ROS manifests itself as a signalling molecule, intervening in the process of regulating the response of plants to abiotic and biotic stress factors. At the same time, ROS and phytohormones are involved in plant growth and development. Abscisic acid and ethylene are considered stress hormones that are involved in the process of seed germination, especially dormancy [92].

The germination process is a crucial stage in the life cycle of plants, where seeds start to germinate and develop into new plants, and is a process that is highly sensitive to various environmental factors, with water deficit or drought stress being the most important stress factors. Seed germination and the post-germination phase are finely regulated by oxidative balance, a process involving lipid peroxidation, reactive oxygen species (ROS) formation, and enzyme activity, all of which have a direct impact on plant viability and subsequent plant development [93–95]. Polyunsaturated fatty acids, the predominant molecules in the membrane structures of plant cells, are vulnerable to ROS attack, and their peroxidation affects both germination capacity and seed longevity [88–90]. One of the most common forms of ROS is hydrogen peroxide, which has a significant cytotoxic effect on cells, being involved in oxidative degradative processes of cell membranes and disrupting essential physiological processes during germination [90,93,94]. In the early stages of imbibition, lipid peroxidation processes are activated and cell membranes undergo structural damage, leading to the accumulation of malondialdehyde (MDA), commonly used as a marker of oxidative stress [96]. The increase in MDA concentration is directly associated with enhanced cellular degradation and abiotic stress, in particular water stress, and this phenomenon contributes to a decrease in germination rate and a progressive reduction in seed viability. Due to all of these aspects, the accumulation of MDA serves as an indicator of irreversible damage to cellular structures, leading to a decrease in germination potential and a shorter seed life [93–95]. The stability of cell membranes is directly influenced by the intensity of the lipid peroxidation process, and the variations in MDA levels observed after seed pretreatment can be correlated with the activity of antioxidant enzymes involved in free radical detoxification, such as superoxide dismutase (SOD), peroxidase (POD), and catalase (CAT) [97,98]. These enzymes play a crucial role in maintaining cellular redox balance, contributing to ROS neutralisation and protecting the structural integrity of cells [95–97]. Under conditions of water stress, plants are forced to develop protective mechanisms to counteract the harmful effects of the reactive oxygen species generated, and antioxidant enzymes such as catalase (CAT), superoxide dismutase (SOD), and peroxidase (POX) play a central role in this protective mechanism. These enzymes, through their specific catalytic properties (SOD—converts superoxide anion into hydrogen peroxide, thus reducing the oxidative potential at the cellular level; CAT—catalyses the decomposition of hydrogen peroxide into water and oxygen, protecting cells from its cytotoxic effects; peroxidase—in the presence of electron donors breaks down hydrogen peroxide, increasing the protection of cellular structures by preventing its accumulation in toxic concentrations), form an antioxidant defence system that maintains the oxidative balance and protects seeds in the critical phases of imbibition and germination, preventing catabolic processes at cellular level and supporting normal plant development [99–101]. Recent studies suggest that modulating the oxidative balance, through targeted interventions that limit ROS accumulation and lipid peroxidation, can significantly improve germination capacity and seed longevity, offering new prospects for improving agricultural productivity and crop sustainability under abiotic stress. In taking all of this into account, an extended experimental study was developed to evaluate the antioxidant properties of fenugreek seed extract in the seedling material based on its ability to modulate the enzymatic activity of protein enzymes involved in phase I, counteracting oxidative stress. The experimental results are presented in the figures below (Figure 4).

From the experimental data, it is observed that the activity of antioxidant enzymes (CAT, SOD, peroxidase) and the concentration of the stress marker (MDA) in *Triticum aestivum* are up-regulated and significantly increased under the influence of abiotic stresses, indicating the activation of the plant defence mechanisms. Under normal conditions, these values are relatively low, while under stress conditions (water, heat, etc.), the values increase to combat lipid peroxidation and ROS accumulation. The process is greatly accelerated by the presence of fenugreek seed extract, which manages under water stress conditions to counteract the harmful effects of reactive oxygen species by activating protective enzymatic mechanisms.

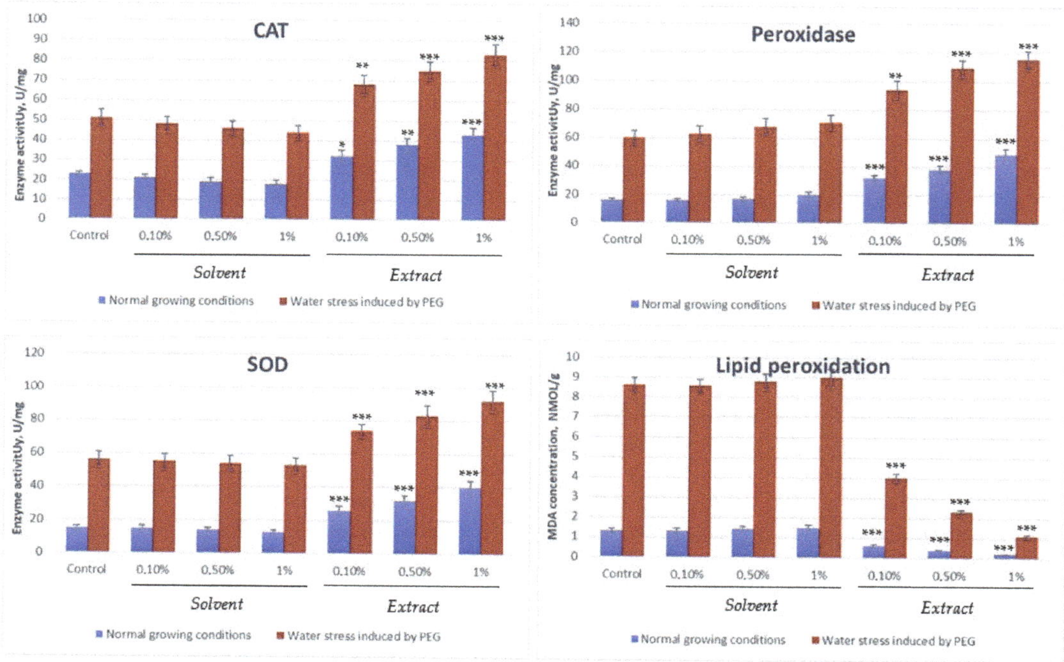

Figure 4. Evaluation of the enzyme activity of enzymes involved in counteracting oxidative stress and the degree of lipid peroxidation (* $p < 0{:}05$, ** $p < 0{:}01$ and *** $p < 0{:}001$ vs. control).

2.4. *Trigonella foenum-graecum* Seed Extract Antifungal Activity

Following the qualitative evaluation of the antifungal activity of the standardised extract in 5 mg/mL gallic acid (GAE) obtained from *Trigonella foenum-graecum* L. (Fenugreek) seeds in 40% ethanol, it can be observed that it is effective, especially at the concentration of 500 mg/mL, on the three tested strains. The extract concentrations (62.5–500 mg/mL) were selected to cover a broad range for in vitro bioactivity screening and to identify the minimum inhibitory concentration (MIC) capable of producing measurable antifungal effects. These concentrations were chosen based on preliminary solubility and diffusion assays, as well as values commonly reported in similar in vitro studies involving crude plant extracts. It is important to note that these values do not directly reflect agronomically applicable doses but rather serve as an initial step to characterise the antifungal potential of *Trigonella foenum-graecum* extract. Further investigations under greenhouse and field conditions are needed to calibrate these concentrations for practical application in crop protection.

According to the results obtained, it can be observed that the most pronounced fungicidal effect was obtained on the *Fusarium graminearum* strain, where the concentration of 500 mg/mL inhibited the development of the strain by 84.43 ± 4.29% compared to the antifungal control Amphotericin B (10 µg/mL) (Table 8, Figure 5b) and with the solvent (64.56 ± 5.88%). For the next three concentrations tested (250 mg/mL, 125 mg/mL, and 62.5 mg/mL), lower inhibitory values were obtained, the associated effect being possibly fungistatic. However, it is important to note that at these concentrations, 40% ethanol was devoid of antifungal activity; the results obtained are entirely associated with the effect of the extract. In the case of the *Aspergillus niger* strain, the most effective antifungal activity was observed at 500 mg/mL, namely 78.01 ± 5.61%, while at 250 mg/mL, 61.78 ± 3.72% inhibition was obtained compared to the Nystatin control (5 µg/mL) (Table 8, Figure 5a). However, the antifungal effect of 40% ethanol was more pronounced compared with the extract at the tested concentrations; the 500 mg/mL sample, where the inhibition was 87.32 ± 5.74% was the significant one. The following concentrations tested in the presence of the extract indicated significantly lower inhibition values, respectively, 37.80 ± 1.54% and 22.62 ± 3.11. In the case of the *Monilinia laxa* strain, the 500 mg/mL concentration of the extract led to an inhibition of 66.23 ± 5.55% compared to the Imidazole control (10 µg/mL) (Table 8, Figure 5c). It can be observed for the tested strains that the activity of the extract is directly proportional to the tested concentrations. The 40% ethanol showed significant percentages of inhibition for tested concentrations, but a synergistic/potentiating effect of the extract activity can be observed, especially in the *Monilinia laxa* strain.

Table 8. Comparative qualitative evaluation of antifungal activity for different concentrations of *Trigonella foenum-graecum* L. seed extract in 40% ethanol against the phytopathogenic strains *Aspergillus niger*, *Fusarium graminearum*, and *Monilinia laxa* (mean ± SD, n = 3).

Concentrations (mg/mL)	Extract	*Aspergillus niger*	*Fusarium graminearum*	*Monilinia laxa*
		Inhibition Percentage (%)	Inhibition Percentage (%)	Inhibition Percentage (%)
500	Fenugreek seed extract conditioned in 40% ethanol	78.01 ± 5.61	84.43 ± 4.29	66.23 ± 5.55
	40% ethanol (solvent)	87.32 ± 5.74	64.56 ± 5.88	57.05 ± 1.99
	p value	<0.05	<0.05	<0.05
250	Fenugreek seed extract conditioned in 40% ethanol	61.78 ± 3.72	47.21 ± 0.66	45.37 ± 5.41
	40% ethanol	66.24 ± 0.41	0.00 ± 0.00	33.64 ± 10.59
	p value	<0.05	<0.05	<0.05
125	Fenugreek seed extract conditioned in 40% ethanol	37.80 ± 1.54	44.82 ± 3.63	28.78 ± 2.63
	40% ethanol	32.50 ± 3.72	0.00 ± 0.00	16.87 ± 3.33
	p value	<0.05	<0.05	<0.05
62.5	Fenugreek seed extract conditioned in 40% ethanol	22.62 ± 3.11	28.77 ± 0.51	20.19 ± 6.80
	40% ethanol	25.30 ± 1.46	0.00 ± 0.00	13.08 ± 5.45
	p value	<0.05	<0.05	<0.05

According to the literature, a series of phytochemical constituents or main compounds identified in *Trigonella foenum-graecum* L., namely saponins, steroidal saponins, flavonoids, phenols, proteins (especially in aqueous and ethanolic extracts), carbohydrates, or alkaloids (especially in aqueous extracts), are responsible for the antifungal activity on *Aspergillus niger* strains [102,103]. As for the effect of Fenugreek extract on the *Fusarium graminearum* strain, the results obtained correspond to those reported in the literature [50], in particular, the presence of a dose–response effect on the antifungal activity. As for *Monilinia laxa*, the literature does not provide results that directly correlate the antifungal effects of Fenugreek extract with this pathogen. Most studies focus on evaluating the effect of volatile oils on strains of the *Monilinia* species [104]; thus, the determinations made in the present study are innovative in terms of the development of new biopesticides necessary to maintain crop safety. After harvest, the biggest challenge facing the agro-tech market is certainly fungal. Fungi, without effective control methods, can result in harvest losses of up to 24% [105]. Most of the losses are due to diseases of commercially important fruits and vegetables that result from pre- or post-harvest infections with fungal pathogens. This is due to their adaptability, which allows them to grow and develop under storage conditions. Thus, the major fungal threats after harvest are moulds and fungi that can infect a wide range of plant species [106].

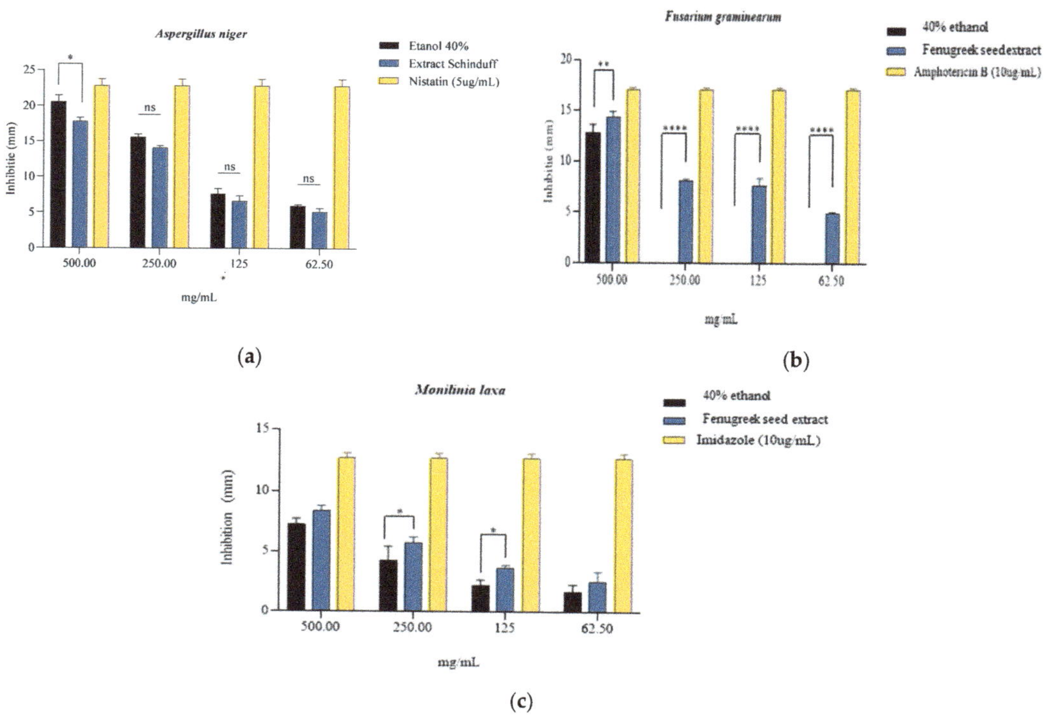

Figure 5. Qualitative antifungal activity expressed by inhibition zones (mm) of the extract obtained from *Trigonella foenum-graecum* L. seeds, with 40% ethanol used as solvent and specific fungistatic controls for strains of (**a**) *Aspergillus niger*, * is $p = 0.0274$; (**b**) *Fusarium graminearum*, ** is $p = 0.0037$, **** = $p < 0.0001$; and (**c**) *Monilinia laxa*, * is $p = 0.0455$; ns = non-significant statistically; $n = 3$.

The quantitative evaluation of antifungal activity was performed at four extract concentrations for each strain, and the minimum inhibitory concentration (MIC) was thus evaluated, noting that in the case of *Fusarium graminearum* (Figure 6b), the reduction in cell viability reported for the solvent was statistically significant ($p < 0.5$) for all concentrations tested. The most significant reduction (viability 15.57 $\pm$ 4.28%) was associated with the concentration of 500 mg/mL extract. In the case of *Aspergillus niger* and *Monilinia laxa* strains, the minimum inhibitory concentration was observed at 500 mg/mL, with no statistically significant differences compared to the activity of the solvent (Figure 6a,c). For the antifungal strain *Monilinia laxa*, lower values of cell proliferation were observed compared to the solvent (positive control). This buffering/protective effect may be due to the presence of phenolic compounds that are associated with the reduction in intracellular oxidative stress (hesperidin, vanillic acid, quercetin, and caffeic acid) generated by the solvent [107–110].

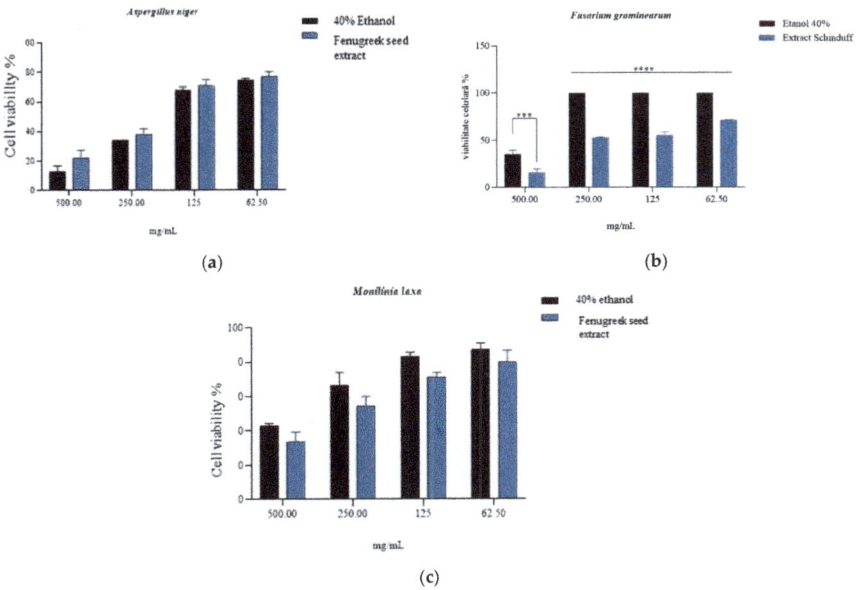

Figure 6. Quantitative antifungal activity of the *Trigonella foenum-graecum* L. (Fenugreek) seed extract in 40% ethanol, compared to that obtained for solvent (40% ethanol) on strains of *Aspergillus niger* (**a**), *Fusarium graminearum* (**b**), and *Monilinia laxa* (**c**), where *** is $p = 0.0002$, **** = $p < 0.0001$; ns = non-significant statistically; $n = 3$.

The MIC was defined as having the lowest concentration of *Trigonella foenum-graecum* seed extract that completely inhibited visible fungal growth after incubation for 7 days, compared to the solvent control. Based on the quantitative evaluation of inhibition and viability assays, the MIC values were determined as 500 mg/mL for *Fusarium graminearum* and >500 mg/mL for *Aspergillus niger* and *Monilinia laxa* (Figure 5a–c). The pronounced inhibitory effect at 500 mg/mL on *F. graminearum* (viability 15.57 ± 4.28%) indicates a fungicidal activity, whereas the partial inhibition and subsequent regrowth observed for *A. niger* and *M. laxa* after subculturing onto PDA without extract support a fungistatic effect at this concentration. Similar fungicidal–fungistatic thresholds have been reported for crude plant extracts rich in saponins and phenolics, which cause irreversible membrane disruption at higher concentrations, while lower concentrations only suppress sporulation and hyphal elongation [111]. These observations reinforce the notion of the concentration-dependent mode of action of the fenugreek extract and justify its potential use as a biofungicide prototype.

To evaluate the effect of the minimum inhibitory concentrations obtained for the development capacity of fungal mycelia, *T. foenum-graecum* extract was incorporated into PDA plates at a final concentration of 500 mg/mL. The plates were seeded with fragments of the active growth zone of the fungal culture with a diameter of 6 mm. The plates were incubated at 27 ± 1 °C for 5–7 days. After incubation, the diameters of the developed colonies were measured. The diameters of the colonies treated with extract were compared with those of the colonies grown on the culture medium without the addition of extract (control). Inhibition (%) was calculated using the formula shown above.

The evaluation of the antifungal effect expressed by inhibiting the development capacity of fungal mycelia was carried out in the presence of 500 mg/mL extract and solvent (40% ethanol) at the same concentrations. The results obtained indicate an inhibition of over 50% for the *Fusarium graminearum* strain (60.1 ± 0.9%), while for 40% ethanol, the inhibition was considerably lower (11.1 ± 1.1%), with significant differences compared to the solvent (Figure 7, $p < 0.0001$). In the case of the *Aspergillus niger* strain and *Monilinia laxa*, the inhibition percentages obtained are similar to those of the solvent, the percentages being below 20%, with only the reduction obtained for *Monilinia laxa* by the solvent being significant (16.5 ± 0.5% extract, 9.06 ± 0.96% solvent, $p < 0.001$).

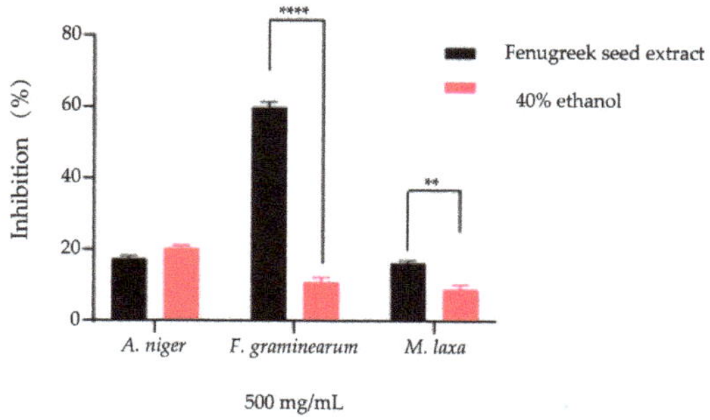

Figure 7. Evaluation of the antifungal effect by inhibition of mycelium growth (%) on PDA medium compared to the positive control (untreated strain) for both the *T. foenum-graecum* seed extract and the solvent (40% ethanol) used at similar concentrations (** $p = 0.0016$, **** $p = < 0.0001$) on the strains *Aspergillus niger*, *Fusarium graminearum*, and *Monilinia laxa*.

The antifungal activity of *Trigonella foenum-graecum* seed extract can be attributed to the synergistic action of its major phytochemical classes, particularly steroidal saponins and flavonoids. Steroidal saponins such as diosgenin and protodiosgenin are amphiphilic molecules that interact with membrane sterols (e.g., ergosterol), forming complexes that disrupt fungal cell membrane integrity and increase permeability, leading to the leakage of cellular contents and eventual cell lysis [112,113]. This mechanism has been reported in several *Fusarium* and *Aspergillus* species treated with saponin-rich extracts [114,115]. Flavonoids such as luteolin, vitexin, and apigenin derivatives also exert antifungal effects by interfering with ergosterol biosynthesis, inhibiting hyphal growth, and inducing oxidative stress through ROS generation. Moreover, phenolic acids (salicylic, caffeic, and ferulic acids) may act synergistically by altering cell wall rigidity via interactions with chitin and β-glucan components, thereby potentiating the activity of membrane-active saponins [116]. This combination of membrane disruption, enzyme inhibition, and oxidative imbalance likely explains the strong inhibitory effect observed against *Fusarium graminearum* in our assays [117].

The pronounced antifungal effect observed against *Fusarium graminearum* compared to *Aspergillus niger* and *Monilinia laxa* may be related to species-specific differences in cell wall architecture and membrane composition. *F. graminearum* exhibits a less compact chitin–glucan structure and higher ergosterol [118], which can increase susceptibility to membrane-active compounds such as steroidal saponins and phenolic acids present in the extract. In contrast, *A. niger* and *M. laxa* possess more robust cell walls, rich in melanin and cross-linked polysaccharides [119,120], providing greater resistance to oxidative and surfactant-like agents. This structural variability likely contributes to the differential sensitivity observed among the tested fungi. Similar patterns have been reported for plant-derived saponins and flavonoids, which preferentially disrupt the cell membranes of *Fusarium* species while exhibiting limited effects on melanized or spore-forming fungi.

3. Materials and Methods

3.1. Method of Obtaining the Vegetable Extract and Physicochemical Characterisation

The plant material used consists of dried fenugreek seeds obtained commercially from the company Ceaiul Casei: country of origin, Romania. The fenugreek seeds were passed three times through purified water and dried in an oven at a temperature of 40 °C for 1 h, the process being followed by their grinding. The fenugreek seed extract was obtained by initially macerating them in 70% ethanol for 10 days at ambient temperature (25 °C). The resulting extraction mixture was separated from the vegetable waste using filter paper. The vegetable waste was taken up again, over which 1000 g of 70% ethyl alcohol was added, with the extraction then being continued for 24 h, at room temperature, away from light, and with occasional shaking. This operation was repeated for another 24 h. The extraction ratio was 1:10 (*w*/*v*). The resulting extract was analysed for gallic acid concentration and concentrated by removing the alcohol in a rotary evaporator, followed by resuming the extraction process in 40% ethanol to achieve a gallic acid concentration of at least 5 mg GAE/mL—*m*/*v*. The resulting extract was stored at 40 °C until the related studies on extract analysis were started.

3.2. *Qualitative Extract Screening*

3.2.1. Highlighting the Presence of Phytochemical Compounds in the Dry Extract of Fenugreek Seeds and Plant Material by the ATR-FT-IR Method

The spectra of the dry extract and vegetal material of fenugreek were obtained using Fourier transform infrared spectroscopy (Thermo Scientific, Waltham, MA, USA, Nicolet iS50 FT-IR Spectrometer with automated beamsplitter exchange) in the mid-infrared range, 4000–500 cm^{-1}. Approximately 5–10 mg of dry extract/vegetal material was placed on the surface of the diamond crystal of the attenuated total reflective (ATR), and then it was pressed gently and analysed directly. A total of 64 spectra were acquired for background and 64 spectra for each sample, at a resolution of 4 cm^{-1}. Detection was performed using a dedicated DLaTGS detector (InfraTec, Dresden, Germany), and KBr as a beamsplitter. The reading of the FT-IR spectra for the extract sample was performed after it had previously been subjected to the drying process in an oven at 30 °C for 72 h [121].

3.2.2. Identification of Bioactive Compounds Using Fourier Transform Ion Cyclotron Resonance High-Resolution Mass Spectrometry (FT-ICR-MS) Analysis

FT-ICR-MS analysis was performed using a high-resolution mass spectrometer equipped with a 15 T superconducting magnet (Solar X-XR, QqqFT-ICR HR, Bruker, Daltonics, Germany). The instrument was calibrated before the analysis using a sodium trifluoroacetate (NaTFA) solution. The sample was prepared by diluting 30 μL of hydroalcoholic fenugreek extract in 30 mL of ultrapure water and adding 10 μL of formic acid. The resulting solution was ultrasonicated for 10 min and introduced into the electrospray ionisation (ESI) source using the direct infusion system at a flow rate of 120 μL/h. The spectra were recorded using positive ionisation in a mass range between 92 and 1500 amu, with 50 scans, with an ion accumulation time of 0.700 s, a time of flight of 0.001 s, and a data acquisition size of 4,194,304. The extract analysis parameters were as follows: capillary voltage of 3700 V, Spray Shield of −450 V, drying gas pressure (N_2) 3.0 bar at 180 °C, and a flow rate of 1.2 L/min.

3.2.3. Total Polyphenol Content Dosage—TPC

The total polyphenol content was determined following the method described by Singleton et al. [122], with minor modifications. The total phenol content (TPC) assay uses the electron transfer method to detect the presence of antioxidants, rather than the capacity. However, since it uses a redox-type reaction, it is often considered an antioxidant capacity assay. It uses the Folin–Ciocalteu Reactant (FCR, Sigma Aldrich, Darmstadt, Germany, product code F9252) which can be reduced by the phenolate anions formed by phenols in a basic medium. The scientific consensus is that molybdenum (Mo) allows FCR to receive an electron from phenols, thus being reduced from Mo^{6+} to Mo^{5+}. During the reaction, the colour shifts from yellow to blue, and it can be followed by measuring the absorbance at 765 nm. The reaction media is composed of a total of 200 μL of methanol and antioxidant extract, to which 2.5 mL of FCR (diluted in a 1:10 ratio with purified water) is added. After 10 min at room temperature, 2 mL of sodium carbonate 7.5% solution is added, and the reaction medium is left at room temperature for 2 h. A calibration curve was produced for concentrations of caffeic acid prepared with methanol, ranging from 25 to 250 μg/mL. For the extract tested, 3 solutions with either 2, 4, or 6 μL of plant extract are prepared in the same way, and are used to find the average antioxidant concentration with the help of the plot formula. The samples were run in triplicate, and then reported as mean and the ±standard deviation (SD) was calculated. The results were expressed in mg caffeic acid equivalents/L extract.

3.2.4. Total Flavonoid Content—TFC

Total Flavonoid Content antioxidant assays are commonly used for plant extracts, as more than half the identified phenolics found in plants are flavonoids. The most common method is based on the aluminium chloride colorimetric assay, where Al (III) acts as a complexing agent and can form yellow-coloured chelates of Al (III)-flavonoids [123]. Flavonoids possess many oxo and hydroxyl groups, which allow them to bind to metals such as Al (III) at a 1:1 ratio. This reaction can be followed by spectrophotometry, as the absorbance measured at 415 nm increases proportionally to the quantity of chelates formed, and, by extension, proportionally to the quantity of antioxidants in the solution. First, a solution of $AlCl_3$ (10%) was made by dissolving 0.9 g of $AlCl_3{\cdot}H_2O$ in 5 mL of H_2O. This solution was used as the reactant for this test. A 1M potassium acetate solution was also made by dissolving 0.591 g of potassium acetate in 5 mL of H_2O. A calibration curve was made with 2.8 mL H_2O, 100 μL of $AlCl_3$ (10%), 100 μL of potassium acetate (1 M), and 1500 μL of a mixture of methanol and increasing volumes of rutin (1 mg/mL). The blank contained 2.8 mL H_2O, 100 μL of $AlCl_3$ (10%), 100 μL of potassium acetate (1 M), and 1500 μL of methanol. For the plant extracts tested, 3 solutions were made with the same composition, swapping the rutin from the calibration curve with 2, 4, and 6 μL of plant extract. After leaving the reaction medium at room temperature for 30 min, the absorbance at 415 nm was measured. The calibration curve was made, and the plot line formula was used to calculate the flavonoid concentration in each extract.

3.2.5. Determination of Reducing Sugar Content (RSC)

For the quantification of reducing sugar content, the 3,5-dinitrosalicylic acid (DNSA) method was used, described by Khatri et al. [51], with some modifications. DNSA was prepared using 5.35 g NaOH, 45.5 g Na and K tartrate, and 1.25 g Na bisulfite, dissolved in distilled water. The resulting mixture was heated to 80 °C for 5 min. Measures of 1.25 g phenol and 1.575 g 3,5-dinitro salicylic acid were added and distilled water was added to make a total volume of 250 mL. For sample preparation, 350 μL of the extract sample was taken, and 650 μL of DNSA reagent was pipetted onto it. The mixture was heated to 90 °C for 15 min. After cooling, 150 μL of the reaction mixture was pipetted into 96-well plates, and the absorbance of the resulting solution was read at 540 nm. In parallel, control samples were also prepared, using 40% ethanol. The calibration curve was represented by D-glucose in different concentrations from 0 to 0.1 mg/mL.

3.2.6. Dosage of Soluble Protein Content

The method used in the determination of soluble proteins was adapted from that described by Bradford et al. [124]. Thus, the supernatant was used, and a volume of 10 μL of the extract sample was taken to which 200 μL of Bradford reagent was added. The sample and standard solutions, together with the reagent, were incubated for 15 min at room temperature. The absorbance was read at λ = 595 nm. The calibration curve was made, starting from a stock solution of bovine serum albumin of 2 mg/mL for concentrations that varied between 1.50 and 0.00 mg/mL.

3.3. Antioxidant Activity

The benefits of antioxidants have become more mainstream, and more and more studies focus on them. Between 1993 and 2003 alone, the number of publications on this subject quadrupled, and this upward trend has continued in recent decades [125]. Several antioxidant assays can be used to determine the antioxidant capacity of a sample, and they are usually based on electron transfer (ET), hydrogen atom transfer (HAT), or a mixture of both [126,127]. ET-based assays determine the capacity of an antioxidant to transfer one of its electrons and reduce the oxidant. The latter changes colour proportionally to the amount of antioxidant found in the sample, thus making it possible to follow the reaction by measuring the absorbance at a specific wavelength. The majority of HAT assays, on the other hand, involve a competitive reaction between the oxidant and the antioxidant, competing with the substrate [128,129].

3.3.1. Antioxidant Activity by DPPH Method

The DPPH (2,2-Diphenyl-1-picrylhydrazyl Radical) assay is one of the most used mixed-mode methods for determining antioxidant capacity [130]. It combines HAT as well as ET methods. However, it could be considered as mostly an ET method, given the fact that the abstraction of the hydrogen atom by the $DPPH^{\bullet}$ free radical from the antioxidant compound (HAT mechanism) happens less easily than the transfer of the electron from the free radical $DPPH^{\bullet}$ towards the antioxidant compound (ET mechanism). During this assay, a solution of $DPPH^{\bullet}$ free radicals is made by dissolving DPPH, commercially available as a powder, into distilled water. Because the chemical structure allows the decolouration of the free radical via mesomeric effects, the free radicals are highly stable and absorb light in the visible spectrum, giving the solution a deep purple colour. This technique has the advantage of being a simple and fast method to find the antioxidant capacity of a solution, and the reactive being stable enough to be commercialised and ready for use. It does, however, require the use of alcoholic solutions as media, which hinders its use for emulsions, as proteins often precipitate in alcohol. Moreover, due to steric hindrance, large antioxidant molecules react more slowly or not at all, not reaching the radical site [131,132]. Methanolic solutions were made using aliquots of 12 different plant extracts and reaching a total volume of 3 mL. The reaction is started by the addition of 1.5 mL of DPPH reactive (0.04%). It is left for 20 min at room temperature and is followed by measuring the decrease in absorbance at 515–528 nm for the 7 probes with increasing extract volume, as well as for the reference, in triplicate (containing only 3 mL of methanolic solution and 1.5 mL of reactive). For each probe, the inhibition percentage is calculated using the following formula:

$$\%I = \frac{Abs\ reference - Abs\ probe}{Abs\ reference} \times 100 \quad (1)$$

A curve is plotted % I f(log (Extract Volume)) and the line formula is used to find the IC_{50} value, the antioxidant concentration for which 50% inhibition of the free radicals is observed.

Determining antioxidant activity using the DPPH method was performed according to the method reported by Madhu [133], with minor modifications.

3.3.2. Antioxidant Activity by the CUPRAC Method

The principle underlying the CUPRAC method consists of the reduction in the cupric complex, neocuproine (Cu (II)-Nc), by antioxidants in the cuprous form (Cu (I)-Nc). The reduction of copper ions was performed according to the method described by Celik et al. [134], as follows. A measure of 60 µL of sample/standard solutions of different concentrations was mixed with 50 µL CuCl2 (10 mM), 50 µL neocuproine (7.5 mM), and 50 µL 1 M ammonium acetate buffer, pH = 7.00. The samples were incubated for 30 min, and the absorbance was measured at 450 nm.

3.3.3. Antioxidant Activity by the FRAP Method

The determination of the antioxidant power of reducing iron was performed using the method described by Corbu et al. [135] and Thaipong et al. 2006 [136], with some modifications. The following solutions were prepared: 300 mM acetate buffer, pH 3.6, 10 mM TPTZ (2,4,6-tripyridyltriazine) stock solution in 40 mM HCl, and 20 mM FeCl3 solution in distilled water. The FRAP reagent was prepared by mixing 300 mM acetic acid–sodium acetate buffer solution, pH 3.6, with 10 mM TPTZ solution and 20 mM $FeCl_3$ solution (10:1:2). The FRAP reagent was kept in a water bath at 37 °C until the analysis was performed. Over 10 µL sample/standard solution, 190 µL FRAP reagent was added and incubated for 30 min at 37 °C. After incubation, the absorbance was read at 593 nm. A 1 mM Trolox stock solution was used to plot the calibration curve, with concentrations ranging from 30 to 250 µM Trolox/mL.

3.3.4. Antioxidant Activity by TEAC (ABTS) Method

Another commonly used antioxidant capacity assay is the ABTS (2,2′-azino-bis (3-ethylbenzothiazoline-6-sulfonic acid)) method. It too has a mixed mode of functioning, combining HAT and ET mechanisms, although the ET mechanism happens faster. The $ABTS^{\bullet+}$ free radical has a blue-green colour and absorbs at 417, 734, and 815 nm. When reduced by antioxidants, it decreases the solution absorbance at 734 nm. The reaction can thus be followed by spectrophotometry, the decrease in absorbance being proportional to the logarithm of the antioxidant volume. By measuring the absorbance at 734 nm, one limits the interference of other compounds found in the extract and the turbidity. The method is simple, fast (the reaction takes less than 30 min), and can be used on a wide pH range for both lipophilic and hydrophilic compounds in both aqueous and organic solvents. Because the free radicals are not commercially available, 7 mM of ABTS diammonium salt was dissolved and oxidised by 2.5 mM potassium persulfate, in distilled water, to form the oxidant species by the emission of one electron from a nitrogen atom. The solution was left for 16 h to generate the free radicals, turning a dark teal colour. The reactive solution was later diluted in methanol until reaching an absorbance of 0.7 at 734 nm. For the extract, 3 probes were made containing 2.900 mL of the ABTS reactive and 0.100 mL of the methanolic extract solutions of increasing concentrations. After 5 min, the absorbance at 734 nm was measured, and the inhibition percentage was calculated as stated above. A plot %I = f(log extract Volume) was later made for each extract, and the IC_{50} value was found.

3.4. Germination Bioassay

This study aimed to evaluate the influence of the extract obtained from fenugreek seeds (*Trigonella foenum-graecum*) in 40% ethanol on wheat seeds. The method used was adapted from the procedures described by Mitelut and Popa [72], Ghayal et al. [137], and Perisoara et al. [138]. The *Trigonella foenum-graecum* extract, as well as the solvent (40% ethanol), were studied at concentrations of 0.10%, 0.50%, 1.00%, and 1.50%, diluted with distilled water, and the results were reported compared to a negative control, in this case, distilled water. Before starting the study, the Petri plates, with a diameter of 100 mm × 200 mm (Corning Petri plates, New York, NY, USA), were disinfected by being sprayed with isopropyl alcohol 70% (Contec IPA) and left to dry overnight. The filter papers (IDL. GMBH, type blue, 90 diameters) were sterilised in a hood equipped with a UV lamp (JOUAN MSC 12, Yerville, France) for 30 min on each side and the wheat seeds (*Triticum aestivum*) were washed with purified water and dried in an incubator (Memmert IPP30, Schwabach, Germany) at a temperature of 30 °C. After completing the washing, sterilisation, and disinfection procedures, 20 wheat seeds (of medium size) were placed in each Petri plate containing the filter paper, over which 5 mL of the extract sample obtained following the dilutions was pipetted. The same procedure is used for testing the negative control and the solvent (40% ethanol), replacing the 5 mL of the extract sample with the negative control/solvent sample. The entire experiment was performed in triplicate, with each plate containing 20 wheat seeds. The Petri plates were incubated in the dark (Memmert IPP30 incubator) for 6 days, at a temperature of 25 ± 1 °C. At the end of the incubation period, germinated seeds were counted, and the length of the roots of the germinated seeds was measured. Based on the results obtained, the following indicators were determined: germination percentage (GP), relative seed germination index (RSG), relative root growth index (RRG), and germination index (Gi), according to the calculation formulas below [79,139]:

$$\mathrm{GP} = \frac{\textit{Number of germinated seeds}}{\textit{Number of total seeds tested}} \times 100 \tag{2}$$

$$\mathrm{RSG} = \frac{\textit{Average number of germinated seeds in the sample}}{\textit{Average number of germinated seeds in the control}} \times 100 \tag{3}$$

$$\mathrm{RRG} = \frac{\textit{Average root length of germinated seeds in the sample}}{\textit{Average root length of germinated seeds in the control}} \times 100 \tag{4}$$

$$\mathrm{Gi} = \mathrm{RSG} \times \mathrm{RRG}/100. \tag{5}$$

According to the literature [76,78], it appears that Gi values, depending on the results obtained, may indicate a phytostimulant or phytotoxic effect of the extract on the seeds studied. For the determination and interpretation of Gi values, the data obtained regarding RRG and RSG were processed [78]. Obtaining a Gi value lower than 50% indicates a strong phytotoxic activity of the extract, a Gi value as close as possible to 0% highlights extreme phytotoxicity, a Gi value between 50 and 80% is associated with moderate phytotoxic activity on the plant, a value above 80% is considered to indicate that the tested extract does not manifest phytotoxic activity, and a Gi value above 100% indicates that the studied extract possesses compounds with phytostimulant properties [76,78]. The experiment ends after 65% of the seeds in the control sample have germinated and/or developed roots of at least 20 mm in length [140,141].

3.5. Catalase, Superoxide Dismutase, Peroxidase, and Lipid Peroxidation Assessment

The method used to sterilise the seeds was the incubation of seeds for 30 min at 50 °C in distilled water. The seeds, thus conditioned, were evenly distributed in Petri dishes of 50 mm diameter on a soapy bed of filter paper which had been previously sterilised by UV irradiation and moistened with distilled water. In the case of samples where the simulation of moderate drought conditions was desired, the saddling material was moistened with PEG 6000 to achieve an osmotic potential of −0.6 Mpa [142]. They were covered with new filter paper, and then the test solution was added by pipette (about 10 mL). Three replicates were performed for each species. The control variant was moistened with distilled water. The test solution was dripped in as many places as possible to wet the paper evenly. The Petri dish was covered with the upper half, where the test variant and the repetition were marked with a marker. The plates were covered with aluminium foil and randomly placed in a constant 25 °C environment. Seeds were checked daily to monitor germination progress while ensuring that the filter paper did not dry out. If necessary, a few drops of the test solution were added to maintain optimal humidity. Germination was considered complete when the root became visible. Seeds treated for 6 days were sampled for the determination of biochemical parameters (catalase, superoxide dismutase, peroxidase, and lipid peroxidation). Catalase activity was investigated according to the method of Aebi [143]. The estimation was performed spectrophotometrically, measuring the decrease in absorbance at 240 nm. The reaction mixture contained 0.01M phosphate buffer (pH 7.0), 2 mM H_2O_2, and cell lysates. The specific activity of catalase is expressed in terms of units/mg protein. A unit is defined as the velocity constant per second. Superoxide dismutase is a metalloenzyme that catalyses the dismutation of superoxide anion into oxygen and hydrogen peroxide. We have been using the spectrometric procedures described by Sigma Aldrich to determine the SOD activity in samples. The method is based on the spectrophotometric evaluation (550 nm absorption spectra) of the inhibition rate of cytochrome C reduction by competing for the superoxide radical with superoxide dismutase [144]. Peroxidase was analysed according to Prochazkova [145]. The 3 mL reaction mixture contained 28 µL guaiacol, 2 mM H_2O_2, 0.1 M phosphate buffer (pH 6.0), and 0.1 mL enzyme. Absorbance was recorded at 470 nm. Lipid peroxidation was evaluated using the method with thiobarbituric acid (TBA). The MDA-TBA adducts formed after the reaction of TBA and MDA from the biological sample were measured using a colorimetric method (λ = 532 nm).

3.6. Determination of Antifungal Activity

3.6.1. Qualitative Screening of the Antifungal Activity of the 40% Ethanol-Conditioned Extract of *Trigonella foenum-graecum* L. Seeds Against Some Pathogenic Fungal Species

Qualitative screening enables the collection of preliminary information on the mode of action, the effect of a potential antifungal product (Fenugreek seed extract) on fungal cultures on the semi-solid substrate, as well as the stability of preliminary concentrations in correlation with the effect obtained. Qualitative screening of the antifungal activity of the *Trigonella foenum-graecum* L. seeds in 40% ethanol was performed according to protocols outlined in the specialised literature [146–148] and the CLSI M38M51S standard [149], using three pathogenic fungal species, namely *Aspergillus niger* and *Fusarium graminearum* obtained from the Microbial Strains Collection of the Faculty of Biology, University of Bucharest, Romania; and *Monilinia laxa* obtained from the Microbial Strains Collection of the Faculty of Agriculture, University of Agronomic Sciences and Veterinary Medicine Bucharest and confirmed by the MALDI-TOF method.

The evaluation was performed using the disc-diffusion method on the semi-solid PDA (Potato Dextrose Agar) medium, liquefied, cooled to 40–45 °C, and poured into Petri dishes. The culture medium was inoculated with the fungal inoculum adjusted to the nephelometric standard 1 McFarland (106 CFU/mL). Sterilised filter paper discs (6 mm in diameter) were dispersed on the surface of the inoculated plates, over which 10 µL of the extract was prepared at concentrations of 500 mg/mL,250 mg/mL, 125 mg/mL, and 62.5 mg/mL. After 5–7 days of incubation at 27 ± 1 °C, the inhibition diameter of the colonies developed on the culture medium was measured. The diameter of inhibition of colonies formed by fungal strains tested in the presence of Fenugreek extract and the solvent represented by 40% ethanol was compared with that of colonies exposed to the antifungal control (Nystatin, Amphotericin B, Imidazole)—positive control. The negative control (absence of extract and antifungal) was represented by DMSO. The determination of the relative percentage of inhibition (PRI) was performed using the formula [150]

$$\text{PRI} = \frac{100(X - Y)}{(Z - Y)} \tag{6}$$

where X = total inhibition area of the extract/solvent tested; Y = total inhibition area of the negative control (DMSO); and Z = total inhibition area of the standard fungicide (positive control). The total inhibition area was calculated using area = πr2, where r = radius of the inhibition zone.

3.6.2. Quantitative Determination of the Antifungal Activity of the *Trigonella foenum-graecum* L. Seeds Extract in 40% Ethanol on Some Pathogenic Fungal Species

The quantitative determination of antifungal activity was performed using the microdilution method according to the protocols in the specialised literature [146–148] and the CLSI M38M51S standard [149], using the pathogenic fungal species *Aspergillus niger* and *Fusarium graminearum*. Using RPMI (Roswell Park Memorial Institute) 1640 medium, in 24-well plates, the fungal inoculum, adjusted to the nephelometric standard 1 McFarland (106 CFU/mL), was seeded. The final tested concentrations of the extract (500 mg/mL, 250 mg/mL, 125 mg/mL, and 62.5 mg/mL) were made at a final volume of 2 mL, directly in the culture medium. After 5–7 days of incubation at a temperature of 27 $\pm$ 1 °C, the samples were read spectrophotometrically at λ = 620 nm to determine the cell viability (%). The results obtained were compared with the positive control (fungistatic agent) and the negative control (fungal culture without treatment) to determine the antifungal activity, using the following formula, according to Javed et al. [151]:

$$\text{Inhibitory concentration (\%)} = (A - B)/A \times 100 \tag{7}$$

where A is the concentration of the negative control (fungal culture without treatment) and B represents the concentration of the positive control (culture treated with extract/ethanol).

4. Statistical Analysis

All studies were performed in triplicate. The results were reported as averages, and the mean standard deviation ($\pm$) and relative standard deviation (RSD) were calculated. The statistical analysis for the enzyme activity was performed using Microsoft Excel 2019. Statistical evaluations for the germination bioassay and antifungal activity were performed using GraphPad Prism 9 (San Diego, CA, USA). Data were analysed using the two-way ANOVA test. The level of significance was set to * $p < 0{:}05$, ** $p < 0{:}01$, and *** $p < 0{:}001$ vs. control.

5. Conclusions

In this study, we obtained an extract from *Trigonella foenum-graecum* seeds, rich in phytochemical compounds known for their antioxidant and antimicrobial properties, which could be effectively utilised in the plant protection industry.

The extract was found to contain significant levels of reducing sugars; proteins; phenolic compounds, including flavonoids and polyphenols, such as rutin, abscisic acid, salicylic acid, and vitexin; and steroidal saponins, the most notable of which are diosgenin and its precursor, protodiosgenin; as well as a compound belonging to the pyridine alkaloids category, trigonelline. The compounds identified are well documented in the literature for their antimicrobial and antioxidant activities.

Both qualitative and quantitative assessments confirmed the antifungal activity of the extract, particularly against *Fusarium graminearum* and *Monilinia laxa* strains at the highest concentration tested (500 mg/mL). These findings are consistent with the observed phytotoxic effect on wheat seeds at the maximum concentration (1.50%), which can be attributed to the high accumulation of biological compounds in the extract. Future research will include studies on plants under greenhouse conditions to establish effective concentrations for field application, taking into account soil dilution, plant surface absorption, and phytotoxicity thresholds.

This work provides valuable insights into obtaining a potential biologically active product ecologically and efficiently, without a negative impact on the environment, with proven antifungal properties. It also offers a promising alternative for protecting cereal and fruit crops.

Supplementary Materials: The following supporting information can be downloaded at https://www.mdpi.com/article/10.3390/plants14213320/s1: Figure S1: Full mass spectra of fenugreek extract using positive ESI ionisation; Figure S2: Molecular peak for Protodiosgenin (C51H84O22)—m/z calculated is 1049.552, ESI-positive; Figure S3: Full mass spectra of Sodium trifluoroacetate using positive ESI ionisation for the calibration of the instrument; Figure S4: Full mass spectra of blank sample using positive ESI ionisation; Figure S5: Molecular peak for Diosgenin (C27H42O3)—m/z calculated is 415.320, ESI-positive; Figure S6: Molecular peak for Trigonelline (C7H7O2)—m/z calculated is 138.054, ESI-positive; Figure S7: Molecular peak for Tigogenine (C27H44O3)—m/z calculated is 417.336, ESI-positive; Figure S8: Molecular peak for Yamogenin (C27H42O3)—m/z calculated is 415.320, ESI-positive; Figure S9: Molecular peak for Vitexin (C21H20O10)—m/z calculated is 433.112, ESI-positive; Figure S10: Molecular peak for Salicylic Acid (C7H6O3)—m/z calculated is 139.038, ESI-positive; Figure S11: Molecular peak for Apigenin 6-C-galactoside 8-C-arabinoside (C26H28O14)—m/z calculated is 565.115, ESI-positive; Figure S12: Molecular peak for Luteoline-7-O glucoside (C21H18O12)—m/z calculated is 463.087, ESI-positive; Figure S13: Molecular peak for Apigenin–7–O–glucoside (C21H20O10)—m/z calculated is 433.112, ESI-positive; Figure S14: Molecular peak for Abscisic Acid (C15H20O4)—m/z calculated is 265.143, ESI-positive; Figure S15: Molecular peak for Rutin (C27H30O16)—m/z calculated is 611.160, ESI-positive; Figure S16: Molecular peak for t-Resveratrol (C14H12O3)—m/z calculated is 228.078, ESI-positive.

Author Contributions: Conceptualization, A.P., B.-M.T., and M.D.E.; methodology, A.P., B.-M.T., M.C., M.D.E., L.P., and E.Ş.I.; software A.P., B.-M.T., M.D.E., M.C., A.-M.F., B.P., D.E.M., R.C.Z., and L.P.; validation, A.P., B.-M.T., M.C., M.D.E., B.P., D.E.M., E.Ş.I., A.-M.F., and L.P.; formal analysis, A.P., B.-M.T., M.C., M.D.E., E.Ş.I., B.P., A.-M.F., R.C.Z., and L.P.; investigation, A.P., B.-M.T., and M.D.E.; resources, A.P., B.-M.T., M.C., M.D.E., B.P., D.E.M., A.-M.F., and L.P.; data curation, A.P., B.-M.T., M.C., M.D.E., E.Ş.I., B.P., D.E.M., A.-M.F., R.C.Z., and L.P.; writing—original draft preparation, A.P., B.-M.T., and M.D.E.; writing—review and editing, A.P., B.-M.T., and M.D.E.; visualisation, S.C., L.P., D.E.M., and R.C.Z. supervision, S.C.; funding acquisition, A.P., B.-M.T., and M.D.E. and S.C. All authors have read and agreed to the published version of the manuscript.

Funding: The research was funded by the University of Agronomic Sciences and Veterinary Medicine in Bucharest and by the laboratories involved in the research.

Institutional Review Board Statement: Not applicable.

Data Availability Statement: The data presented in this study and supplymentary material and are available in article.

Acknowledgments: We would like to express our thanks and gratitude to The University of Agronomical Sciences and Veterinary Medicine of Bucharest for its financial and logistical support, and also to all those involved in the completion of this work.

Conflicts of Interest: The authors declare no conflicts of interest.

References

1. Youssef, K.; de Oliveira, A.G.; Tischer, C.A.; Hussain, I.; Roberto, S.R. Synergistic Effect of a Novel Chitosan/Silica Nanocomposites-Based Formulation against Gray Mold of Table Grapes and Its Possible Mode of Action. *Int. J. Biol. Macromol.* **2019**, *141*, 247–258. [CrossRef]
2. Hussien, A.; Ahmed, Y.; Al-Essawy, A.H.; Youssef, K. Evaluation of Different Salt-Amended Electrolysed Water to Control Postharvest Moulds of Citrus. *Trop. Plant Pathol.* **2018**, *43*, 10–20. [CrossRef]
3. Porteous-Alvarez, A.J.; Maldonado-González, M.M.; Mayo-Prieto, S.; Lorenzana, A.; Paniagua-García, A.I.; Casquero, P.A. Green Strategies of Powdery Mildew Control in Hop: From Organic Products to Nanoscale Carriers. *J. Fungi* **2021**, *7*, 490. [CrossRef]
4. González-Fernández, R.; Prats, E.; Jorrín-Novo, J.V. Proteomics of Plant Pathogenic Fungi. *J. Biomed. Biotechnol.* **2010**, *2010*, 932527. [CrossRef] [PubMed]
5. El-Baky, N.A.; Amara, A.A.A.F. Recent Approaches towards Control of Fungal Diseases in Plants: An Updated Review. *J. Fungi* **2021**, *7*, 900. [CrossRef] [PubMed]
6. Buttar, Z.A.; Cheng, M.; Wei, P.; Zhang, Z.; Lv, C.; Zhu, C.; Ali, N.F.; Kang, G.; Wang, D.; Zhang, K. Update on the Basic Understanding of *Fusarium graminearum* Virulence Factors in Common Wheat Research. *Plants* **2024**, *13*, 1159. [CrossRef]
7. Yang, F.; Jacobsen, S.; Jørgensen, H.J.L.; Collinge, D.B.; Svensson, B.; Finnie, C. *Fusarium graminearum* and Its Interactions with Cereal Heads: Studies in the Proteomics Era. *Front. Plant Sci.* **2013**, *4*, 41579. [CrossRef] [PubMed]
8. Desjardins, A.E.; Proctor, R.H. Molecular Biology of *Fusarium* Mycotoxins. *Int. J. Food Microbiol.* **2007**, *119*, 47–50. [CrossRef]
9. Balsells-Llauradó, M.; Silva, C.J.; Usall, J.; Vall-llaura, N.; Serrano-Prieto, S.; Teixidó, N.; Mesquida-Pesci, S.D.; de Cal, A.; Blanco-Ulate, B.; Torres, R. Depicting the Battle between Nectarine and *Monilinia laxa*: The Fruit Developmental Stage Dictates the Effectiveness of the Host Defenses and the Pathogen's Infection Strategies. *Hortic. Res.* **2020**, *7*, 167. [CrossRef]
10. Bellingeri, M.; Quilot-Turion, B.; Lino, L.O.; Bevacqua, D. The Crop Load Affects Brown Rot Progression in Fruit Orchards: High Fruit Densities Facilitate Fruit Exposure to Spores but Reduce the Infection Rate by Decreasing Fruit Growth and Cuticle Cracking. *Front. Ecol. Evol.* **2018**, *5*, 306567. [CrossRef]
11. Dini, M.; Raseira, M.d.C.B.; Corre, M.N.; Signoret, V.; Quilot-Turion, B. Influence of Fruit Wounding on Subsequent *Monilinia laxa* Infection of Nectarines. *Agronomy* **2023**, *13*, 1235. [CrossRef]
12. Samson, R.A.; Houbraken, J.; Thrane, U.; Frisvad, J.C.; Andersen, B. *Food and Indoor Fungi*, 2nd ed.; CBS-KNAW Fungal Biodiversity Centre: Utrecht, The Netherlands, 2010; p. 390.
13. Ichim, E.; Marutescu, L.; Popa, M.; Cristea, S. Antimicrobial efficacy of some plant extracts on bacterial ring rot pathogen, clavibacter michiganensis ssp. sepedonicus. *EuroBiotech J.* **2017**, *1*, 85–88. [CrossRef]
14. Reshi, Z.A.; Ahmad, W.; Lukatkin, A.S.; Javed, S.B. From Nature to lab: A review of secondary metabolite biosynthetic pathways, environmental influences, and in vitro approaches. *Metabolites* **2023**, *13*, 895. [CrossRef]
15. Ramakrishna, K.; Chakravarthi, G.; Rai, S.; Singh, P.; Mishra, S.; Mishra, D.; Tiwari, N. Biochemical characterization of plant secondary metabolites. In *Secondary Metabolites and Biotherapeutics*, 1st ed.; Kumar, A., Kumar, S., Eds.; Academic Press: Kolkata, India, 2024; Volume 1, pp. 39–61.
16. Suteu, D.; Rusu, L.; Zaharia, C.; Badeanu, M.; Daraban, G.M. Challenge of utilization vegetal extracts as natural plant protection products. *Appl. Sci.* **2020**, *10*, 8913. [CrossRef]
17. Deresa, E.M.; Diriba, T.F. Phytochemicals as alternative fungicides for controlling plant diseases: A comprehensive review of their efficacy, commercial representatives, advantages, challenges for adoption, and possible solutions. *Heliyon* **2023**, *9*, e13810. [CrossRef] [PubMed]
18. Zabka, M.; Pavela, R.; Gabrielova-Slezakova, L. Promising antifungal effect of some Euro-Asiatic plants against dangerous pathogenic and toxinogenic fungi. *J. Sci. Food Agric.* **2011**, *91*, 492–497. [CrossRef] [PubMed]
19. Cosoveanu, A.; Cabrera, R.; Giménez Mariño, C.; Iacomi, B.M.; González-Coloma, A. Antifungal activity of plant extracts against pre and postharvest pathogens. *Sci. Pap. Ser. A Agon.* **2013**, *LVI*, 206–211.
20. Chon, B.; Xu, S.; Lee, Y.J. Compensation of Strong Water Absorption in Infrared Spectroscopy Reveals the Secondary Structure of Proteins in Dilute Solutions. *Anal. Chem.* **2021**, *93*, 2215–2225. [CrossRef]
21. Barth, A. Infrared Spectroscopy of Proteins. *Biochim. Biophys. Acta-Bioenerg.* **2007**, *1767*, 1073–1101. [CrossRef]
22. Sene, C.F.B.; McCann, M.; Wilson, R.; Grinter, R. Fourier-Transform Raman and Fourier-Transform Infrared Spectroscopy (an Investigation of Five Higher Plant Cell Walls and Their Components). *Plant Physiol.* **1995**, *106*, 1623–1631. [CrossRef]
23. Portaccio, M.; Errico, S.; Chioccarelli, T.; Cobellis, G.; Lepore, M. Fourier-Transform Infrared Microspectroscopy (FT-IR) Study on Caput and Cauda Mouse Spermatozoa. *Proceedings* **2019**, *42*, 19. [CrossRef]
24. El-Bahy, G. FTIR and Raman spectroscopic study of Fenugreek (*Trigonella foenum graecum* L.) seeds. *J. Appl. Spectrosc.* **2005**, *72*, 111–116. [CrossRef]
25. Akbari, S.; Abdurahman, N.H.; Yunus, R.M.; Fayaz, F. Microwave-assisted extraction of saponin, phenolic and flavonoid compounds from *Trigonella foenum-graecum* seed based on two level factorial design. *J. Appl. Res. Med. Aromat. Plants* **2019**, *14*, 100212. [CrossRef]

26. Silva, S.D.; Feliciano, R.P.; Boas, L.V.; Bronze, M.R. Application of FTIR-ATR to Moscatel dessert wines for prediction of total phenolic and flavonoid contents and antioxidant capacity. *Food Chem.* **2014**, *150*, 489–493. [CrossRef] [PubMed]
27. Coates, J. Interpretation of infrared spectra: A practical approach. In *Encyclopedia of Analytical Chemistry*; Meyers, R.A., Ed.; John Wiley and Sons: Chichester, UK, 2006; pp. 1–23.
28. Houghton, P.J.; Raman, A. Laboratory Handbook for the Fractionation of Natural Extracts. In *Laboratory Handbook for the Fractionation of Natural Extracts*; Springer Science Business Media: London, UK, 1998; p. 199.
29. Azmir, J.; Zaidul, I.S.M.; Rahman, M.M.; Sharif, K.M.; Mohamed, A.; Sahena, F.; Omar, A.K.M. Techniques for extraction of bioactive compounds from plant materials: A review. *J. Food Eng.* **2013**, *117*, 426–436. [CrossRef]
30. Chatterjee, S.; Variyar, P.S.; Sharma, A. Effect of γ-irradiation on the antioxidant activity of fenugreek (*Trigonella foenum-graecum*) seed extract. *J. Environ. Agric. Food Chem.* **2011**, *10*, 2798–2805.
31. Dai, J.; Mumper, R. Plant phenolics: Extraction, analysis and their antioxidant and anticancer properties. *Molecules* **2010**, *15*, 7313–7352. [CrossRef]
32. Rahmani, M.; Hamel, L.; Toumi-Benali, F.; Dif, M.M.; Moume, F.; Rahmani, H. Determination of antioxidant activity, phenolic quantification of four varieties of fenugreek *Trigonella foenum graecum* L. seed extract cultured in west Algeria. *J. Mater. Environ. Sci.* **2018**, *9*, 1656–1661.
33. Turker, I.; Dastan, S.; Isleroglu, H. Extraction of phenolic compounds from fenugreek seeds using different extraction techniques. In Proceedings of the III. International Agricultural, Biological & Life Science Conference, Edirne, Turkey, 1–3 September 2021.
34. Lohvina, H.; Sándor, M.; Wink, M. Effect of ethanol solvents on total phenolic content and antioxidant properties of seed extracts of fenugreek (*Trigonella foenum-graecum* L.) varieties and determination of phenolic composition by HPLC-ESI-MS. *Diversity* **2021**, *14*, 7. [CrossRef]
35. Jokic, S.; Bucić-Kojić, A.; Planinić, M.; Velić, D.; Tomas, S.; Bilic, M.; Bešvir, Ž. The effect of solvent and temperature on extraction yield of phenolic compounds from soybeans, antioxidant activity and colour of extracts. In Proceedings of the 5th International Congress FLOUR-BREAD '09, 7th Croatian Congress of Cereal Technologists, Opatja, Croatia, 21–23 October 2009.
36. Norziah, M.H.; Fezea, F.A.; Bhat, R.; Ahmad, M. Effect of extraction solvents on antioxidant and antimicrobial properties of fenugreek seeds (*Trigonella foenum-graecum* L.). *Int. Food Res. J.* **2015**, *22*, 1261–1271.
37. Amid, B.T.; Mirhosseini, H. Effect of different purification techniques on the characteristics of heteropolysaccharide-protein biopolymer from durian (*Durio zibethinus*) seed. *Molecules* **2012**, *17*, 10875–10892. [CrossRef] [PubMed]
38. Kaczorová, D.; Karalija, E.; Dahija, S.; Bešta-Gajević, R.; Parić, A.; Ćavar Zeljković, S. Influence of Extraction Solvent on the Phenolic Profile and Bioactivity of Two *Achillea* Species. *Molecules* **2021**, *26*, 1601. [CrossRef]
39. Rivas-García, L.; Crespo-Antolín, L.; Forbes-Hernández, T.Y.; Romero-Márquez, J.M.; Navarro-Hortal, M.D.; Arredondo, M.; Llopis, J.; Quiles, J.L.; Sánchez-González, C. Bioactive Properties of Tagetes erecta Edible Flowers: Polyphenol and Antioxidant Characterization and Therapeutic Activity against Ovarian Tumoral Cells and Caenorhabditis elegans Tauopathy. *Int. J. Mol. Sci.* **2024**, *25*, 280. [CrossRef] [PubMed]
40. Syed, Q.A.; Rashid, Z.; Ahmad, M.H.; Shukat, R.; Ishaq, A.; Muhammad, N.; Rahman, H.U.U. Nutritional and therapeutic properties of fenugreek (*Trigonella foenum-graecum*): A review. *Int. J. Food Prop.* **2020**, *23*, 1777–1791. [CrossRef]
41. Yilmaz, Y.; Toledo, R.T. Oxygen radical absorbance capacities of grape/-wine industry byproducts and effect of solvent type on the extraction of grape seed polyphenols. *J. Food Compos. Anal.* **2006**, *19*, 8. [CrossRef]
42. Premanath, R.; Sudisha, N.; Devi, N.L.; Aradhya, S.M. Antibacterial and anti-oxidant activities of fenugreek (*Trigonella foenum graecum* L.) leaves. *Res. J. Med. Plant* **2011**, *5*, 695–705. [CrossRef]
43. Sudina, G.F.; Mirzoeva, O.K.; Pushkarevr, M.A.; Korshunova, G.; Sumbatyan, N.V.; Varfolomeev, S.D. Caffeic acid phenethyl ester as a lipoxygenase inhibitor with antioxidant properties. *FEBS Lett.* **1993**, *329*, 21–24. [CrossRef]
44. Sakhria, M.; Hichem, A.; Hafsia, B.; Abdelfettah, E.; Najla, H. Phytochemical study and protective effect of *Trigonella foenum-graecum* (Fenugreek seeds) against carbon tetrachloride-induced toxicity in liver and kidney of male rat. *Biomed. Pharmacother.* **2016**, *88*, 19–26.
45. Pasha, I.; Shabbir, M.A.; Haider, M.A.; Afzal, B.; Chughtai, M.F.; Ahmad, S.; Manzoor, M.S. Biochemical Evaluation of *Trigonella foenum graecum* (Fenugreek) with Special Reference to Phenolic Acids. *Pak. J. Sci. Ind. Res. Ser. B Biol. Sci.* **2017**, *60*, 154–161.
46. Singh, P.; Bajpai, V.; Gond, V.; Kumar, A.; Tadigoppula, N.; Kumar, B. Determination of Bioactive Compounds of Fenugreek (*Trigonella foenum-graecum*) Seeds Using LC-MS Techniques. *Mol. Biol.* **2020**, *2107*, 377–393.
47. Bandara, K.R.V.; Padumadasa, C.; Peiris, D.C. Potent antibacterial, antioxidant and toxic activities of extracts from *Passiflora suberosa* L. leaves. *PeerJ Life Environ.* **2018**, *6*, e4804. [CrossRef] [PubMed]
48. Amborabé, B.-E.; Fleurat-Lessard, P.; Chollet, J.-F.; Roblin, G. Antifungal effects of salicylic acid and other benzoic acid derivatives towards Eutypa lata: Structure–activity relationship. *Plant Phys. Biochem.* **2002**, *40*, 1051–1060. [CrossRef]
49. Hernández-Guzmán, C.; Hernández-Montiel, L.G.; Velázquez-Lizarraga, A.E.; Ríos-González, L.J.; Huerta-Ochoa, S.; Cervantes-Güicho, V.d.J.; Morales-Martínez, T.K.; Mejía-Ruíz, C.H.; Reyes, A.G. Improved antioxidant, antifungal and herbicidal activities through the bioconversion of diosgenin by *Yarrowia lipolytica* P01a. *Plants* **2024**, *13*, 2629. [CrossRef]

50. Haouala, R.; Hawala, S.; El-Ayeb, A.; Khanfir, R.; Boughanmi, N. Aqueous and organic extracts of *Trigonella foenum-graecum* L. inhibit the mycelia growth of fungi. *J. Environ. Sci.* **2008**, *20*, 1453–1457. [CrossRef]
51. Khatri, D.; Chhetri, S.B.B. Reducing Sugar, Total Phenolic Content, and Antioxidant Potential of Nepalese Plants. *BioMed Res. Int.* **2020**, *2020*, 7296859. [CrossRef]
52. Dawood, M.G.; El-Rorkiek, K.G.A.; El-Awadi, M.E.-S.; Sadak, M.S. Physiological effect of aqueous seed extract of fenugreek on productivity and grain quality of wheat plant. *Asian J. Appl. Sci.* **2020**, *13*, 107–113. [CrossRef]
53. Jeandet, P.; Formela-Luboińska, M.; Labudda, M.; Morkunas, I. The Role of Sugars in Plant Responses to Stress and Their Regulatory Function during Development. *Int. J. Mol. Sci.* **2022**, *23*, 5161. [CrossRef]
54. Patel, J.; Dhruve, J.; Talati, J. Nutraceutical and molecular characterization of different fenugreek (*Trigonella foenum-graecum* L.) genotypes. *World J. Pharm. Pharm. Sci.* **2015**, *4*, 1267–1287.
55. Ghevariya, H.; Vatukiya, V.; Mistry, N.; JJain, N. Comparative evaluation of bioactive compounds and antioxidant properties of fenugreek (*Trigonella foenum-graecum* L.) seed, stem, leaf, and microgreens. *Third Concept Int. J. Ideas* **2023**, *37*, 62–75.
56. Valko, M.; Leibfritz, D.; Moncol, J.; Cronin, M.T.D.; Mazur, M.; Telser, J. Free radicals and antioxidants in normal physiological functions and human disease. *Int. J. Biochem. Cell Biol.* **2007**, *39*, 44–84. [CrossRef] [PubMed]
57. Radi, R. Oxygen radicals, nitric oxide, and peroxynitrite: Redox pathways in molecular medicine. *PNAS J.* **2018**, *115*, 5839–5848. [CrossRef] [PubMed]
58. Sies, H.; Jones, D.P. Reactive oxygen species (ROS) as pleiotropic physiological signalling agents. *Nat. Rev. Mol. Cell Biol.* **2020**, *21*, 363–383. [CrossRef] [PubMed]
59. Murphy, M.P. How mitochondria produce reactive oxygen species. *Biochem J.* **2009**, *417*, 1–13. [CrossRef]
60. Turrens, J.F. Mitochondrial formation of reactive oxygen species. *J. Physiol.* **2003**, *552*, 335–344. [CrossRef]
61. Bahorun, T.; Soobrattee, M.A.; Luximon-Ramma, V.; Aruoma, O.I. Free radicals and antioxidants in cardiovascular health and disease. *Internet J. Med. Update* **2006**, *1*, 25–41. [CrossRef]
62. Butterfield, D.A. Amyloid beta-peptide (1-42)-induced oxidative stress and neurotoxicity: Implications for neurodegeneration in Alzheimer's disease brain. A review. *Free Radic. Res.* **2002**, *36*, 1307–1313. [CrossRef]
63. Mahajan, A.; Tandon, V.R. Antioxidants and rheumatoid arthritis. *J. Indian Rheumatol. Ass.* **2004**, *12*, 139–142.
64. Meyer, C.H.; Sekundo, W. Nutritional supplementation to prevent cataract formation. *Dev. Ophthalmol.* **2005**, *38*, 103–119. [PubMed]
65. Hracsko, Z.; Orvos, H.; Novak, Z.; Pal, A.; Varga, I.S. Evaluation of oxidative stress markers in neonates with intra-uterine growth retardation. *Redox. Rep.* **2008**, *13*, 11–16. [CrossRef]
66. Manjamalai, A.; Berlin Grace, V.M. Antioxidant activity of essential oils from *Wedelia chinensis* (osbeck) in vitro and in vivo lung cancer bearing C57BL/6 mice. *Asian Pac. J. Cancer Prev.* **2012**, *13*, 3065–3071. [CrossRef]
67. Bari, M.W.; Islam, A.; Islam, M.M.; Sultana, M.J.; Afroz, R.; Khan, M.M.R.; Parul, S.S.; Swaraz, A.M.; Hossain, M.I.; Islam, M.A. Determination of in vitro antioxidant activity and in vivo antineoplastic effects against Ehrlich ascites carcinoma of methanolic extract of *Sphagneticola calendulacea* (L.) Pruski. *Heliyon* **2021**, *7*, e07228. [CrossRef]
68. Lee, K.J.; Oh, Y.C.; Cho, W.K.; Ma, J.Y. Antioxidant and Anti-Inflammatory Activity Determination of One Hundred Kinds of Pure Chemical Compounds Using Offline and Online Screening HPLC Assay. *Evid.-Based Complement. Altern. Med.* **2015**, *2015*, 165457. [CrossRef]
69. Floegel, A.; Kim, D.; Chung, S.; Koo, S.I.; Chun, O.K. Comparison of ABTS/DPPH assays to measure antioxidant capacity in popular antioxidant-rich US foods. *J. Food Compos. Anal.* **2011**, *24*, 1043–1048. [CrossRef]
70. Kaya, E.; Akbas, P. Determination of the Antimicrobial, Antioxidant Activities and Fatty Acid Composition of *Peganum harmala* Seed. *Braz. Arch. Biol. Tech.* **2022**, *66*, e23220237.
71. Bukhari, S.B.; Bhanger, M.I.; Memon, S. Antioxidative Activity of Extracts from Fenugreek Seeds (*Trigonella foenum-graecum*). *Pak. J. Anal. Environ. Chem.* **2008**, *9*, 78–83.
72. Mitelut, A.C.; Popa, M.E. Seed germination bioassay for toxicity evaluation of different composting biodegradable materials. *Rom. Biotechnol. Lett.* **2011**, *16*, 12–129.
73. Zucconi, F.; Monaco, A.; Forte, M.; Bertoldi, M. Phytotoxins during the stabilization of organic matter. In *Composting of Agricultural and Other Wastes*; Gasser, J.K.R., Ed.; Elsevier Applied Science: London, UK; New York, NY, USA, 1985; pp. 73–88.
74. Handreck, K.; Black, N. *Growing Media for Ornamental Plants and Turfs*, 3rd ed.; Wales University Press: Kensington, NSW, Australia, 2002; pp. 136–151.
75. Oncel, I.; Keles, Y.; Ustun, A.S. Interactive effects of temperature and heavy metal stress on the growth and some biochemical compounds in wheat seedlings. *Environm. Pollut.* **2000**, *107*, 315–320. [CrossRef]
76. Ravindran, B.; Mupambwa, H.A.; Silwana, S.; Mnkeni, P.N.S. Assessment of Nutrient Quality, Heavy Metal and Phytotoxic Properties of Chicken Manure on Selected Commercial Vegetable Crops. *Heliyon* **2017**, *3*, e00493. [CrossRef]
77. Araujo, A.S.F.; Monteiro, R.T.R. Plant bioassays to assess the toxicity of textile sludge compost. *Sci. Agri.* **2005**, *62*, 286–290. [CrossRef]

78. Emino, E.R.; Warman, P.R. Biological Assay for Compost Quality. *Compost. Sci. Util.* **2004**, *12*, 342–348. [CrossRef]
79. Luo, Y.; Liang, J.; Zeng, G.; Mo, D.; Li, G.; Zhang, D. Seed germination test for toxicity evaluation of compost: Its roles, problems and prospects. *Waste Manag.* **2018**, *71*, 109–114. [CrossRef]
80. Haugland, E.; Brandsaeter, L.O. Experiments on bioassay sensitivity in the study of allelopathy. *J. Chem. Ecol.* **1996**, *22*, 1845–1859. [CrossRef]
81. Hoffman, M.L.; Weston, L.A.; Snyder, J.C.; Regnier, E.E. Allelopathic influence of germinating seeds and seedlings of cover crops on weed species. *Weed Sci.* **1996**, *44*, 579–584. [CrossRef]
82. Mominul Islam, A.K.M.; Kato-Noguchi, H. Phytotoxic Activity of *Ocimum tenuiflorum* Extracts on Germination and Seedling Growth of Different Plant Species. *Sci. World J.* **2014**, *2014*, 676242. [CrossRef]
83. Azizi, G.; Kondori, M.J.; Siah-Marguee, A.; Alimoradi, L. Bioassay study of fenugreek extract's allelopathic effects on thegermination and growth of several crops and weeds. *J. Plant Breed. Crop. Sci.* **2011**, *3*, 229–239.
84. Leather, G.R.; Einhellig, F.A. Bioassay of naturally occurring allelochemicals for toxicity. *J. Chem. Ecol.* **1988**, *14*, 1821–1828. [CrossRef] [PubMed]
85. Xuan, T.D.; Tsuzuki, E.; Hiroyuki, T.; Mitsuhiro, M.; Khanh, T.D.; Chung, I.M. Evaluationon phytotoxicity of neem (*Azadirachta indica.* A. Juss) to crops and weeds. *Crop Prot.* **2004**, *23*, 335–345. [CrossRef]
86. Fuentes, A.; Llorens, M.; Saez, J.; Aguilar, M.I.; Ortuno, J.F.; Meseguer, V.F. Phytotoxicity and heavy metals speciation of stabilised sewage sludges. *J. Hazard. Mater.* **2004**, *108*, 161–169. [CrossRef]
87. Roy, S.; Asaduzzaman, M.; Pramanik, H.; Prodhan, K. Effect of banana plant extracts on germination and seedling growth of some vegetable crops. *Bangladesh J. Crop Sci.* **2006**, *17*, 235–242.
88. Cesaro, A.; Belgiorno, V.; Guida, M. Compost from organic solid waste: Quality assessment and European regulations for its sustainable use. *Resour. Conserv. Recycl.* **2015**, *94*, 72–79. [CrossRef]
89. Reigosa, M.J.; Souto, X.C.; Gonzalez, L. Effect of phenolic compounds on the germination of six weeds species. *Plant Growth Regul.* **1999**, *28*, 83–88. [CrossRef]
90. Dai, L.; Wu, L.; Zhou, X.; Jian, Z.; Meng, L.; Xu, G. Effects of water extracts of Flaveria bidentis on the seed germination and seedling growth of three plants. *Sci. Rep.* **2022**, *12*, 17700. [CrossRef]
91. Kucera, B.; Cohn, M.A.; Leubner-Metzger, G. Plant hormone interactions during seed dormancy release and germination. *Seed Sci. Res.* **2005**, *15*, 281–307. [CrossRef]
92. El-Maarouf-Bouteau, H.; Sajjad, Y.; Bazin, J.; Langlade, N.; Cristescu, S.M.; Balzergue, S.; Baudouin, E.; Bailly, C. Reactive oxygen species, abscisic acid and ethylene interact to regulate sunflower seed germination. *Plant Cell Environ.* **2015**, *38*, 364–374. [CrossRef]
93. Ma, Z.; Marsolais, F.; Bykova, N.V.; Igamberdiev, A.U. Nitric Oxide and Reactive Oxygen Species Mediate Metabolic Changes in Barley Seed Embryo during Germination. *Front. Plant Sci.* **2016**, *7*, 138. [CrossRef]
94. Park, M.R.; Hasenstein, K.H. Oxygen dependency of germinating Brassica seeds. *Life Sci. Space Res.* **2016**, *8*, 30–37. [CrossRef]
95. Doria, E.; Pagano, A.; Ferreri, C.; Larocca, A.V.; Macovei, A.; Araújo, S.d.S.; Balestrazzi, A. How does the seed pre-germinative metabolism fight against imbibition damage? Emerging roles of fatty acid cohort and antioxidant defence. *Front. Plant Sci.* **2019**, *10*, 1505. [CrossRef] [PubMed]
96. Gomes, P.M.; Garcia, Q.S. Reactive oxygen species and seed germination. *Biologia* **2013**, *68*, 351–357. [CrossRef]
97. Del Río, L.A.; Sandalio, L.M.; Palma, J.M.; Bueno, P.; Corpas, F.J. Reactive oxygen species and reactive nitrogen species in peroxisomes. Production, scavenging, and role in cell signaling. *Plant Physiol.* **2006**, *141*, 330–335. [CrossRef]
98. Gill, S.S.; Tuteja, N. Reactive oxygen species and antioxidant machinery in abiotic stress tolerance in crop plants. *Plant Physiol. Biochem.* **2010**, *48*, 909–930. [CrossRef] [PubMed]
99. Mittler, R. Oxidative stress, antioxidants and stress tolerance. *Trends Plant Sci.* **2002**, *7*, 405–410. [CrossRef]
100. Apel, K.; Hirt, H. Reactive oxygen species: Metabolism, oxidative stress, and signal transduction. *Annu. Rev. Plant Biol.* **2004**, *55*, 373–399. [CrossRef] [PubMed]
101. Hasanuzzaman, M.; Hossain, M.A.; Fujita, M. Plant responses and tolerance to abiotic oxidative stress: Antioxidant defenses is a key factor. *HERO J.* **2012**, 261–315.
102. Kulkarni, M.; Hastak, V.; Jadhav, V.; Date, A.A. Fenugreek Leaf Extract and Its Gel Formulation Show Activity Against Malassezia Furfur. *Assay Drug Dev. Technol.* **2020**, *18*, 45–55. [CrossRef]
103. Tan, L.F.; Yap, V.L.; Rajagopal, M.; Wiart, C.; Selvaraja, M.; Leong, M.Y.; Tan, P.L. Plant as an Alternative Source of Antifungals against *Aspergillus* Infections: A Review. *Plants* **2022**, *11*, 3009. [CrossRef]
104. lvarez-García, S.; Moumni, M.; Romanazzi, G. Antifungal Activity of Volatile Organic Compounds from Essential Oils against the Postharvest Pathogens *Botrytis cinerea*, *Monilinia fructicola*, *Monilinia fructigena*, and *Monilinia laxa*. *Front. Plant Sci.* **2023**, *14*, 1274770. [CrossRef]
105. Oerke, E.C.; Dehne, H.W. Safeguarding Production-Losses in Major Crops and the Role of Crop Protection. *Crop Prot.* **2004**, *23*, 275–285. [CrossRef]

106. Pétriacq, P.; López, A.; Luna, E. Fruit Decay to Diseases: Can Induced Resistance and Priming Help? *Plants* **2018**, *7*, 77. [CrossRef]
107. Cizmarova, B.; Hubkova, B.; Bolerazska, B.; Marekova, M.; Birková, A. Caffeic acid: A brief overview of its presence, metabolism, and bioactivity. *Bioact. Compd. Health Dis.* **2020**, *3*, 74–81.
108. Ingole, A.S.; Kadam, M.; Dalu, A.P.; Kute, S.M.; Mange, P.R.; Theng, V.; Lahane, O.R.; Nikas, A.P.; Kawal, Y.V.; Nagrik, S.U.; et al. A Review of the Pharmacological Characteristics of Vanillic Acid. *J. Drug Deliv. Ther.* **2021**, *11*, 200–204. [CrossRef]
109. Anand David, A.V.; Arulmoli, R.; Parasuraman, S. Overviews of Biological Importance of Quercetin: A Bioactive Flavonoid. *Pharmacogn. Rev.* **2016**, *10*, 84–89. [CrossRef]
110. Pyrzynska, K. Hesperidin: A Review on Extraction Methods, Stability and Biological Activities. *Nutrients* **2022**, *14*, 2387. [CrossRef]
111. Shakeel, A.; Jewel, J.N.; Uzma, J.; Aabida, G.; Zafar, H.; Nasheeman, A. Saponins, the Unexplored Secondary Metabolites in Plant Defense: Opportunities in Integrated Pest Management. *Plants* **2025**, *14*, 861. [CrossRef]
112. Abdelrahman, M.; Sudisha, J. Saponins Versus Plant Fungal Pathogens. *Bioact. Mol. Plant Def.* **2020**, 37–45. [CrossRef]
113. Cho, J.; Hyemin Choi, J.; Lee, M.S.K.; Ho, Y.S.; Dong, G. The Antifungal Activity and Membrane-Disruptive Action of Dioscin Extracted from Dioscorea Nipponica. *BBA Biomembr.* **2013**, *1828*, 1153–1158. [CrossRef] [PubMed]
114. Chapagain, B.P.; Zeev, W.; Leah, T. In Vitro Study of the Antifungal Activity of Saponin-Rich Extracts against Prevalent Phytopathogenic Fungi. *Ind. Crops Prod.* **2007**, *26*, 109–115. [CrossRef]
115. Nguyen, D.H.; Vo, T.N.N.; Le, N.T.T.; Thi, D.P.N.; Thi, T.T.H. Evaluation of Saponin-Rich/Poor Leaf Extract-Mediated Silver Nanoparticles and Their Antifungal Capacity. *Green Process. Synth.* **2020**, *9*, 429–439. [CrossRef]
116. Castano-Duque, L.; Lebar, M.D.; Mack, B.M.; Lohmar, J.M.; Carter-Wientjes, C. Investigating the Impact of Flavonoids on Aspergillus Flavus: Insights into Cell Wall Damage and Biofilms. *J. Fungi* **2024**, *10*, 665. [CrossRef]
117. Dembińska, K.; Ambika, H.S.; Pejchalová, M.; Richert, A.; Swiontek, B.M. The Application of Natural Phenolic Substances as Antimicrobial Agents in Agriculture and Food Industry. *Foods* **2025**, *14*, 1893. [CrossRef]
118. Trang, T.-M.; Ponts, N.; Bernillon, S.; Waffo-Téguo, P.; Rouger, C.; Richard-Forget, F.; Atanasova, V. Integrated Metabolomic and Transcriptomic Analysis to Elucidate the Antifungal and Antimycotoxin Mechanisms of Natural Stilbenoids against *Fusarium graminearum*. *Microbiol. Res.* **2025**, 128370. [CrossRef]
119. Kamaldeep, C.; Hausner, G.; Harris, S.D. Regulation of Melanin Production in Fungi. *Front. Fungal Biol.* **2025**, *6*, 1621764. [CrossRef] [PubMed]
120. Rodríguez-Pires, S.; Melgarejo, P.; De Cal, A.; Espeso, E.A. Proteomic Studies to Understand the Mechanisms of Peach Tissue Degradation by *Monilinia laxa*. *Front. Plant Sci.* **2020**, *11*, 569243. [CrossRef] [PubMed]
121. Perisoara, A.; Marinas, I.C.; Geana, E.I.; Constantin, M.; Angheloiu, M.; Pirvu, L.; Cristea, S. Phytostimulation and Synergistic Antipathogenic Effect of *Tagetes erecta* Extract in Presence of Rhizobacteria. *Horticulturae* **2022**, *8*, 779. [CrossRef]
122. Singleton, V.L.; Orthofer, R.; Lamuela-Raventós, R.M. Analysis of Total Phenols and Other Oxidation Substrates and Antioxidants by Means of Folin-Ciocalteu Reagent. *Methods Enzymol.* **1999**, *299*, 152–178.
123. Formagio, A.S.N.; Kassuya, C.A.L.; Neto, F.F.; Volobuff, C.R.F.; Iriguchi, E.K.K.; Vieira, M.d.C.; Foglio, M.A. The Flavonoid Content and Antiproliferative, Hypoglycaemic, Anti-Inflammatory and Free Radical Scavenging Activities of *Annona dioica* St. Hill. *BMC Complement. Altern. Med.* **2013**, *13*, 14. [CrossRef]
124. Bradford, M.M. A rapid and sensitive method for the quantitation of microgram quantities of protein utilizing the principle of protein-dye binding. *Anal. Biochem.* **1976**, *72*, 248–254. [CrossRef] [PubMed]
125. Lorca, M.; Vega-Muñoz, A.; Acosta, A.; Cabezas, D.; Díaz, K.; Mella, J.; Sabadini, G.; Salazar-Sepúlveda, G.; Contreras-Barraza, N.; Mellado, M. Past and Present of the Antioxidant Studies in Chile: A Bibliometric Study from 2000 to 2024. *Antioxidants* **2025**, *14*, 985. [CrossRef]
126. Knez, E.; Kadac-Czapska, K.; Grembecka, M. Evaluation of Spectrophotometric Methods for Assessing Antioxidant Potential in Plant Food Samples—A Critical Approach. *Appl. Sci.* **2025**, *15*, 5925. [CrossRef]
127. Huang, D.; Ou, B.; Prior, R.L. The chemistry behind antioxidant capacity assays. *J. Agric. Food Chem.* **2005**, *53*, 1841–1856. [CrossRef]
128. Kotha, R.R.; Tareq, F.S.; Yildiz, E.; Luthria, D.L. Oxidative Stress and Antioxidants—A Critical Review on In Vitro Antioxidant Assays. *Antioxidants* **2022**, *11*, 2388. [CrossRef]
129. Tumilaar, S.G.; Hardianto, A.; Dohi, H.; Kurnia, D. A Comprehensive Review of Free Radicals, Oxidative Stress, and Antioxidants: Overview, Clinical Applications, Global Perspectives, Future Directions, and Mechanisms of Antioxidant Activity of Flavonoid Compounds. *J. Chem.* **2024**, 5594386. [CrossRef]
130. Blois, M.S. Antioxidant determinations by the use of a stable free radical. *Nature* **1958**, *181*, 1199–1200. [CrossRef]
131. Mishra, K.; Ojha, H.; Chaudhury, N.K. Estimation of antioxidant activity of plant extracts using DPPH assay: A review. *Free. Radic. Antioxid.* **2012**, *2*, 24–30.
132. Prior, R.L.; Wu, X.; Schaich, K. Standardized methods for the determination of antioxidant capacity and phenolics in foods and dietary supplements. *J. Agric. Food Chem.* **2005**, *53*, 4290–4302. [CrossRef] [PubMed]

133. Madhu, G.; Bose, V.C.; Aiswaryaraj, A.S.; Maniammal, K.; Biju, V. Defect Dependent Antioxidant Activity of Nanostructured Nickel Oxide Synthesized through a Novel Chemical Method. *Colloids Surf. A Physicochem. Eng. Asp.* **2013**, *429*, 44–50. [CrossRef]
134. Çelik, S.E.; Özyürek, M.; Güçlü, K.; Apak, R. Determination of Antioxidants by a Novel On-Line HPLC-Cupric Reducing Antioxidant Capacity (CUPRAC) Assay with Post-Column Detection. *Anal. Chim. Acta* **2010**, *674*, 79–88. [CrossRef] [PubMed]
135. Corbu, V.M.; Gheorghe, I.; Marinaș, I.C.; Geană, E.I.; Moza, M.I.; Csutak, O.; Chifiriuc, M.C. Demonstration of *Allium sativum* Extract Inhibitory Effect on Biodeteriogenic Microbial Strain Growth, Biofilm Development, and Enzymatic and Organic Acid Production. *Molecules* **2021**, *26*, 7195. [CrossRef]
136. Thaipong, K.; Boonprakob, U.; Crosby, K.; Cisneros-Zevallos, L.; Hawkins Byrne, D. Comparison of ABTS, DPPH, FRAP, and ORAC Assays for Estimating Antioxidant Activity from Guava Fruit Extracts. *J. Food Compos. Anal.* **2006**, *19*, 669–675. [CrossRef]
137. Ghayal, N.; Biware, M.; Gharpure, P. Phytotoxic Effects of Leaf Leachates of Invasive Weeds *Cosmos sulphureus* and *Xanthium strumarium* on Agricultural Crops. *Biosci. Biotechnol. Res. ASIA* **2018**, *15*, 821–832. [CrossRef]
138. Perişoară, A.; Pîrvu, L.; Ene, D.M.; Pavel, M.; Olariu, L.; Cristea, S. The influence of two types of extracts obtained from *Tagetes erecta* flowers on radish seeds. *Sci. Pap. Ser. B Hortic.* **2024**, *68*, 486.
139. Milon, A.R.; Chang, S.W.; Ravindran, B. Biochar amended compost maturity evaluation using commercial vegetable crops seedlings through phytotoxicity germination bioassay. *J. King Saud Univ. Sci.* **2022**, *34*, 101770. [CrossRef]
140. EPA. *Ecological Effects Test Guidelines OPPTS 850.4200, Seed Germination/Root Elongation Toxicity Test*; EPA: Washington, DC, USA, 1996.
141. Alamri, S.; Siddiqui, M.; Al-Khaiskani, M.; Ali, H. Response of salicylic acid on seed germination and physio-biochemical 587 changes of wheat under salt stress. *Acta Sci. Agric.* **2018**, *2*, 36–42.
142. Kaya, M.; Okçu, G.; Atak, M.; Çikili, Y.; Kolsarici, Ö. Seed treatments to overcome salt and drought stress during germination in sunflower (*Helianthus annuus* L.). *EJA* **2006**, *24*, 291–295. [CrossRef]
143. Aebi, H. Catalase in vitro. *Methods Enzymol.* **1984**, *105*, 121–126.
144. McCord, J.M.; Fridovich, I. Superoxide dismutase: An enzymic function for erythrocuprein (hemocuprein). *J. Biol. Chem* **1969**, *244*, 6049–6055. [CrossRef] [PubMed]
145. Prochazkova, D.; Wilhelmová, N.; Vysloužilová, M. Antioxidant enzyme activities in relation to stress resistance of plants. *Biol. Plant* **2001**, *44*, 345–350.
146. Al-Hussainy, A.D.L. Evaluation of the Antimicrobial (Antibacterial and Antifungal) Activity of Ethanolic Extracts of Some Medical Plants. *J. Nat. Sci. Res.* **2015**, *5*, 87–95.
147. Sindhusha, V.B.; Rajasekar, A. Preparation and Evaluation of Antimicrobial Property and Anti-Inflammatory Activity of Fenugreek Gel Against Oral Microbes: An Invitro Study. *Cureus* **2023**, *15*, e47659. [CrossRef]
148. Omezzine, F.; Bouaziz, M.; Daami-Remadi, M.; Simmonds, M.S.J.; Haouala, R. Chemical Composition and Antifungal Activity of *Trigonella foenum-graecum* L. Varied with Plant Ploidy Level and Developmental Stage. *Arab. J. Chem.* **2017**, *10*, S3622–S3631. [CrossRef]
149. *CLSI M38M51S*; Performance Standards for Antifungal Susceptibility Testing of Filamentous Fungi. Clinical and Laboratory Standards Institute: Wayne, PA, USA, 2022; pp. 1–18.
150. Dharajiya, D.; Patel, P.; Patel, M.; Moitra, N. In Vitro Antimicrobial Activity and Qualitative Phytochemical Analysis of *Withania Somnifera* (L.) Dunal Extracts. *Int. J. Pharm. Sci. Rev. Res.* **2014**, *27*, 349–354.
151. Javed, S.; Shoaib, A.; Mahmood, Z.; Mushtaq, S.; Iftikhar, S. Analysis of Phytochemical Constituents *of Eucalyptus citriodora* L. Responsible for Antifungal Activity against Post-Harvest Fungi. *Nat. Prod. Res.* **2012**, *26*, 1732–1736. [CrossRef] [PubMed]

plants

MDPI

Article

Antifungal Activity of *Artemisia capillaris* Essential Oil Against Alternaria Species Causing Black Spot on Yanbian Pingguoli Pear in China

Zu-Xin Kou [1,†], Yue Dang [1,†], Li Liu [2], Xue-Hong Wu [2,*] and Yu Fu [1,*]

1 Department of Chemistry, Yanbian University, Yanji 133002, China; 15841950399@163.com (Z.-X.K.); 13500813498@163.com (Y.D.)
2 Department of Plant Pathology, China Agricultural University, Beijing 100193, China; liulill1124@163.com
* Correspondence: wuxuehong@cau.edu.cn (X.-H.W.); fuyu@ybu.edu.cn (Y.F.)
† These authors contributed equally to this work.

Abstract: Black spot is currently one of the most widespread diseases affecting Yanbian Pingguoli pears (*Pyrus pyrifolia* cv. 'Pingguoli'), resulting in significant economic losses for fruit farmers. It is mainly caused by infestation by the fungal group of *Alternaria* species. To date, no research has reported the presence of *Alternaria* species and the pathogen of black spot disease on Yanbian Pingguoli pears in China. This study isolated, identified, and performed molecular profiling of 124 *Alternaria* strains collected from 15 major growing areas of Yanbian Pingguoli pear (more than 5000 trees). Moreover, the study evaluated the ability of *Artemisia capillaris* essential oil (AcEO) to suppress the mycelial expansion of *Alternaria* pathogens and conducted comprehensive chemical profiling. Overall, 124 pathogenic fungi were identified as *Alternaria tenuissima* (67 isolates, 54.0%) and *A. alternate* (57 isolates, 46.0%). AcEO showed a strong inhibitory effect on the two *Alternaria* species, with a minimal inhibitory concentration (MIC) value equivalent to 5.0 μL/mL. Eucalyptol, 2,2-Dimethyl-3-methylenebicyclo [2.2.1] heptane, (-)-alcanfor, and β-copaene were identified as the predominant bioactive components of AcEO. AcEO demonstrated concentration-dependent inhibition of the mycelial growth of *A. tenuissima* and *A. alternata*. These findings position AcEO as a promising candidate for developing sustainable fungicides to combat *Alternaria*-induced crop losses.

Keywords: *Artemisia capillaris*; essential oil; Pingguoli pear; antifungal activity

1. Introduction

Pingguoli pear (*Pyrus pyrifolia* cv. 'Pingguoli') belongs to Maloideae of the family Rosaceae. As the pear's shape is oblate, the background color is yellow and green, the sun gives it a red glow and, from far away, it looks like an apple on the tree, it is called Pingguoli. There are many advantages of the Pingguoli pear, including large fruit, a small nucleus, fine and crisp flesh, high sugar content, fragrant odor, and resistance during storage, making it widely loved by the public [1–3]. Yanbian Pingguoli pear, one such excellent pear variety, has been cultivated for more than 100 years in the east of China. At present, its planting scale is 12,000 hectares across more than 10 thousand households and fruit farmers, with an annual output of more than 90 thousand tons, producing an output value of nearly CNY 100 million. Yanbian Pingguoli pear production has become a pillar industry in the east of China, which is Asia's largest apple and pear production base.

While Pingguoli pears can be stored at cooler temperatures for up to six months, the incidence of black spot is as high as 37% at the later stage of storage [4,5]. Black spot disease

https://doi.org/10.3390/plants14203146

is currently one of the most widely observed diseases in Pingguoli pear, and is mainly caused by infestation with the fungal group of *Alternaria* species [6,7], such as Japanese pear [8], European pear [9], Asian pear [10], sandy pear [11] and Korla fragrant pear [12]. It mainly affects the new leaves, new shoots, and young fruits of Pingguoli pear and can invade through the flowers and fruits, causing a large number of diseased fruits and early fruit drop. When the disease is serious, it can even lead to fruit tree defoliation, fruit rot, shedding, and the death of new tips, thus affecting the yield of the fruit tree and resulting in huge economic losses [4,5]. Meanwhile, toxins produced by the pathogenic fungus of black spot on Pingguoli pear can trigger human esophageal cancer development, causing food safety issues [13]. The continuous use of chemical fungicides as the main means of control, such as iprodione, tebuconazole, and mancozeb, has gradually resulted in the increasing resistance of pathogens, reduced sensitivity, environmental pollution, the accumulation of pesticide residues in food and an associated risk of contracting cancer [14]. Therefore, there is an urgent need to develop an effective and environmentally friendly alternative to these traditional fungicides to control black spot on Pingguoli pear.

Plant essential oils, also called volatile oils, are oily liquids containing a variety of volatile organic compounds extracted from the roots, stems, leaves, flowers, fruits, and other parts of plants [15]. There are more than 3000 kinds of plant essential oils, which play a wide range of roles in insecticidal, anti-inflammatory, anti-oxidant, anti-cancer, and anti-microbial activities [16–18]. The chemical profiling of plant essential oils reveals a complex matrix of bioactive constituents, primarily comprising hydrocarbon skeletons (terpenoid frameworks) and their oxygenated functional groups (phenolic aldehydes/esters). These compounds encompass terpenes, alcohols, amines, aldehydes, ethers, oxides, ketones, esters, amides, phenols, and heterocycles, forming intricate combinations [19]. Generally, around 60–70% of the total essential oil constituents comprise two to three major compounds [20]. The composition of secondary metabolites within essential oils is influenced by a multitude of factors, including geography, photoperiod, abiotic and edaphic factors, seasons, microbial diversity, the specific plant part utilized, developmental stage, and the method of extraction [21].

Several studies have reported that various plant essential oils showed high efficiency in controlling *Alternaria* species. Perveen et al. [22] found that the *Ocimum basilicum* L. essential oil successfully constrained the growth and conidia germination of *A. alternata*. The highest concentration of basil essential oil (10%) provided maximum growth reduction (88%) in the fungus. Kamsu et al. [23] tested a Massep (*Ocimum gratissimum* L.) essential oil concentration of 2400 μL/L and found it was effective in inhibiting *A. solani* at a rate of 71.81%. Chen et al. [24] showed a 50.80% reduction in *A. alternata* incidence in inoculated blueberries by day 4 post-*Artemisia argyi* essential oil treatment. *Artemisia capillaris* Thunb. is a traditional medicine with suggested hepatoprotective effects, which belongs to the *Artemisia asteraceae* family [25]. Although previous studies have suggested that AcEO has demonstrated antimicrobial effects [26], limited experimental evidence exists regarding its effect on the *Alternaria* pathogenic fungi of black spot disease. The objective of this study was to evaluate the antifungal potential of AcEO against *Alternaria* pathogens associated with black spot disease on Yabian Pingguoli pears. This is expected to provide a new avenue for the development and utilization of *Artemisia capillaris* as a novel biofungicide.

2. Results

2.1. Sample Collection and Isolation of Alternaria Isolates

During the period spanning 2020 through 2022, specimens of *Alternaria* were cultured from Yanbian Pingguoli pear leaves displaying distinct black spot lesions. These samples originated from 15 Pingguoli pear cultivation sites, encompassing over 5000 trees (Table 1).

Notably, both *A. tenuissima* and *A. alternata* species prevailed across all investigated orchards. *A. tenuissima* isolates were not detected in the fruit farm of Longjing County, and *A. alternata* isolates were not detected in Xicheng Town in Helong County and Liangshui Town in Tumen County.

Table 1. Regional distribution and phylogenetic affiliations of *Alternaria* spp. isolates collected from symptomatic Yanbian Pingguoli pear foliage exhibiting characteristic black spot symptoms in China.

Geographic Districts of Yanbian	Number of Fields	Number of *Alternaria* Isolates	
		A. tenuissima	*A. alternata*
Longjing County, Zhixin Town	1	1 (1) [a]	4 (2)
Longjing County, Fruit Farm	1	0 (0)	6 (2)
Longjing County, Shanhe Town	1	1 (1)	11 (3)
Helong County, Bajiazi Town	1	4 (2)	9 (3)
Helong County, Xicheng Town	1	5 (2)	0 (0)
Helong County, Longcheng Town	1	3 (2)	2 (1)
Tumen County, Yueqing Town	1	14 (3)	1 (1)
Tumen County, Liangshui Town	1	4 (2)	0 (0)
Yanji City, Yilan Town	2	14 (4)	3 (1)
Yanji City, Chaoyangchuan Town	1	5 (2)	7 (3)
Yanji City, Xiaoying Town	1	5 (2)	1 (1)
Hunchun County, Yingan Town	3	11 (5)	13 (5)
Total	15	67 (26)	57 (22)
Ratio	–	54.0%	46.0%

[a] The parenthetical values denote the corresponding *Alternaria* strains employed in pathogenicity tests.

2.2. Morphological Characterization of Alternaria Isolates

The morphological analysis of 67 *A. tenuissima* strains revealed initial colony pigmentation ranging from grayish green to olive brown on potato dextrose agar (PDA). Morphometric characterization demonstrated unbranched conidiophores (16.9 μm–54.3 μm in length) terminating in phialides producing conidia (2.9 μm–7.1 μm wide). Conidia exhibited ovoid to obovoid morphology, with dimensions spanning 14.8 μm–28.4 μm (length) and 4.2 μm–10.3 μm (width). On potato carrot agar (PCA), conidial chains displayed 12 conidia per linear arrangement, featuring 1–4 transverse septa and 0–2 intercalary cells, with the entire structure embedded within lateral branches (Figure 1a–c).

The morphological characterization of 57 *A. alternata* isolates revealed initial colony pigmentation transitioning from tan to dark brown on potato dextrose agar (PDA). The pathogen exhibited filamentous conidial chains containing 8–12 conidia per linear arrangement. Conidiophores measured (10.3 μm–48.1 μm (length)) × (4.9 μm–13.4 μm (width)), terminating in phialides that produced ovoid to ellipsoidal conidia ((15.9–31.2) μm × (5.5–12.2) μm). Conidial structures displayed 1–5 transverse septa and 0–2 longitudinal septa, with intercalary cells embedded within septate conidia (Figure 1d–f). These phenotypic traits, including colony morphology, conidial dimensions, and septation patterns, were morphologically congruent with *A. tenuissima* and *A. alternata* reference strains as documented by Simmons [27].

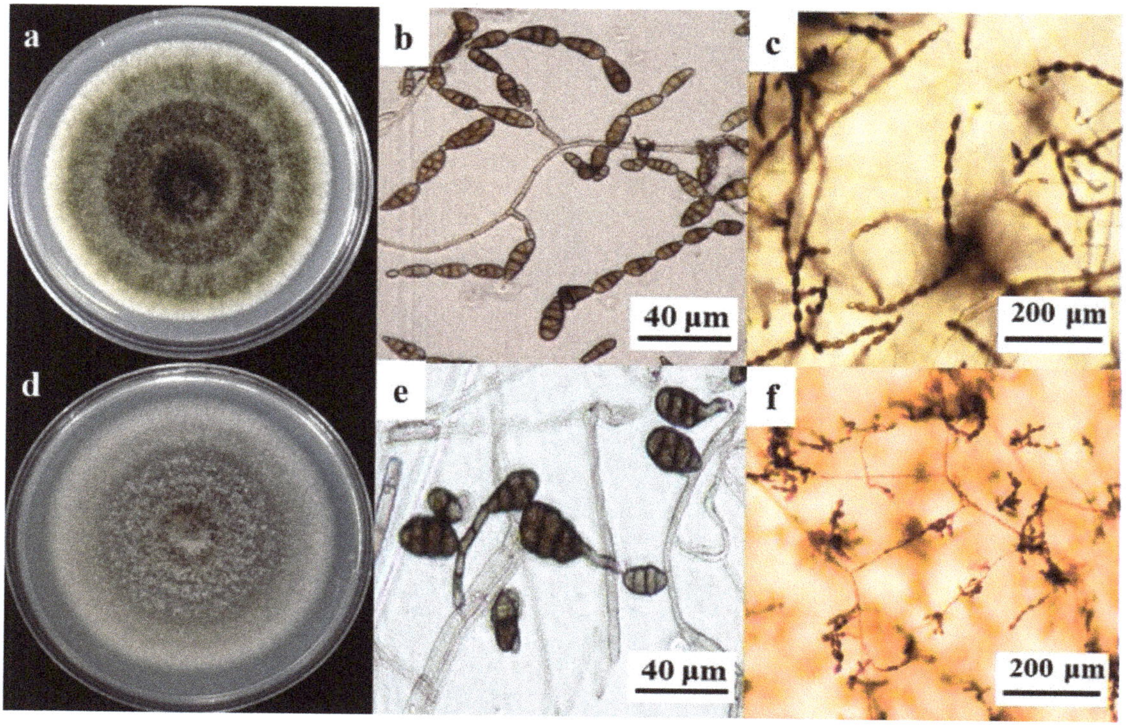

Figure 1. Comparative morphological analysis of *A. tenuissima* and *A. alternata*. (**a**) Colony phenotypes: *A. tenuissima* exhibited tan-colored colonies with radial rugae on PDA. (**b**) Conidial ontogeny: Ovoid conidia ((15.9–31.2) × (5.5–12.2) µm) displaying 1–5 transverse septa. (**c**) Asexual reproductive structures: Linear conidial chains (8–12 conidia per chain) embedded within lateral branches. (**d**) Colony morphometrics: *A. alternata* formed olive-brown colonies with zonate growth patterns. (**e**) Conidial characteristics: Ellipsoidal conidia ((14.8–28.4) × (4.2–10.3) µm) bearing 0–2 intercalary cells. (**f**) Sporulation architecture: Curved conidial chains (5–9 conidia per chain) terminating in phialide clusters.

2.3. Molecular Characterization of Alternaria Isolates

The molecular identification of 124 *Alternaria* isolates was conducted via PCR amplification of the ITS region (570 bp) using consensus primers ITS1/ITS4. BLASTn analysis revealed ≥99% nucleotide homology with *A. tenuissima* (KR867207, AF347032) and *A. alternata* (AF347031, MG744379, KP124372.1), along with high similarity to other *Alternaria* spp. in GenBank. Phylogenetic reconstruction based on histone H3 gene sequences (546 bp for Group I; 440 bp for Group II) demonstrated clustering patterns: Group I (n = 67) exhibited ≥99% identity to *A. tenuissima* reference sequences (JX495167, JX495168, AF404634), and Group II (n = 57) showed ≥98% homology with *A. alternata* strains (AF404624, MG744388, MK085979.1). The complete nucleotide datasets (n = 124) have been deposited in GenBank (Supplementary Table S1). Unrooted phylogenetic trees constructed from ITS and histone H3 data (Figures 2 and 3) revealed distinct clades corresponding to the two genotypic groups identified in pathogenicity trials (n = 48 isolates).

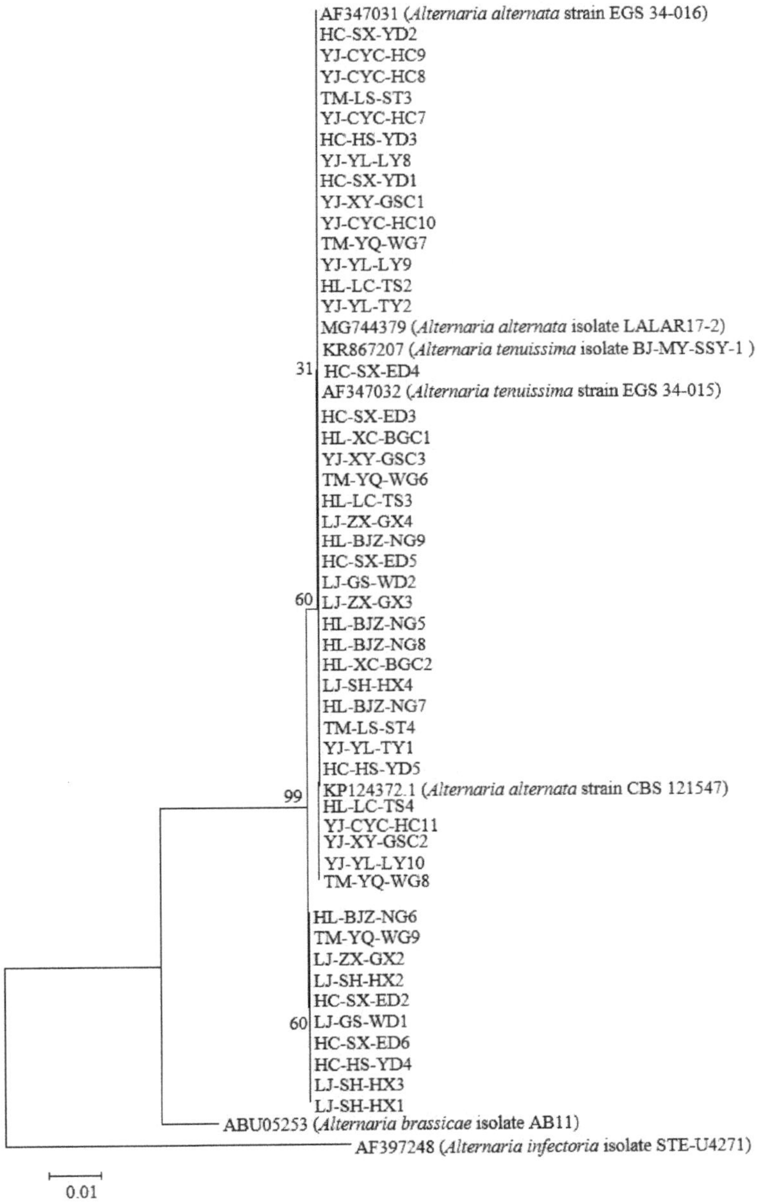

Figure 2. Phylogenetic reconstruction was performed using 48 *Alternaria* pathogenic ITS sequences curated from the National Center for Biotechnology Information (NCBI) nucleotide database. Maximum likelihood analysis generated a topology supported by 1000 bootstrap replicates. The tree was rooted with *A. brassicae* AB11 (GenBank: ABU05253) and *A. infectoria* STE-U4271 (GenBank: AF397248), two phylogenetically distinct taxa representing distinct host adaptation lineages.

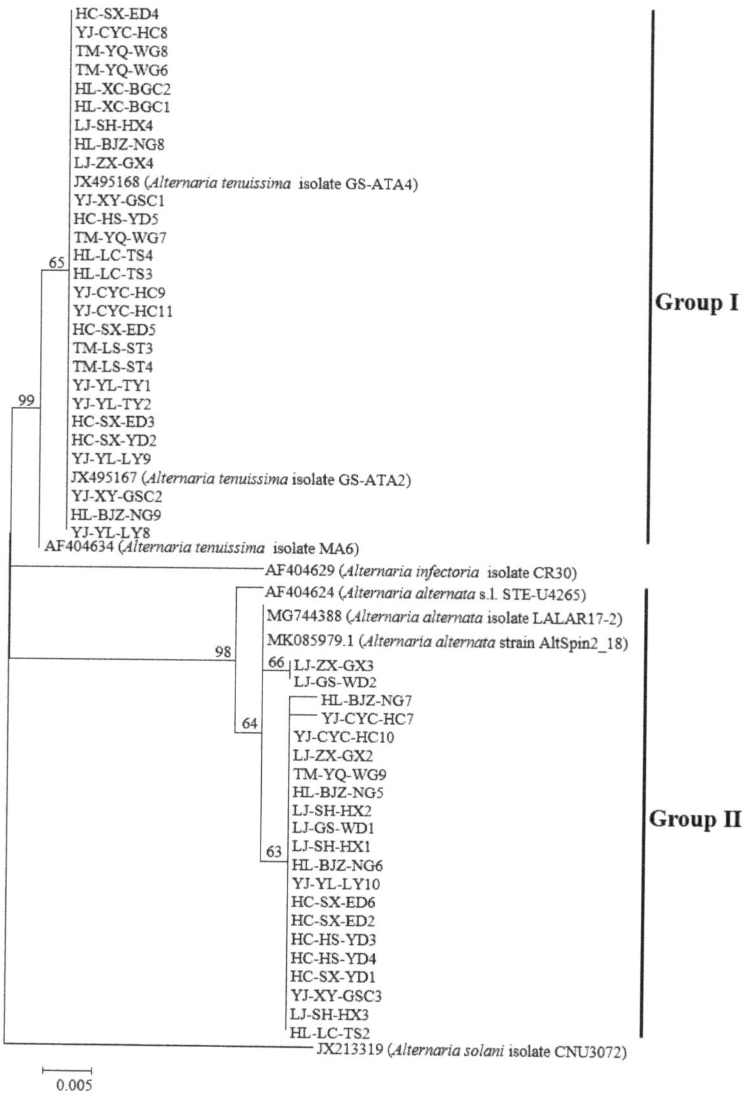

Figure 3. Phylogenetic reconstruction of histone H3 gene sequences (546 bp) resolved 48 *Alternaria* pathogenic strains into two major clades. Bootstrap support (1000 replicates) and Bayesian posterior probabilities (2M generations) demonstrated (1) Group I (*A. tenuissima*) with ≥99% support and (2) Group II (*A. alternata*) with ≥98% support. Outgroup taxa (*A. solani* isolate CNU3072 (JX213319) and *A. infectoria* isolate CR30 (AF404634)) were chosen to root the tree at the *Alternaria* crown group.

2.4. Pathogenicity of Representative Alternaria Isolates

Following the administration of needle puncture wounds to both the left and right sides of a healthy leaf, the left side was co-inoculated with *Alternaria* isolates. The right puncture site was maintained as a mock-inoculated control, as shown in the areas indicated by the red circles in Figure 4. The manifestation of conspicuous necrotic lesions was observed at five days post-treatment. After two weeks, when Yanbian Pingguoli pear leaves were challenged with *Alternaria* isolates, excised tissues exhibited distinct concentric ring patterns and necrotic spots (Figure 4). Pathogen recovery and molecular identification from symptomatic tissues further validated Koch's postulates. The pathogenicity assay demonstrated that *A. tenuissima* induced higher disease incidence (51.7–100.0%) and severity (30.5–50.4) on detached pear leaves compared to *A. alternata* (31.2–100.0% and 18.2–29.5). The differences in disease parameters between the two fungi were statistically significant ($p < 0.05$) (Table 2).

Figure 4. Pathogenicity of 48 *Alternaria* isolates on detached and healthy Yanbian Pingguoli pear leaves (6 cm long and 3.5 cm wide) naturally grown in Pingguoli pear orchards at the end of May. (**a**): Adaxial side of the leaf inoculated with *A. tenuissima*. (**b**): Abaxial side of the leaf inoculated with *A. tenuissima*; (**c**): Adaxial side of the leaf inoculated with *A. alternata*. (**d**): Abaxial side of the leaf inoculated with *A. alternata*.

Table 2. Disease incidence and disease index of detached leaves of Yanbian Pingguoli pear caused by 48 *Alternaria* isolates.

Alternaria Species	Isolate Number	Disease Incidence (%) [Mean ± Standard Deviation (Range)]	Disease Index [Mean ± Standard Deviation (Range)]
A. tenuissima	26	71.2 ± 13.5 (51.7–100.0) ^{a}	57.3 ± 6.9 (30.5–50.4) ^{a} *
A. alternata	22	48.2 ± 8.6 (31.2–100.0) ^{b}	19.8 ± 3.7 (18.2–29.5) ^{b} *

* The mean ± standard deviation (SD) values (a,b) were calculated from triplicate experiments using Alternaria isolates specific to each Yanbian Pingguoli pear cultivar. Symptom severity (DS) was quantified following a 14-day incubation period under controlled conditions (28 °C, 90% RH). Incidence rates and disease indices, labeled with superscript letters, demonstrated statistically distinct variations ($p < 0.05$). For disease incidence, single-factor variance analysis (ANOVA) with Dunnett-T3 multiple comparisons revealed significant differences, whereas disease index data were processed using an LSD post hoc test for multiple comparisons.

2.5. Chemical Composition of AcEO Determined by Gas Chromatography–Mass Spectrometry (GC-MS)

The chemical compositions of the AcEOs were determined using GC-MS chromatographic analyses, in which 40 volatile compounds were identified (Table 3). The structural formulas of various components in AcEO are shown in Supplementary Table S2. The spectral identification (>99% confidence) was conducted through the comparison of experimental mass spectra with the NIST mass spectral library (version 20), complemented by chromatographic co-elution validation using reference standards synthesized in-house. The primary chemical constituents of the AcEO predominantly included eucalyptol (60.40%), 2,2-Dimethyl-3-methylenebicyclo [2.2.1] heptane (5.03%), (-)-alcanfor (3.86%), and β-copaene (3.25%).

Table 3. GC-MS analysis characterizing the volatile profile of *Artemisia capillaris* essential oil.

No.	Compound	Retention Time (min)	Molecular Formula	CAS	Relative Content (>0.05%)
1	Bicyclo [3.1.0]hex-2-ene	9.68	$C_{10}H_{16}$	2867-05-2	0.13
2	(1R)-2,6,6-Trimethylbicyclo[3.1.1]hept-2-ene	9.99	$C_{10}H_{16}$	7785-70-8	1.46
3	α-Pinene	10.17	$C_{10}H_{16}$	80-56-8	0.95
4 *	**2,2-Dimethyl-3-methylenebicyclo[2.2.1]heptane**	**10.71**	$\mathbf{C_{10}H_{16}}$	**5794-04-7**	**5.03**
5	1-Isopropyl-4-meth-ylenebicyclo[3.1.0]hexane	11.85	$C_{10}H_{16}$	3387-41-5	1.97
6	Myrcene	12.49	$C_{10}H_{16}$	123-35-3	1.79
7	1-Octen-3-ol	13.65	$C_9H_{16}O$	3391-86-4	1.31
8 *	**Eucalyptol**	**15.20**	$\mathbf{C_{10}H_{18}O}$	**470-82-6**	**60.40**
9	Cis-3,7-Dimethyl-1,3,6-octatriene	15.32	$C_{10}H_{16}$	3338-55-4	1.03
10	Crithmene	15.75	$C_{10}H_{16}$	99-85-4	1.96
11	Isoterpinolene	16.56	$C_{10}H_{16}$	586-63-0	0.2
12	Isothujone	17.62	$C_{10}H_{16}O$	471-15-8	1.73
13	Thujone	18.06	$C_{10}H_{16}O$	546-80-5	0.46
14	(4E,6Z)-2,6-Dimethyl-2,4,6-octatriene	18.43	$C_{10}H_{16}$	7216-56-0	0.84
15 *	**(-)-Alcanfor**	**19.30**	$\mathbf{C_{10}H_{16}O}$	**76-22-2**	**3.86**
16	L(-)-Borneol	20.76	$C_{10}H_{18}O$	464-45-9	2.42
17	Terpinen-4-ol	20.89	$C_{10}H_{18}O$	562-74-3	0.75
18	Terpineol	21.33	$C_{10}H_{18}O$	98-55-5	0.52
19	(Z)-piperitol	21.73	$C_{10}H_{18}O$	16721-38-3	0.07
20	Capillin	23.12	$C_{12}H_{18}O$	495-74-9	2.26
21	1,7,7-Trimethylbicyclo[2.2.1]hept-2-yl acetate	24.32	$C_{12}H_{20}O_2$	92618-89-8	0.31
22	(1S,3S,5S)-1-Isopropyl-4-methylenebicyclo[3.1.0]hexan-3-yl acetate	24.52	$C_{12}H_{20}O_2$	139757-62-3	0.42
23	γ-Elemene	25.93	$C_{15}H_{24}$	29873-99-2	0.08
24	α-copaene	27.48	$C_{15}H_{24}$	3856-25-5	0.09
25 *	**β-copaene**	**29.15**	$\mathbf{C_{15}H_{24}}$	**18252-44-3**	**3.25**
26	(-)-Isogermacrene D	29.86	$C_{15}H_{24}$	317819-80-0	0.10
27	(-)-Caryophyllene	30.11	$C_{15}H_{24}$	87-44-5	0.14
28	α-caryophyllene	30.27	$C_{15}H_{24}$	6753-98-6	0.10
29	(-)-Germacrene D	31.25	$C_{15}H_{24}$	23986-74-5	1.64
30	1,2,3,4,4a,5,6,7-octahydro-4a,5-dimethyl-3-(1-methylethenyl)-1-Naphthalenol	31.35	$C_{15}H_{24}O$	61847-19-6	0.16
31	bicyclogermacrene	31.60	$C_{15}H_{24}$	24703-35-3	0.05
32	ε-Muurolene	31.75	$C_{15}H_{24}$	30021-46-6	0.07
33	δ-Cadinene	32.34	$C_{15}H_{24}$	483-76-1	0.39
34	α-Panasinsene	33.11	$C_{15}H_{24}$	56633-28-4	0.64
35	(+)-Phyllocladene	34.50	$C_{20}H_{32}$	469-86-3	1.27
36	Gitoxigenin	35.71	$C_{23}H_{34}O_5$	545-26-6	0.21
37	Muurolol	36.29	$C_{15}H_{26}O$	19912-62-0	0.17
38	Neointermedeol	36.90	$C_{15}H_{26}O$	5945-72-2	0.53
39	cis-p-mentha-2,8-dien-1-ol	37.50	$C_{10}H_{16}O$	425394-92-9	0.14
40	Bi-1-cycloocten-1-yl	42.04	$C_{16}H_{26}$	61468-42-6	0.55

* The components with high content % have been highlighted.

2.6. Antifungal Activity of the AcEO Against Alternaria spp.

As illustrated in Figure 5, incubation at 28 °C for 7 days resulted in the complete colony colonization of *A. tenuissima* YJ-XY-GSC1 and *A. alternata* YJ-XY-GSC3 on agar plates, exhibiting characteristic black flocculent mycelial morphology. Exposure to AcEO induced progressive radial constriction of fungal colonies, with hyphal extension rates demonstrating statistically significant attenuation compared to untreated controls ($p < 0.01$). Dose–response analysis revealed concentration-dependent suppression kinetics, where complete growth inhibition (MIC = 5.0 μL/mL) was achieved through carvacrol-mediated mechanisms. This concentration-dependent inhibition pattern suggests AcEO's potential as a natural fungicide against these phytopathogens.

Scanning electron microscopy (SEM) analysis revealed dose-dependent ultrastructural modifications in *Alternaria* spp. hyphae. As depicted in the micrographs (Figure 6), radial constriction and surface corrugation became pronounced in *A. tenuissima* YJ-XY-GSC1 and *A. alternata* YJ-XY-GSC3 under AcEO gradients (0.5–5.0 μL/mL), with progressive loss of cytoplasmic continuity at 2.5 μL/mL (1/2 MIC). The control groups (*A. tenuissima* YJ-XY-GSC1 and *A. alternata* YJ-XY-GSC3) displayed intact hyphal networks with uniform cylindrical geometry. The groups with 1/2 MIC of AcEO exhibited pronounced wrinkles and concavity. Notably, this study corroborated emerging evidence that plant-derived essential oils disrupt fungal integrity through the damaged fungal morphology, as previously reported [24,28].

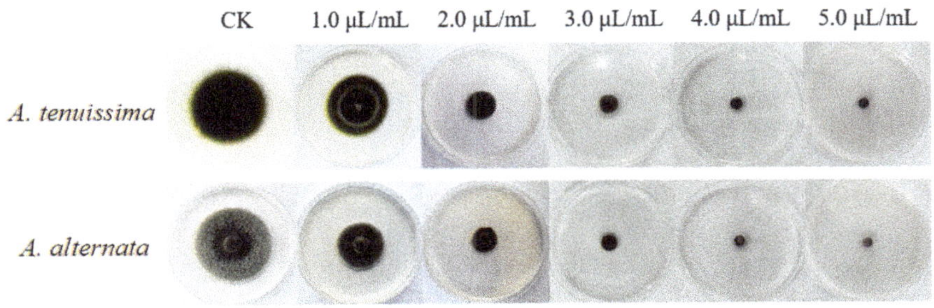

Figure 5. Antifungal inhibitory effects of AcEO on mycelial growth of *A. tenuissima* YJ-XY-GSC1 and *A. alternata* YJ-XY-GSC3 in PDA medium supplemented with gradient concentrations of AcEO (0, 1.0, 2.0, 3.0, 4.0, and 5.0 μL/mL).

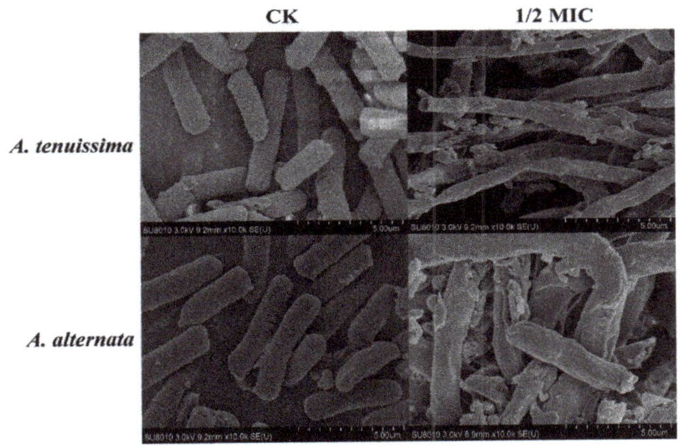

Figure 6. SEM pictures of *A. tenuissima* YJ-XY-GSC1 and *A. alternata* YJ-XY-GSC3 treated with varying AcEO concentrations (0 and 1/2MIC).

3. Discussion

Pingguoli pear was introduced to Yanbian Korean autonomous prefecture in Jilin province in China in 1921 by the Korean cultivator Fandou Cui, and originates from North Korea. After several generations of meticulous cultivation and continuous breeding, it was officially named 'Pingguoli pear' in 1958. Today, this fruit is mainly produced in the Yanbian Korean autonomous prefecture in Jilin province, particularly in three counties; namely, Longjing, Helong, and Yanji. Additionally, there are some distribution centers in Tumen, Hunchun, and Wangqing counties, and in other cities. From 1997 to 2004, due to extensive cultivation across provinces or autonomous regions such as Liaoning, Inner Mongolia, and Gansu, Pingguoli pear production surged while prices plummeted into a state of depression. Coupled with limited transportation radius and storage capacity for Pingguoli pears along with increasing production costs over time, fruit farmers witnessed declining profits year after year, leading many to cut down their orchards or repurpose them for alternative uses. Consequently, this resulted in a significant reduction in Pingguoli pear growing areas. Due to the custom of gifting apple pears among Chinese Koreans, the primary growing regions remain within the Yanbian Korean autonomous prefecture of Jilin province, while some are also cultivated in Gansu province. In 2023, Harbin's tourism boom resulted in a significant increase in the demand for 'frozen autumn pears'. These pears are typically obtained from Huagai pear, autumn pear, white pear, and Jianba pear (while the Yanbian area freezes Yanbian Pingguoli pear). Due to their growing popularity, Yanbian Pingguoli pears have attracted more and more attention.

Pear black spot disease, caused by *Alternaria* species, is one of the most serious diseases affecting pear varieties such as Japanese pear [8], European pear [9], Asian pear [10], Korla fragrant pear [29], and sandy pear [11]. Market analysis indicates that Pingguoli pear derivatives are predominantly commercialized through fresh and dried consumption. The pathogenic *Alternaria* strain exhibits robust resistance to cold and is capable of surviving in various environments, including Pingguoli pear trees, leaves, fruits, and soil. Commonly, the Pingguoli pear is storable and can be stored for 6 months at 0 °C, while concurrently exhibiting a high incidence of black spot during the late storage period, reaching up to 37% [30]. This highlights the significant impact of *Alternaria* species on Pingguoli pear production throughout all the storage stages; thus, the fungus has become a pivotal factor impeding the development of the Yanbian Pingguoli pear industry. Previous reports have suggested that *A. alternata* is the dominant etiological factor in black spot disease affecting Pingguoli pears in Gansu province in China [30,31]. To date, no reports have been released announcing that the *Alternaria* species causes black spot on Yanbian Pingguoli pears in China. Therefore, this investigation reveals novel phytopathogenic characteristics of *Alternaria* species on Yanbian pear black spot outbreaks from Jilin Province, China. The molecular characterization of 124 obtained fungi confirms that the selected universal primers ITS1 and ITS4 enabled attribution of the observed disease to *Alternaria*. Furthermore, the partial coding sequence of the histone H3 gene was used to confirm that the *Alternaria* species causing black spot were *A. tenuissima* and *A. alternata*. The phylogenetic analyses in this study are in agreement with other studies, which have shown a clear separation of *A. alternata* and *A. tenuissima* from the *Alternaria* species complex by the partial coding sequence of the histone 3 gene [32–35]. This study provides the first evidence of *A. tenuissima* and *A. alternata* co-infecting Yanbian Pingguoli, with pathogenic validation through Koch's postulates and multigene phylogenetic analyses. The observed black spot symptoms, characterized by irregular necrotic lesions with chlorotic halos, mark a novel host–pathogen interaction in China.

Artemisia species have been used as food additives and in traditional herbal medicines, particularly for treating diseases such as cancer, inflammation, malaria, hepatitis, and microbial infections [36]. Of these species, *Artemisia capillaris* is known as the Chinese drug 'Yin Chen Hao' ('Injinho' in Korean medicinal terminology), and is distributed broadly in the north-east area of China. Its traditional effects are to clear away dampness and heat, promote gallbladder function and reduce jaundice [37]. Many investigations have demonstrated that the essential oil of *A. capillaris* exhibits significant biological activity [37]. Our GC-MS findings indicate that the most abundant constituents of AcEO are eucalyptol (60.40%), 2,2-Dimethyl-3-methylenebicyclo [2.2.1] heptane (5.03%), (-)-alcanfor (3.86%), and β-copaene (3.25%). There are significant differences compared to previous reports [38]. The observed phytochemical variability is mediated by environmental parameters including soil mineral composition, solar radiation intensity, growth stage, and nutrient availability gradients.

In this study, we also investigated the effectiveness of AcEO volatiles as a control for *A. alternata* and *A. tenuissima* infestation during Yanbian Pingguoli pear growth. The results showed that the 5.0 μL/mL AcEO treatment completely inhibited the mycelial growth and morphology of *A. alternata* and *A. tenuissima*. Implementing AcEO in Pingguoli commercial practices may decrease chemical fungicide dependency, simultaneously mitigating antifungal resistance development and ecological contamination. The observed phenomena demonstrate synergistic alignment with a study conducted by Chen et al. [24]. The antifungal activity of essential oils (Castor, Jasmine, Clove, Sesame, Neem, Coconut, Henna, Black seed, and Mint) against *A. alternata* was evaluated through radial growth inhibition assays [39]. Extracts were tested at concentrations ranging from 1% to 6% using in vitro inhibition protocols. Singh et al. [28] provided compelling evidence of a significant inhibition of *A. tennussima* growth by *Eucalyptus globulus* essential oil. Various studies have affirmed the antifungal potential of essential oils obtained from a variety of plants on *Alternaria* isolates. Allagui et al. [40] investigated the potential of the essential oil derived from *Cinnamomum verrum* and *Syzygium aromaticum*. Experimental results revealed that these plant-derived extracts exhibited notable efficacy in inhibiting the mycelial growth of *A. alternata*, with MICs ranging from 0.31 to 0.45 mg/mL and 0.37 to 0.57 mg/mL, respectively, as determined through in vitro assays. Despite their proven effectiveness, essential oils face three critical limitations in food system applications: (i) a tendency to volatilize under thermal processing (>40% loss at 60 °C), (ii) aqueous insolubility (<1 mg/mL solubility in PBS), and (iii) chemical instability against oxidative stress (half-life < 24 h). Kamsu et al. [23] confirmed that Massep (*Ocimum gratissimum* L.) essential oil nanoemulsion significantly inhibited the mycelial growth of *A. solani* compared to pure essential oil. Further studies should focus on the preparation of a nanoemulsion using AcEO to maximize the efficacy in controlling pathogens causing black spot in Pingguoli.

4. Materials and Methods

4.1. Fungal Isolation

The growing area was mainly concentrated in the five cities of Longjing, Helong, Tumen, Hunchun, and Yanji, in Yanbian Korean autonomous prefecture in China. From 2020 to 2022, the pathogenic-infected foliage of Yanbian Pingguoli pear was systematically harvested and then maintained in a 4-degree refrigerated microbial incubator for strain separation. The *Alternaria* isolates causing black spot on Yanbian Pingguoli pear were screened using the single spore separation method, as reported by Fu et al. [35]. Briefly, symptomatic tissues were cleaned with sterile water, disinfected with alcohol and 5% sodium hypochlorite, and cut into small pieces (1–2 mm), which were incubated on potato dextrose agar (PDA) for 7 days. The isolates were selected and purified for subsequent experiments.

4.2. Morphological and Molecular Characterization

Purified isolates were measured for conidial morphological parameters, rostrate projection dimensions, morphological silhouette, and transverse septation frequency. The isolates were cultured under controlled conditions including solid medium (potato carrot agar (PCA) poured into 90 mm Petri dishes) and incubation duration (168 h). The morphometric parameters of conidial chains and branching patterns were determined using a Leica M165C stereomicroscope connected to the Image-Pro plus 6.0 image analysis and processing software (Leica, Wetzlar, Germany). The phenotypes of the isolates were systematically evaluated against reference strains of *A. alternata*, *A. tenuissima*, and other species of *Alternaria* [27] through standardized protocols.

Purified isolates were used to extract DNA following the Two-liquid Plant DNA extraction Kit's protocol (O' BioLab, Beijing, China). PCR amplification targeted the rDNA-ITS fragments with the primer pair ITS1 (5′-TCC GTA GGT GAA CCT GCG G-3′) and ITS4 (5′-TCC TCC GCT TAT TGA TAT GC-3′). The cycling protocol included an initial melting step (94 °C, 5 min), followed by 35 cycles of 94 °C for 40 s (melting), 58 °C for 40 s (annealing), and 72 °C for 1 min (elongation), concluding with a final extension at

72 °C for 10 min. This protocol was optimized to ensure the efficient amplification of conserved regions across diverse fungal taxa [41]. PCR amplification targeted the histone H3 coding region with the primer pair H3-1a (5′-ACT AAG AG ACC GCC CGC AGG-3′) and H3-1b (5′-GCG GGC GAG CTG G1 GTC CTT-3′). The cycling protocol included an initial melting step (96 °C, 2 min), followed by 30 cycles of 96 °C for 15 s (melting), 55 °C for 30 s (annealing), and 75 °C for 35 s (elongation), concluding with a final extension at 72 °C for 2 min. This protocol was designed to maximize yield while minimizing nonspecific amplification artifacts [42]. Sequencing services for the PCR products were conducted through Beijing Tianyi Huiyuan Biotechnology Co., Ltd. Raw sequence data were processed using the DNAMAN bioinformatics software (Version 5.0) to generate consensus sequences, which were subsequently compared against the non-redundant nucleotide database (nt) via the BLASTn algorithm to retrieve homologous sequences from the NCBI library. Maximum likelihood (ML) analysis of concatenated nucleotide datasets from ITS and histone H3 loci was performed to infer phylogenetic relationships among 124 *Alternaria* strains using Clustal_W v1.83 [43] (for multiple sequence alignment) and the MEGA 5 software (release 5.2.2; accessible at http://www.megasoftware.net/) (for tree construction). For the haplotype comparison, phylogenetic analysis of concatenated ITS and histone H3 sequences demonstrated bootstrap support values (1000 replicates) for internal nodes. *A. brassicae isolate* AB11, *A. infectoria* isolate STE-U4271, *A. solani* isolate CNU3072, and *A. infectoria* isolate CR30 were used as outgroups.

4.3. Pathogenicity Tests

In order to confirm Koch's postulates, 48 *Alternaria* isolates with their representation were assessed with relatively minor modifications, as reported by Pryor and Michailides [44]. Briefly, 20 holes were pricked in the recipient leaves with a sterilized prick needle, and a drop of sterile water was placed on one side where the prick was made as a blank. Next, 48 *Alternaria* isolates covered with PDA Petri dishes were pressed out with a hole puncher with a diameter of 5 mm and placed on the side where the holes were pierced with the prick needle. Healthy, detached Yanbian Pingguoli pear leaves (6 cm long and 3.5 cm wide) that naturally grew in Pingguoli pear orchards at the end of May were selected as receptors. Thirty symptomatic leaves were assigned to individual isolation groups and maintained under controlled conditions (25 °C, 90% relative humidity) with a 12 h photoperiod for 14 days. Symptom progression was monitored daily, with diseased tissues subjected to pathogen re-isolation. Disease severity (DS) was quantified using a modified 5-point scale [44]: 0 (no lesions), 1 (<1 mm), 2 (1–5 mm), 3 (5–10 mm), and 4 (>10 mm). The disease index (DI) was calculated as $DI = (\Sigma [n \times DS])/N \times 25$, where n indicates lesion frequency per grade and N represents total inoculation points [32]. The LSD post hoc analysis revealed statistically significant differences ($p < 0.05$) among experimental conditions.

4.4. Extraction of AcEO and GC-MS Analysis

Fresh *Artemisia capillaris* samples were collected from a sunny slope in Yanji City, Jilin Province, China. They were authenticated by an expert from the College of Agriculture, Yanbian University. The whole plants were air-dried in the shade and cut into 5 cm segments. The AcEOs were extracted according to the patent CN117778103B [45]. After 3 h, the AcEOs were collected and stored in brown sample bottles at 4 °C in darkness until further analysis.

GC-MS was performed on a Shimadzu QP 2010 Ultra system (Shimadzu, Tokyo, Japan) equipped with a DB-5MS column (30 m × 0.25 mm ID, 0.25 μm film thickness). The analytical configuration included splitless injection, a helium carrier gas flow (0.98 mL/min), and an injection volume of 2.0 μL. Ionization was achieved via electron impact (70 eV) with the source temperature set at 280 °C. The oven program started at 45 °C (4 min), increased to 280 °C at 6 °C/min, and finalized with a 15 min isothermal hold. Full-scan acquisitions (35–500 m/z) were acquired at 0.56 scans/s. Compound identification utilized spectral matching against the NIST 20 mass spectral library and was validated through comparative literature analysis [26,46,47].

4.5. Antifungal Activity of the AcEO

The antifungal activities of AcEO were tested against the mycotoxigenic strains *A. tenuissima* YJ-XY-GSC1 and *A. alternata* YJ-XY-GSC3, using the contact assay on PDA. For the AcEO, 5 concentrations were used (1.0 μL/mL, 2.0 μL/mL, 3.0 μL/mL, 4.0 μL/mL, and 5.0 μL/mL), which were dissolved in PDA with 1% DMSO. Fungal cultures (7 days old) were inoculated onto 4 mm diameter disks and cultured at 28 °C until complete mycelial coverage of negative controls was achieved. Each concentration included five technical replicates to ensure statistical validity. MIC was determined as the lowest AcEO concentration that completely suppressed fungal growth after a 7-day incubation period.

SEM was used to examine the impacts of AcEO on the morphology of *A. tenuissima* YJ-XY-GSC1 and *A. alternata* YJ-XY-GSC3. Following treatment with various concentrations of AcEO (specifically 0 and 0.25 MIC), the mycelia of *A. tenuissima* YJ-XY-GSC1 and *A. alternata* YJ-XY-GSC3 were gathered. Mycelial samples were immobilized in 2.5% glutaraldehyde for 12 h at 4 °C. Post-fixation, they were washed three times in 0.01 M phosphate buffer (PBS, pH 7.4). Dehydration was performed through an ascending ethanol series (50–100%, 10 min per step). Critical point drying with liquid CO_2 was applied, followed by sputter-coating with gold. Morphological analysis was conducted under 3.0 kV accelerating voltage using a SU8010 scanning electron microscope (Hitachi, Tokyo, Japan).

5. Conclusions

This study characterized 124 fungal strains associated with black spot in Yanbian pear cultivars in China. Phylogenetic analysis revealed the dominant presence of *A. tenuissima* (54.0%, n = 67) and *A. alternata* (46.0%, n = 57). Our results also indicated that AcEO exhibits potent inhibitory activity against *A. tenuissima* and *A. alternate*, achieving complete suppression of mycelial growth at an MIC of 5.0 μL/mL.

The high volatility, poor water solubility, and susceptibility to degradation limited the application of essential oils. Further studies should focus on the preparation of a nanoemulsion using AcEO to maximize the efficacy in controlling pathogens causing black spot on Pingguoli. Additionally, the chemical components of plant essential oils are highly diverse. The components that truly have antibacterial and antifungal effects may only be one or a few of them. Moreover, there may be antagonistic or synergistic effects among various components. Further studies should also focus on the mechanisms of action and on extending the application of AcEO and the active monomer substances. This study provides new insights into alternative natural gaseous fungicides for black spot management, which may help to reduce economic losses.

6. Patents

Fu, Y. A raw material steaming device for plant essential oil production, 2023 (Patent No. CN117778103B).

Supplementary Materials: The following supporting information can be downloaded at: https://www.mdpi.com/article/10.3390/plants14203146/s1, Table S1: *Alternaria* isolates used for phylogenetic analysis; Table S2: Structural formulas of various components in the essential oil from *Artemisia capillaris*.

Author Contributions: Conceptualization, Y.D. and L.L.; methodology, Y.F. and X.-H.W.; resources writing—original draft preparation, Z.-X.K. and Y.D.; writing—review and editing, Y.F.; supervision, Y.F. and X.-H.W. All authors have read and agreed to the published version of the manuscript.

Funding: This work was supported by the National Natural Science Foundation of China (No. 32360708) and the Science and Technology Research Project of Education Department of Jilin Province (No. JJKH20230628KJ).

Data Availability Statement: The original contributions presented in this study are included in the article and Supplementary Materials. Further inquiries can be directed to the corresponding authors.

Conflicts of Interest: The authors declare no conflicts of interest.

References

1. Ma, J.N.; Wang, S.L.; Zhang, K.; Wu, Z.G.; Hattori, M.; Chen, G.L.; Ma, C.M. Chemical components and antioxidant activity of the peels of commercial apple-shaped pear (fruit of *Pyrus pyrifolia* cv. pingguoli). *J. Food Sci.* **2012**, *77*, C1097–C1102. [CrossRef]
2. Chen, S.J.; Li, Y.C.; Bi, Y.; Yin, Y.; Ge, Y.H.; Wang, Y. Solvent effects on the ultrastructure and chemical composition of cuticular wax and its potential bioactive role against *Alternaria alternata* in Pingguoli pear. *J. Integr. Agric.* **2014**, *13*, 1137–1145. [CrossRef]
3. Dai, J.; Hu, Y.; Si, Q.; Gu, Y.F.; Xiao, Z.Q.; Ge, Q.; Sha, R.Y. Antioxidant and hypoglycemic activity of sequentially extracted fractions from Pingguoli pear fermentation broth and identification of bioactive compounds. *Molecules* **2022**, *27*, 6077. [CrossRef]
4. Li, Y.C.; Bi, Y.; Cao, J.K. Method for extraction, isolation and bioassay of antifungal compounds from *Pyrus pyrifolia* cv Pingguoli. *J. Gansu Agric. Univ.* **1999**, *4*, 354–357. (In Chinese)
5. Li, Y.C.; Bi, Y. Extraction, isolation and identification of preformed antifungal compounds from the peel of Pingguoli pear. *Sci. Technol. Food Ind.* **2006**, *2*, 49–52. (In Chinese)
6. Prusky, D. Pathogen quiescence in postharvest diseases. *Annu Rev. Phytopathol.* **1996**, *34*, 413–434. [CrossRef] [PubMed]
7. Li, Y.C.; Bi, Y. Histopathological study on latent infection of *Alternaria* sp. in *Pyrus bretchneideri*. *J. Gansu Agric. Univ.* **2005**, *4*, 516–520. (In Chinese)
8. Adachi, Y.; Tsuge, T. Coinfection by different isolates of *Alternaria alternata* in single black spot lesions of Japanese pear leaves. *Phytopathology* **1994**, *84*, 447–451. [CrossRef]
9. Tanahashi, M.; Nakano, T.; Akamatsu, H.; Kodama, M.; Otani, H.; Osaki-Oka, K. *Alternaria alternata* apple pathotype (*A. mali*) causes black spot of European pear. *Eur. J. Plant Pathol.* **2016**, *145*, 787–795. [CrossRef]
10. Itai, A.; Igori, T.; Fujita, N.; Egusa, M.; Kodama, M.; Murayama, H. Ethylene analog and 1-methylcyclopropene enhance black spot disease development in *Pyrus pyrifolia* Nakai. *Hortscience* **2012**, *47*, 228–231. [CrossRef]
11. Zhang, P.F.; Zhai, L.F.; Zhang, X.K.; Huang, X.Z.; Hong, N.; Xu, W.X.; Wang, G.P. Characterization of *Colletotrichum fructicola*, a new causal agent of leaf black spot disease of sandy pear (*Pyrus pyrifolia*). *Eur. J. Plant Pathol.* **2015**, *143*, 651–662. [CrossRef]
12. Song, B.; Xu, B.Q.; Zhu, X.F.; He, T.M. Genomic analysis of *Alternaria alternata*, the causal agent of black spot disease in Korla fragrant pears (*Pyrus sinkiangensis* Yü) in China. *Can. J. Plant Sci.* **2025**, *105*, 1–14. [CrossRef]
13. Aichinger, G.; Grgic, D.; Beisl, J.; Crudo, F.; Warth, B.; Varga, E.; Marko, D. N-acetyl cysteine alters the genotoxic and estrogenic properties of Alternaria toxins in naturally occurring mixtures. *Emerg. Contam.* **2022**, *8*, 30–38. [CrossRef]
14. Guerreiro, B.; Ventura, C.; Louro, H.; Silva, M.J. P01-05 Investigating the genotoxic effects of the Alternaria toxin tenuazonic acid in human cells. *Toxicol. Lett.* **2024**, *399*, S73–S74. [CrossRef]
15. Jyotsna, B.; Patil, S.; Surya Prakash, Y.; Rathnagiri, P.; Kavi Kishor, P.B.; Jalaja, N. Essential oils from plant resources as potent insecticides and repellents: Current status and future perspectives. *Biocatal. Agric. Biotechnol.* **2024**, *61*, 103395. [CrossRef]
16. Costa, W.K.; da Cruz, R.C.D.; Carvalho, K.D.; de Souza, I.A.; Correia, M.T.D.; de Oliveira, A.M.; da Silva, M.V. Insecticidal activity of essential oil from leaves of *Eugenia stipitata* McVaugh against *Aedes aegypti*. *Parasitol. Int.* **2024**, *98*, 102820. [CrossRef]
17. Becer, E.; Altundag, E.M.; Güran, M.; Vatansever, H.S.; Ustürk, S.; Hanoglu, D.Y.; Baser, K.H.C. Composition and antibacterial, anti-inflammatory, antioxidant, and anticancer activities of *Rosmarinus officinalis* L. essential oil. *S. Afr. J. Bot.* **2023**, *160*, 437–445. [CrossRef]
18. Lee, T.H.Y.; Yeom, W.; Kim, H.; Beuchat, L.R.; Ryu, J.H. Influence of diluent on antimicrobial activity of cinnamon bark essential oil vapor against *Staphylococcus aureus* and *Salmonella enterica* on a laboratory medium and beef jerky. *Food Microbiol.* **2023**, *115*, 104335. [CrossRef]
19. Pophof, B.; Stange, G.; Abrell, L. Volatile organic compounds as signals in a plant-herbivore system: Electrophysiological responses in olfactory sensilla of the moth *Cactoblastis cactorum*. *Chem. Senses* **2005**, *30*, 51–68. [CrossRef]
20. Bakkali, F.; Averbeck, S.; Averbeck, D.; Idaomar, M. Biological effects of essential oils—A review. *Food Chem. Toxicol.* **2008**, *46*, 446–475. [CrossRef]

21. Viteri, L.O.; Gonza'lez, M.J.; Silva, P.B.; Gomes, J.M.; Svacina, T.; Costa, L.T.M.; Valarezo, E.; Mantilla-Afanador, J.G.; Herrera, O.M.; Aguiar, R.W.S.; et al. GABA and octopamine receptors as potential targets for fumigant actions of *Bursera graveolens* essential oil against *Callosobruchus maculatus* and *Callosobruchus chinensis*. *J. Xenobiotics* **2025**, *15*, 1–16. [CrossRef] [PubMed]
22. Perveen, K.; Bokhari, N.A.; Al-Rashid, S.A.I.; Al-Humaid, L.A. Chemical composition of essential oil of *Ocimum basilicum* L. and its potential in managing the Alternaria rot of tomato. *J. Essent. Oil Bear. Plants* **2020**, *23*, 1428–1437. [CrossRef]
23. Kamsu, F.P.N.; Dikongue, F.J.N.; Ngouana, V.; Tchinda, E.S.; Jiogue, M.B.; Ambata, H.T.A.; Tchameni, S.N.; Sameza, M.L.; Jazet, P.M.D. Effectiveness of Massep (*Ocimum gratissimum* L.) essential oil and its nanoemulsion toward *Sclerotium rolfsii*, *Phytophthora infestans* and *Alternaria solani*, pathogens associated with tomato rot diseases. *Biocatal. Agric. Biotechnol.* **2023**, *47*, 102591. [CrossRef]
24. Chen, Y.Y.; Zhang, Y.D.; Wang, L.; Li, M.Q.; Wang, Y.; Liang, J.; Wang, D.L.; Zhang, H.W. Investigating the antifungal mechanism of *Artemisia argyi* essential oil against *Botrytis cinerea*, *Alternaria alternata*, and *Penicillium digitatum* and its application in extending blueberry shelf life. *Int. J. Food Microbiol.* **2025**, *439*, 111262. [CrossRef]
25. Chen, Y.Y.; Zhang, S.H.; Qu, L.P. The protective effect of *Artemisia capillaris* Thunb. extract against UVB-induced apoptosis and inflammation through inhibiting the cGAS/STING pathway. *J. Photochem. Photobiol. B Biol.* **2024**, *258*, 112989. [CrossRef] [PubMed]
26. Cha, J.D.; Jeong, M.R.; Jeong, S.I.; Moon, S.E.; Kim, J.Y.; Kil, B.S.; Song, Y.H. Chemical composition and antimicrobial activity of the essential oils of *Artemisia scoparia* and *A. capillaris*. *Planta Med.* **2005**, *71*, 186–190. [CrossRef]
27. Simmons, E.G. Alternaria themes and variations. *Mycotaxon* **1995**, *55*, 55–163. [CrossRef]
28. Singh, K.; Deepa, N.; Chauhan, S.; Tandon, S.; Verma, R.S.; Singh, A. Antifungal action of 1,8 cineole, a major component of *Eucalyptus globulus* essential oil against *Alternaria tenuissima* via overproduction of reactive oxygen species and downregulation of virulence and ergosterol biosynthetic genes. *Ind. Crop Prod.* **2024**, *214*, 118580. [CrossRef]
29. Chen, X.F.; Teng, L.P.; Dan, H.X.; Xiong, R.C. Occurrence of black leaf spot caused by *Alternaria alternata* on Korla fragrant pear in Xinjiang of China. *J. Plant Pathol.* **2020**, *102*, 265. [CrossRef]
30. Li, Y.C.; Bi, Y.; An, L.Z. Occurrence and latent infection of Alternaria rot of pingguoli pear (*Pyrus bretschneideri* Rehd. cv. Pingguoli) fruits in Gansu, China. *J. Phytopathol.* **2007**, *155*, 56–60.
31. Yin, Y.; Bi, Y.; Chen, S.J.; Li, Y.C.; Wang, Y.; Ge, Y.H.; Ding, B.; Li, Y.; Zhang, Z. Chemical composition and antifungal activity of cuticular wax isolated from Asian pear fruit (cv. Pingguoli). *Sci. Hortic.* **2011**, *129*, 577–582. [CrossRef]
32. Fu, Y.; Zhang, X.F.; Liu, S.J.H.; Hu, K.L.; Wu, X.H. Characterization of *Alternaria* species associated with black spot of strawberry in Beijing municipality of China. *Can. J. Plant Pathol.* **2020**, *4*, 235–242. [CrossRef]
33. Ma, G.P.; Bao, S.W.; Zhao, J.; Sui, Y.; Wu, X.H. Morphological and molecular characterization of *Alternaria* species causing leaf blight on watermelon in China. *Plant Dis.* **2021**, *105*, 60–70. [CrossRef] [PubMed]
34. Shi, X.J.; Zeng, K.X.; Wang, X.Y.; Liang, Z.H.; Wu, X.H. Characterization of *Alternaria* species causing leaf spot on Chinese cabbage in Shanxi province of China. *J. Plant Pathol.* **2021**, *103*, 283–293. [CrossRef]
35. Sun, X.Z.; Wang, C.Y.; Gao, X.; Wu, X.H.; Fu, Y. Characterization of *Alternaria* species associated with black spot of strawberry in Dandong, China. *Agronomy* **2023**, *13*, 1014. [CrossRef]
36. Masudaa, Y.; Asadaa, K.; Satohb, R.; Takadaa, K.; Kitajima, J. Capillin, a major constituent of *Artemisia capillaris* Thunb. flower essential oil, induces apoptosis through the mitochondrial pathway in human leukemia HL-60 cells. *Phytomedicine* **2015**, *22*, 545–552. [CrossRef]
37. Liang, M.; Xiao, X.; Chen, M.; Guo, Y.; Han, W.T.; Min, Y.H.; Jiang, X.W.; Yu, W.H. *Artemisia capillaris* Thunb. Water extract alleviates metabolic dysfunction-associated Steatotic liver disease Disease by inhibiting miR-34a-5p to activate Sirt1-mediated hepatic lipid metabolism. *J. Ethnopharmacol.* **2025**, *338*, 119030. [CrossRef]
38. Gao, Q.H.; Zhao, X.; Yin, L.; Zhang, Y.B.; Wang, B.; Wu, X.L.; Zhang, X.H.; Fu, X.Y.; Sun, W.H. The essential oil of *Artemisia capillaris* protects against CCl_4-induced liver injury in vivo. *Rev. Bras. Farmacogn.* **2016**, *26*, 369–374. [CrossRef]
39. Gadhi, M.A.; Nizamani, Z.A.; Jatoi, G.H.; Abro, M.A.; Keerio, A.U.; Poussio, G.B.; Qiu, D. In-vitro efficacy of bio-control agent and essential oils against leaf blight of chickpea caused by *Alternaria alternata*. *Acta Ecol. Sin.* **2020**, *40*, 166–171. [CrossRef]
40. Allagui, M.B.; Moumni, M.; Romanazzi, G. Antifungal activity of thirty essential oils to control pathogenic fungi of postharvest decay. *Antibiotics* **2024**, *13*, 28. [CrossRef]
41. White, T.J.; Bruns, T.; Lee, S.; Taylor, J. Amplification and direct sequencing of fungal ribosomal RNA genes for phylogenetics. *PCR Protoc.* **1990**, *18*, 315–322.
42. Glass, N.L.; Donaldson, G.C. Development of primer sets designed for use with the PCR to amplify conserved genes from filamentous ascomycetes. *Appl. Environ. Microbiol.* **1995**, *61*, 1323–1330. [CrossRef]
43. Thompson, J.D.; Higgins, D.G.; Gibson, T.J. Clustal W: Improving the sensitivity of progressive multiple sequence alignment through sequence weighting, position-specific gap penalties and weight matrix choice. *Nucleic Acids Res.* **1994**, *22*, 4673–4680. [CrossRef]
44. Pryor, B.M.; Michailides, T.J. Morphological, pathogenic, and molecular characterization of *Alternaria* isolates associated with Alternaria late blight of pistachio. *Phytopathology* **2002**, *92*, 406–416. [CrossRef]

45. Fu, Y. A Raw Material Steaming Device for Plant Essential Oil Production. No. CN117778103B. 9 August 2023.
46. Zheng, T.Y. Study on Antifungal Mechanism of *Artemisia sieversiana* Essential oil on Pathogenic Fungi Causing Pingguoli pear Black Spot. Master's Thesis, Yanbian University, Yanbian, China, 2024.
47. Oses, R.; Ferrando, M.; Bruna, F.; Retamales, P.; Navarro, M.; Fernández, K.; Vera, W.; Larrazábal, M.J.; Neira, I.; Paredes, A.; et al. Exploring the bioactive potential and chemical profile of *Schinus molle* essential oil: An integrated in Silico and in vitro evaluation. *Plants* **2025**, *14*, 2449. [CrossRef] [PubMed]

plants

MDPI

Article

Eco-Friendly Crop Protection: *Argyrantemum frutescens*, a Source of Biofungicides

Eduardo Hernández-Álvarez [1], Samuel Rodríguez-Sabina [2], Noelia Labrador-García [1], Javier Hernández Pérez [1], Carolina P. Reyes [3], María Ángeles Llaría-López [4], Ignacio A. Jiménez [1] and Isabel L. Bazzocchi [1,*]

1 Instituto Universitario de Bio-Orgánica Antonio González, Departamento de Química Orgánica, Universidad de La Laguna, Avenida Astrofísico Francisco Sánchez 2, 38206 La Laguna, Tenerife, Spain; alu0100947311@ull.edu.es (E.H.-Á.); alu0101028057@ull.edu.es (N.L.-G.); alu0101230672@ull.edu.es (J.H.P.); ignadiaz@ull.edu.es (I.A.J.)

2 Departamento de Botánica, Ecología y Fisiología Vegetal, Universidad de La Laguna, Avenida Astrofísico Francisco Sánchez, 38206 La Laguna, Tenerife, Spain; srodrisa@ull.edu.es

3 Instituto Universitario de Bio-Orgánica Antonio González, Departamento de Bioquímica, Microbiología, Biología Celular y Genética, Universidad de La Laguna, Avenida Astrofísico Francisco Sánchez 2, 38206 La Laguna, Tenerife, Spain; cpreyes@ull.edu.es

4 Área de Gestión del Medio Natural y Seguridad, Cabildo Insular de Tenerife, C/Las Macetas s/n, Pabellón Insular Santiago Martín, 38108 La Laguna, Tenerife, Spain; mllaria@tenerife.es

* Correspondence: ilopez@ull.edu.es; Tel.: +34-922318594

Abstract: Plant-derived biopesticides are emerging as a promising and popular alternative for promoting cleaner and safer agricultural practices. The present work aims to explore *Argyranthemum frutescens* (*Asteraceae*) as a source of botanical pesticides and to validate this through a cultivation process. To this task, a bioassay-guided fractionation of the ethanolic root extracts from both wild and cultivated *A. frutescens* on phytopathogenic fungi of *Botrytis cinerea*, *Fusarium oxysporum*, and *Alternaria alternata* was conducted. This approach led to the identification of polyacetylenes with higher potency than commercial fungicides. Specifically, compounds **3** (capillin) and **5** (frutescinone) showed more than 90% growth inhibition at 0.05 mg/mL concentration on *B. cinerea*, while compounds **2** (capillinol) and **3** were also more active than positive controls, Fosbel-Plus and Azoxystrobin, against *F. oxysporum*. The structures of the isolated polyacetylenes (**1–6**, **9**, and **10**) and alkamides (**7**, **8**, and **11**) were determined through spectroscopic analysis, and the absolute configuration of stereocenter C1 of compounds **1**, **2**, **4** and **9** was determined by NMR-spectroscopy with (*R*)-(-)-α-methoxy-phenylacetic as a chiral derivatizing agent, and biogenetic considerations. Overall, this study supports the potential of polyacetylenes as promising agrochemical lead compounds against phytopathogens, and validates *A. frutescens* cultivation as a viable source of biopesticides.

Keywords: *Argyranthemum frutescens*; cultivation; phytopathogens; biofungicides; polyacetylenes

1. Introduction

Plant diseases result in a variety of direct and quantifiable economic consequences for crop management [1]. In particular, phytopathogenic fungi represents a significant threat to global food security, ecosystems, and human health [2]. Among the phytopathogenic fungi, *Botrytis cinerea*, *Fusarium oxysporum* and *Alternaria alternata*, are considered high-risk plant pathogens affecting crops worldwide [3]. In fact, *A. alternata* is one of the most common and cosmopolitan species causing disease on economically important crops due to its environmental adaptability to UV light, low temperature, and water stress conditions [4].

https://doi.org/10.3390/plants14070985

Dark sunken lesions are usually the expression of *Alternaria* spp. infections on roots, tubers, stems, and fruits in a wide range of vegetable- and fruits-producing plants. The *Alternaria* toxins as natural contaminants in food are a serious risk to human and animal health because of their known toxicity, since their administration/ingestion has been shown to be cytotoxic, fetotoxic, and teratogenic in animal; mutagenic in microbial and mammalian cell systems; and tumorigenic in rats [5]. *B. cinerea* is a widespread plant pathogen with a necrotrophic lifestyle, causing overwhelming diseases in more than 1400 plant species, especially fruit crops, resulting in significant economic losses worldwide [6]. The pathogen causes rotting of fruits at both preharvest and postharvest stages, and aside from causing "gray mold" on the mature fruits, the fungus infects leaves, flowers, and seeds. *B. cinerea* produces two major phytotoxins involved in the infection caused by this fungus, the sesquiterpene botrydial and the polyketide botcinin acid [7]. Many *Fusarium* species, such as *F. oxysporum*, are pathogenic, causing crop diseases during crop production and spoilage of agricultural products in both commercial and smallholder farms. The damage caused by this fungus is through their direct attack of crops in the fields and by the production of allergenic compounds and mycotoxins, causing yield loss and increases in food insecurity and food prices [8].

To minimize the crop losses, synthetic fungicides are widely used, despite their over-application or misuse having raised serious concerns, including their impact on the environment, and the effect on human health and livestock [9]. These harmful effects, combined with the induction of resistance development in pests [10], have promoted the introduction of sustainable, effective, and environmentally friendly alternatives to the use of chemicals in crops protection [11]. In this regard, biopesticides have demonstrated the potential to be used to protect crops against phytopathogens, reducing the use of conventional synthetic fungicides as a vital step towards sustainable crop production [12]. Plant-derived pesticides have great advantages, being effective against various pests, biodegradable, abundant, and less expensive [13].

Moreover, the discovery of biopesticides from a plant species could lead to the overexploitation of natural habitats [14], and therefore, there is a need to find sustainable sources of biomass to prevent an ecological crisis. In this regard, the cultivation of a target species as a source of biopesticides arises as a sustainable strategy to ensure the availability of biomass [15].

The *Asteraceae* family is the most extensive within the angiosperms, and it is distributed worldwide [16]. Species of this botanical family have extensive pharmaceutical applications, including antioxidant, anti-inflammatory, and antibacterial [17], but also pesticide properties [13,18]. Recent studies reported that plants belonging to this family have an excellent ability to synthesize nanoparticles in non-toxic ways having numerous applications [19]. The phytochemical profile of plants in the *Asteraceae* family reveals a multitude of bioactive compounds, such as terpenes, saponins, lignans, and polyphenolic compounds, which are responsible for their wide range of properties [13,20]. In particular, the genus *Argyranthemum* Webb ex Sch.Bip., the largest genus of flowering plants endemic to the Macaronesian archipelagos, is present in all the major habitat zones in Macaronesia, ranging from coastal to subalpine habitats, and comprises twenty-four species [21]; three of them are endemic to Madeira, one to Selvagens and twenty to the Canary Islands. *Argyranthemum frutescens* (L.) Sch. Bip. subsp. *frutescens* named as "magarza" is a perennial plant originating in the Canary (Spain) and Madeira (Portugal) Islands [22] used in stomach and buccal tonics and for asthma treatment [23]. Moreover, its potential as a biopesticide has not been investigated.

The present research provides a comprehensive account of the isolation and structural elucidation of eight polyacetylenes and three alkamides from the ethanolic extracts of the roots of *A. frutescens*, wild as well as cultivated, through a bioassay-guided fractionation targeting three strains of phytopathogenic fungi, *A. alternata*, *B. cinerea* and *F. oxysporum*. Evaluation of the isolated metabolites pointed out the polyacetylenes as promising components to be used as biopesticides. The preliminary structure–activity relationship is also discussed.

2. Results and Discussion

2.1. Bioassay-Guided Fractionation

In recent decades, the use of bioassay-guided fractionation techniques have been rising as the main tool to decipher in the discovery and development of bioactive compounds from natural resources. They are used in the early stages of compounds isolation/purification/identification, serving as a starting point for application-based development [24].

Previous studies on *A. frutescens* conducted three decades ago reported the isolation of acetylenes and an alkamide from various parts of the plant [25], as well as the antimicrobial and cytotoxic activities of the acetone extract from the roots and its isolated metabolites [26]. In this study, as part of our research program aimed at discovering plant-derived biopesticides, we carried out a bioassay-guided fractionation of the ethanolic extracts from both wild-type and cultivated *A. frutescens* roots against three phytopathogenic fungi affecting crops worldwide (Figure 1, and Table S1 in Supporting Materials). Growth inhibition percent (%GI) was determined by a test on the mycelium stage [27] of the phytopathogens: *Botrytis cinerea*, *Fusarium oxysporum*, and *Alternaria alternata*. Samples exhibiting %GI higher than 20% at 1 mg/mL were also evaluated at 0.5 and 0.1 mg/mL doses. Commercial antifungal agents, Fosbel-Plus (35% Fosetil Al and 35% Mancozeb) and Ortiva PC (Azoxystrobin 250 g/L, 22.8% p/p) in the case of *B. cinerea* (since Fosbel-Plus was inactive at the lower concentrations assayed), were used as positive controls for comparative purposes in the current analysis, whereas ethanol was used as a negative control.

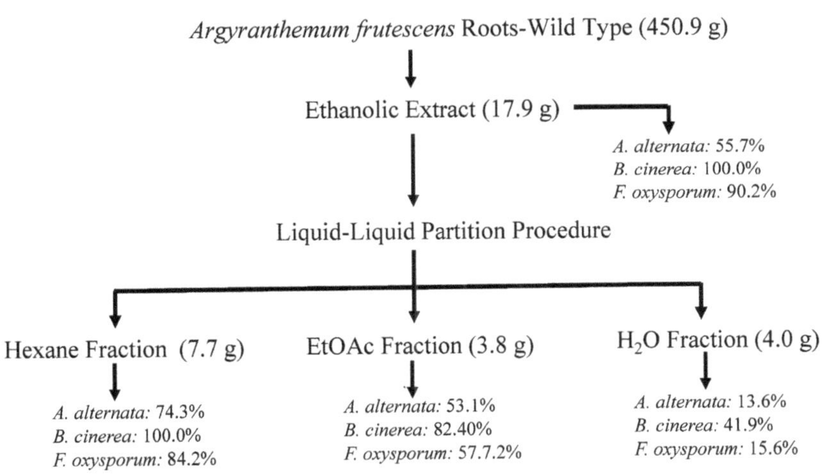

Figure 1. Flowchart of bioassay-guided fractionation of wild *A. frutescens* roots on the mycelium stage of *Alternaria alternata*, *Fusarium oxysporum*, and *Botrytis cinerea*. Values represent the percentage of growth inhibition (%GI) at 1 mg/mL concentration. Fosbel-Plus (%GI 93.5, 83.3, and 93.4, respectively) was used as a positive control.

Thus, the air-dried and powdered roots of wild-type *A. frutescens* plant (450.9 g) were extracted by maceration with 96% EtOH (5 L, 24 h, three times) at room temperature

(22 ± 4 °C). This procedure yielded 17.9 g of residue after the solvent was removed under reduced pressure at 40 °C (4.0% *w*/*w*), and an aliquot of this extract was assayed on phytopathogenic fungi under study. The EtOH extract from the wild-type plant showed remarkable activity at 1 mg/mL concentration with an inhibition growth (%GI) of 55.7%, 100.0% and 90.2% on *A. alternata*, *B. cinerea* and *F. oxysporum*, respectively, showing some degree of activity on *B. cinerea* (47.0%) even at 0.1 mg/mL (Figure 1 and Table S1 in Supporting Materials). These promising results encouraged us to achieve a bioassay-guided fractionation. Therefore, the EtOH extract was successively partitioned into hexane (Hx), EtOAc, and H_2O fractions by liquid–liquid partition, and the fractions were further evaluated. The results revealed that the two organic fractions exhibited activity against the tested fungi, highlighting the Hx fraction with GI values ranging from 74.3 to 100.0% at 1 mg/mL, more potent than the commercial antifungal Fosbel-Plus on *B. cinerea* (GI 100 vs. 83.3%) and slightly lower against *F. oxysporum* (GI 84.2 vs. 93.4%) at 1 mg/mL concentration. By contrast, the aqueous fraction was inactive (%GI < 10 at 1 mg/mL) against two of the phytopathogens, with moderate effect on *B. cinerea* (GI 41.9% at 1 mg/mL).

These promising results regarding the activity of the wild-type *A. frutescens* roots on the assayed phytopathogenic fungus encouraged us to search for a sustainable source of this plant species. In this regard, cultivation of a species would not only provide the possibility of managing the production and preservation of the species but also lead to the production of uniform raw materials, and the possibility of production in all seasons according to the climatic diversity of the country [15].

Therefore, the species *A. frutescens* was cultivated as described in detail in the Experimental section. The air-dried and powdered roots (500.0 g) were extracted by maceration with 96% EtOH at room temperature (22 ± 4 °C), yielding 40.3 g of residue (8.1% *w*/*w*) after the solvent was removed. Following the same procedure as that for the wild-type plant, this EtOH extract was evaluated on the three phytopathogenic fungi, and results indicated some degree of activity against the assayed fungi (67.9% on *A. alternata*, 93.4% on *F. oxysporum*, and 58.6% on *B. cinerea* at 1 mg/mL concentration), although it was lower than the EtOH extract from the wild-type plant on *F. oxysporum* (Figure 2).

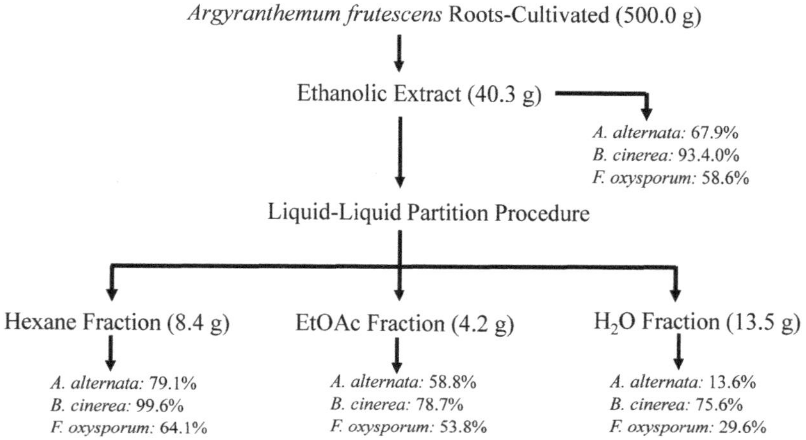

Figure 2. Flowchart of bioassay-guided fractionation of cultivated *A. frutescens* on the mycelium stage of *Alternaria alternata*, *Fusarium oxysporum*, and *Botrytis cinerea*. Values represent the percentage of growth inhibition (%GI) at 1 mg/mL concentration. Fosbel-Plus (GI 93.5, 83.3, and 93.4%, respectively) was used as a positive control.

This ethanolic extract was subjected to fractionation through a liquid–liquid partition and then to biological evaluation, revealing that the hexane fraction showed the highest efficacy with GI of 79.1, 99.6 and 64.1% at 1 mg/mL concentration against *A. alternata*, *B. cinerea*, and *F. oxysporum*, respectively, similar to the extract from the wild-type plant (Figure 2).

Consequently, the hexane fractions emerge as the most promising ones to advance in the bio-guided fractionation process, and those were selected to proceed with the bio-guided fractionation, investigating the metabolites responsible for bioactivity. Thus, the Hex fractions were chromatographed over column chromatography (hexane/EtOAc of increasing polarity, 10:0 to 0:10 as eluent), and subfractions were combined based on their TLC similarity, yielding subfractions A1–A8 and B1–B7 from the wild-type and cultivated plant materials, respectively. Each fraction was assayed against the phytopathogenic fungi, *A. alternata* (Figure 3), *B. cinerea* (Figure 4), and *F. oxysporum* (Figure 5).

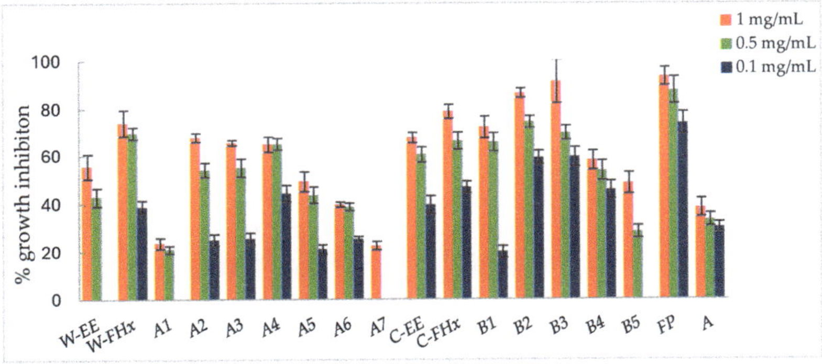

Figure 3. Antifungal effects (% growth inhibition, %GI) of *A. frutescens* roots against *Alternaria alternata*. Wild-type and cultivated plants: ethanolic extracts (W-EE and C-EE), hexane fractions (W-FHx and C-FHx), and subfractions (A1–A7 and B1–B5). Subfractions with a GI higher than 20% at 1 mg/mL were assayed at lower concentrations (0.5 and 0.1 mg/mL). Subfractions with GI < 20% have been omitted. Results are expressed as percentage relative to the negative control. Fosbel-Plus and Azoxystrobin were used as positive controls. Data are presented as mean ± SD (standard deviation, $n = 8$); $p < 0.05$.

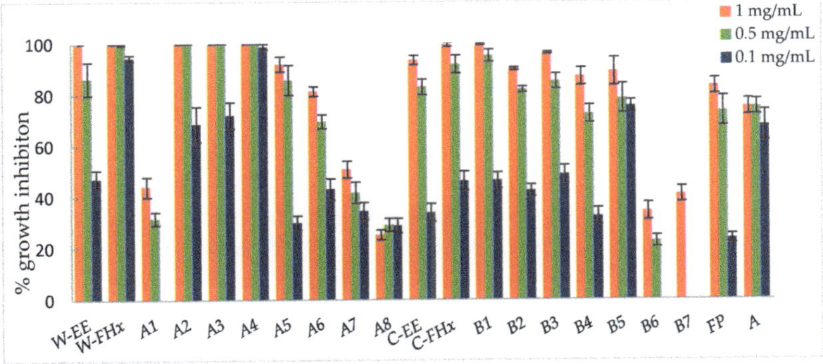

Figure 4. Antifungal effects (% growth inhibition, %GI) of *A. frutescens* roots against *Botrytis cinerea*. Wild-type and cultivated plants: ethanolic extracts (W-EE and C-EE), hexane fractions (W-FHx and C-FHx), and subfractions (A1–A8 and B1–B7). Subfractions with a GI higher than 20% at 1 mg/mL were assayed at lower concentrations (0.5 and 0.1 mg/mL). Subfractions with GI < 20% have been omitted. Results are expressed as percentage relative to the negative control. Fosbel-Plus and Azoxystrobin were used as positive controls. Data are presented as mean ± SD (standard deviation, n = 8); $p < 0.05$.

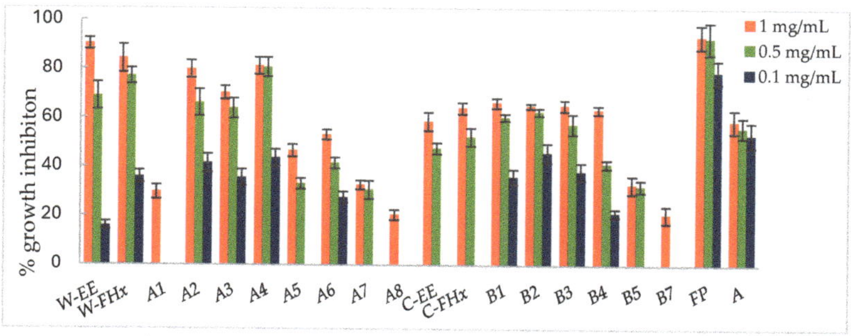

Figure 5. Antifungal effects (% growth inhibition, %GI) of *A. frutescens* roots against *Fusarium oxysporum*. Wild-type and cultivated plants: ethanolic extracts (W-EE and C-EE), hexane fractions (W-FHx and C-FHx), and subfractions (A1–A8 and B1–B7). Subfractions with a GI higher than 20% at 1 mg/mL were assayed at lower concentrations (0.5 and 0.1 mg/mL). Subfractions with GI < 20% have been omitted. Results are expressed as percentage relative to the negative control. Fosbel-Plus and Azoxystrobin were used as positive controls. Data are presented as mean ± SD (standard deviation, n = 8); $p < 0.05$.

Subfractions A2–A4 exhibited percentage inhibition values > 60% at 1 mg/mL against the three fungi (Figures 3–5, Table S1 in Supporting Materials). *B. cinerea* was the most sensitive fungus, with subfractions A2–A4 reaching a 100% of growth inhibition at 0.5 mg/mL, significantly higher than when positive controls, Fosbel-Plus (GI 73.5%) and Azoxystrobin (GI 75.0%) were used. Even subfraction A4 showed a GI of 98.7% at 0.1 mg/mL (Figure 4), the lowest assayed concentration. Regarding *F. oxysporum*, subfraction A4 exhibited a GI of 80.7% at 0.5 mg/mL (Figure 5), whereas *A. alternata* was revealed to be the least sensitive fungus to these subfractions, with the most active subfraction being A4 which showed a GI 65.1% at 0.5 mg/mL (Figure 3), slightly lower than the hexane fraction.

Regarding the results of the subfractions from the cultivated plant against the phytopathogenic fungi (Figures 3–5, Table S2 in Supporting Materials), subfractions B1–B5 exhibited similar behavior pattern as the subfractions from the wild-type plant. Thus, *B. cinerea* was the most sensitive fungus, with %GI ranging from 87.1 to 99.8 at 1 mg/mL, higher than the two positive controls (Fosbel-Plus, GI 83.3% and Azoxystrobin, 75.1%) at the same concentration. Also, these subfractions had an excellent activity at 0.5 mg/mL on this fungus (GI from 72.4 to 95.2%). Moreover, subfractions B1–B3 exhibited a significant degree of growth inhibition on *A. alternata* (72.4–91.5%) and *F. oxysporum* (65.1–66.3%) at 1 mg/mL. Moreover, subfractions B6 and B7 were inactive (%GI < 10%) and poorly active on *A. alternaria* and *F. oxysporum*, respectively, and moderately active on *B. cinerea* (%GI 34.3 and 40.9, respectively).

The results indicated that the antifungal activity of the EtOH extract from the cultivated plant is progressively enriched in the hexane fraction and its respective subfractions by liquid–liquid partitioning and column chromatography. This points to a selective concentration of bioactive compounds responsible for activity. Moreover, the results of the wild-type plant extract indicated that the subfractions were from slightly to moderately less active than the hexane fraction on *A. alternata* and *F. oxysporum*, whereas on *B. cinerea* the potency remained at 100% from the EtOH extract to subfractions. Therefore, the cultivated plant exhibited an efficient bioassay-guided isolation process, whereas the wild-type plant demonstrated partial efficiency, suggesting that the antifungal activity arises from interactions among multiple compounds involving potential synergistic effects.

2.2. Metabolites Identification from Active Subfractions

Based on the bioactivity, subfractions A2–A6 and B1–B5 from the wild-type and cultivated plants, respectively, were selected for successive purification steps by column chromatography to identify the components in those bioactive subfractions. This procedure yielded compounds **1–11** (Figure 6), the structure elucidation of which was achieved by means of spectroscopic techniques and comparison with previously reported data.

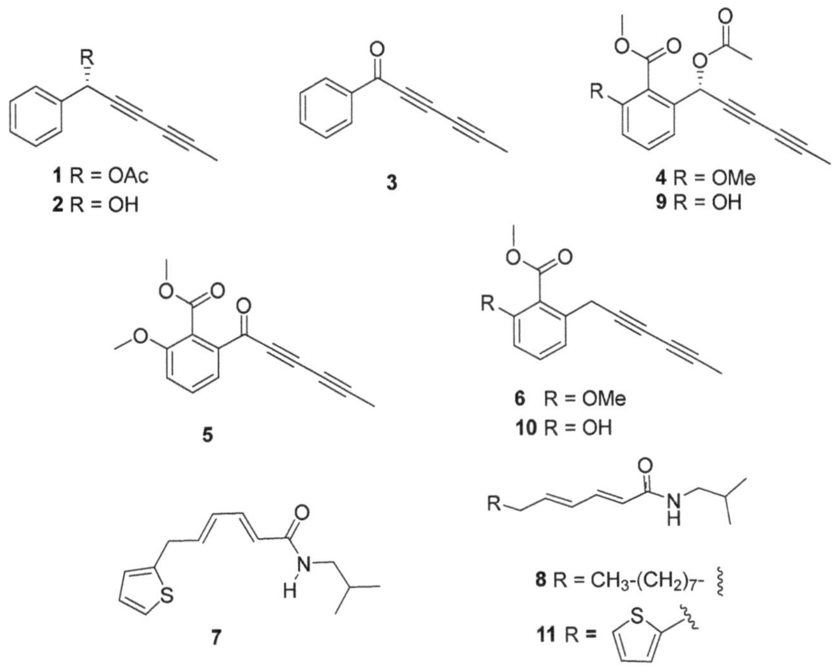

Figure 6. Structure of compounds isolated (**1–11**) from wild-type (**1–8**) and cultivated (**1**, **2**, **4**, **6**, and **8–11**) *Argyranthemum frutescens* roots.

Thus, the isolated compounds were identified as: 1-phenylhexa-2,4-diyn-1-yl acetate (capillinol acetate, **1**) [26], 1-phenylhexa-2,4-diyn-1-ol (capillinol, **2**) [28], capillin (**3**) [28], frutescinol acetate (**4**) [26], frutescinone (**5**) [26], frutescin (**6**) [26], *N*-isobutyl-6-(2-thienyl)-2*Z*,4*Z*-hexadienamide (**7**) [29], 2*E*,4*E*-tetradecadienoic acid isobutyl amide [30] (**8**), 3′-*O*-demethyl frutescinol acetate (**9**) [26], 3′-*O*-demehthyl frutescin [31] (**10**), and *N*-isobutyl-6-(2-thienyl)-2*E*,4*E*-hexadienamide [32] (**11**). Moreover, metabolites **1–8** were isolated from the wild-type plant, whereas metabolites **1**, **2**, **4**, **6** and **8–11** were identified in the cultivated plant.

Although the spectroscopic data and relative stereochemistry of compound **2** (capillinol) were previously reported [28], its absolute configuration remained unresolved. Thus, we report the first determination of the absolute configuration of the secondary alcohol on C1 of this compound by Riguera's method [33], a variation of Mosher's method, based on the use of (*R*)-(-)-α-methoxyphenylacetic (MPA) as the chiral reagent and barium (II) salt as the chelating agent. This methodology offers the advantage that it requires a small amount of sample; only one of the esters, either (*R*) or (*S*)-MPA, needs to be prepared. The absolute configuration was assigned by comparing the chemical shift of substituent G2 (H-2′ and H-6′) before (δ_H 7.4691) and after (δ_H 7.3804) saturation with $Ba(ClO_4)_2$ ($\Delta\delta Ba = +0.0887$) in the ^{1}H NMR spectra. Applying Riguera's model (Figure 7), the most shielded substituent aligns facing the phenyl moiety of the chiral ester, thereby dictating the absolute configuration of the secondary alcohol as 1*S*. Moreover, the absolute configuration of this stereocenter in compounds **1**, **4** and **9** was assumed based on biosynthetic considerations, since they contained the same polyacetylene core and specific rotation sign as compound **2**.

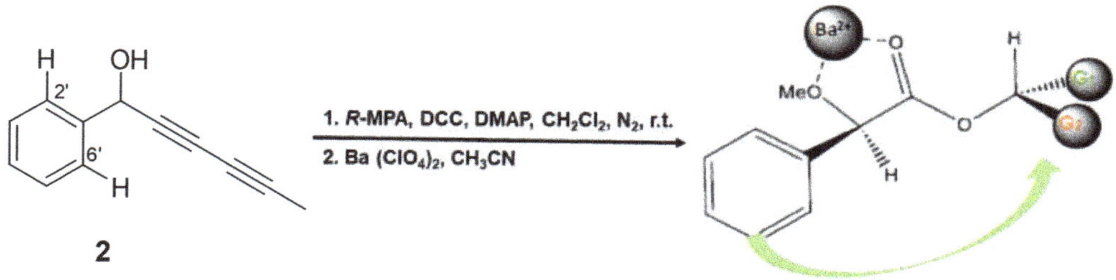

Figure 7. Synthesis of the capillinol (*R*)-α-methoxyphenylacetate and saturation with barium (II) salt as the chelating agent. *R*-MPA: acid, DCC: *N*, *N*′-dicyclohexylcarbodiimide; DMAP: 4-(dimethylamine)pyridine; r.t.: room temperature.

2.3. *Antifungal Activity Assays of the Isolated Compounds*

The effects on fungi viability of the eleven compounds (**1–11**) isolated from both wild-type and cultivated *A. frutescens* roots were evaluated individually against the three phytopathogenic fungi (Table 1). Compounds with an inhibition growth higher than 20% at 0.1 mg/mL were assayed at lower concentrations (0.05 and 0.01 mg/mL). Fosbel-Plus and Ortiva PC were used as positive controls.

Table 1. Antifungal effects (% growth inhibition) of compounds **1–11** at 0.1–0.01 mg/mL from *Argyranthemum frutescens* roots against *Alternaria alternata*, *Botrytis cinerea*, and *Fusarium oxysporum*.

Sample	*A. alternata*			*B. cinerea*			*F. oxysporum*		
	0.1	0.05	0.01	0.1	0.05	0.01	0.1	0.05	0.01
1	34.6 ± 3.3	31.7 ± 2.2	22.6 ± 2.5	76.3 ± 5.4	58.4 ± 6.54	23.6 ± 7.6	36.2 ± 4.2	34.0 ± 3.2	14.4 ± 3.0
2	66.3 ± 2.3	35.3 ± 3.4	24.7 ± 4.6	95.8 ± 3.2	60.0 ± 5.4	NA	83.6 ± 3.8	44.5 ± 4.9	18.2 ± 3.8
3	25.5 ± 3.8	NA	ND	100.0 ± 0.0	92.7 ± 7.1	39.9 ± 6.8	94.9 ± 1.4	84.4 ± 11.7	NA
4	56.6 ± 4.1	55.4 ± 2.4	37.9 ± 3.2	61.8 ± 8.9	60.9 ± 7.4	21.1 ± 5.9	23.9 ± 5.2	17.4 ± 2.5	NA
5	48.1 ± 4.2	40.4 ± 4.9	13.3 ± 5.2	90.8 ± 4.1	90.8 ± 2.3	70.6 ± 4.5	45.6 ± 6.3	31.8 ± 4.7	NA
6	42.6 ± 6.8	28.2 ± 4.2	NA	85.1 ± 3.2	73.2 ± 2.6	32.2 ± 3.1	23.9 ± 2.4	NA	ND
7	39.3 ± 2.3	38.4 ± 1.9	13.8 ± 2.9	39.7 ± 4.6	37.4 ± 5.6	20.2 ± 5.8	19.5 ± 1.5	ND	ND
8	NA	ND	ND	NA	ND	ND	18.9 ± 2.5	ND	ND
9	71.3 ± 2.7	45.3 ± 2.9	33.4 ± 6.6	90.2 ± 5.5	80.6 ± 2.2	57.4 ± 3.0	41.7 ± 1.9	12.7 ± 5.9	ND
10	29.3 ± 2.1	10.7 ± 3.3	NA	70.1 ± 6.6	70.0 ± 6.1	23.7 ± 4.1	20.6 ± 2.0	18.7 ± 8.1	ND
11	40.8 ± 2.9	NA	ND	42.5 ± 2.5	38.5 ± 4.5	17.2 ± 5.2	43.4 ± 0.5	13.9 ± 4.7	ND
FP	74.2 ± 4.8	65.5 ± 4.0	38.1 ± 3.8	23.6 ± 4.8	21.9 ± 3.8	13.5 ± 3.6	78.8 ± 4.7	44.4 ± 4.8	28.7 ± 2.5
A	30.4 ± 2.1	31.2 ± 4.0	33.4 ± 1.9	67.8 ± 4.1	50.0 ± 7.3	32.6 ± 6.1	53.2 ± 5.1	51.6 ± 5.1	32.2 ± 6.2

The values are expressed as percentage of growth inhibition (%GI) relative to control growth in the absence of inhibitory agents (negative control). Compounds with a GI higher than 20% at 0.1 mg/mL were assayed at lower concentrations (0.05 and 0.01 mg/mL). NA: Non-active (% GI < 10%). ND: not determined. FP: Fosbel-Plus and A: Azoxystrobin were used as positive controls. The data shown are the average of eight independent experiments ± SD (standard deviation).

The results showed that compounds **2**, **4** and **9** exhibited good activity against *A. alternata* with %GI values ranging from 56.6% to 71.3%, highlighting compound **9** with slightly lower potency than the positive control, Fosbel-Plus (%GI = 74.2%) at the concentration of 0.1 mg/mL. Moreover, compounds **1**, **5**, **6**, **7**, and **11** exhibited moderate activity against this fungus with GI ranging from 34.6 to 48.1% at 0.1 mg/mL, whereas compounds **3** and **10** showed low potency, and compound **8** was inactive (%GI < 10). Regarding *B. cinerea*, all the assayed compounds, except for compund **8**, showed some degree of antifungal activity. In fact, compounds **1–3**, **6**, **9**, and **10** (%GI from 70.1 to 100%), exhibited higher potency than the commercial fungicide, Azoxytrobin (%GI 67.8) at 0.1 mg/mL, highlighting compounds

3 (capillin) and 5 (frutescinone) with %GI 92.7 and 90.8, respectively, at 0.05 mg/mL, and this last one keeping a %GI 70.6 at the lowest assayed concentration. This fungus was the most sensitive one to the assayed compounds, in concordance with the results of the subfractions from the wild-type and cultivated plants. *Fusarium oxysporum* was also sensitive to the compounds and compounds **2** (%GI = 44.5) and **3** (%GI = 84.4) were also more active than the positive controls (Fosbel Plus, %GI = 44.4 and Azoxystrobin, %GI 51.6) at the concentration of 0. 05 mg/mL.

Polyacetylenes are a kind of naturally occurring metabolites broadly present in the *Apiaceae*, *Araliaceae*, and *Asteraceae* families, which have attracted considerable attention owing to their multiple pharmacological effects, including antitumor, anti-inflammatory, antifungal, and antibacterial activities [34,35]. Polyacetylenes are also considered to be phytoalexins, which play an important ecological role in plants' response to disease states, microbial attacks or abiotic stresses [36]. In fact, some polyacetylenes have been reported to exhibit insecticidal activity and can be used as biological control agents [34], such as atractylodin from *Atractylodes lancea* that shows repellent effects against *Tribolium castaneum*, a worldwide pest of stored products, particularly food grains [37]. Regarding polyacetylenes isolated herein, capillin has been reported to elicit the human leukemia HL-60 cell apoptosis [38] and to inhibit cell proliferation on various human tumor cell lines and the carcinogenic Epstein–Barr virus [39]. In addition, capillin and capillinol hold great promise as anti-diabetic drug candidates [40]. However, to our knowledge, the present work is the first one that reports on their potential as biopesticides.

The results in the present study pointed out not only the potential of *A. frutescens* as a source of polyacetylenes with a potent fungicide activity on high-risk plant pathogens but also validate the cultivated plant species as having similar phytochemical and bioactivity profiles that the wild-type plant.

2.4. SAR Analysis

Taking into consideration the obtained %GI values (Table 1) of the assayed compounds on phytopathogenic fungi, the influence of substitution patterns and some structure-activity requirements can be established for the two types of isolated compounds, polyacetylenes (**1–6**, **9** and **10**) and alkamides (**7**, **8** and **11**) (Figure 8). (1) Considering the polyacetylene compounds, the results of the analysis revealed the following trends: (a) The type of the functional group at the propargyl C-1 position seems critical for the activity. Thus, the comparison of the activities of polyacetylenes **1–3** and **4–6** whose only structural difference is the substituent at C-1, showed that the most effective group at this position was the ketone group (**3** and **5**), followed by hydroxyl or methylene (**2** and **6**), and acetyl group (**1** and **4**) for *B. cinerea* and *F. oxysporum*, whereas the results for *A. alternata* showed a different behavior, with a hydroxyl group (**2**) being preferable to a ketone (**3**) or acetyl moiety (**1**). (b) The number of substituents on the aromatic ring modulated the bio-fungicide profile. Thus, compounds with a monosubstituted aromatic ring were more potent than those trisubstituted (**1** vs. **4**, and **3** vs. **5**) on *B. cinerea* and *F. oxysporum*. Furthermore, a carboxy-methyl ester at C-2′ and a methyl ether at C-3′ strongly affect the activity, since their substitution by a hydrogen atom enhances the activity (**5** vs. **3** and **4** vs. **1**). Once again, results for *A. alternata* were contrary, with the trisubstituted polyacetylenes (**4** and **5**) being more active than the monosubstituted ones (**1** and **3**), suggesting that oxidation of the aromatic ring had a favorable effect on the activity against this fungus. (2) Concerning the compounds with an alkamide skeleton, the replacement of a thiophene ring by an alkyl side chain moiety (**8** vs. **11**) significatively decreased the potency.

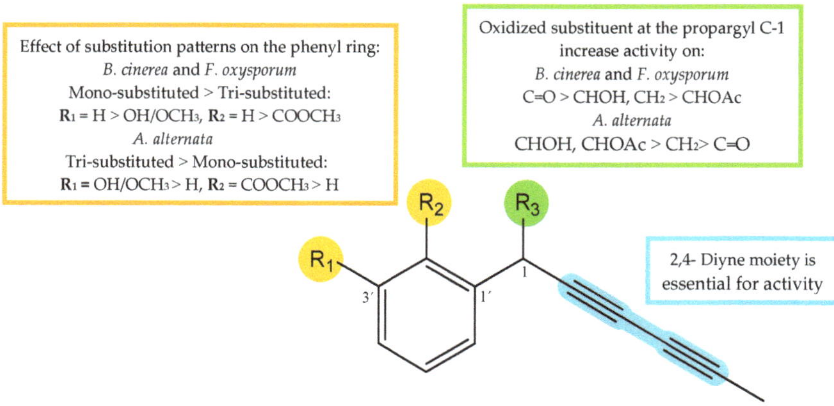

Figure 8. Structure–activity relationship of most active compounds, polyacetylenes (**1–6**, **9**, and **10**).

3. Materials and Methods

3.1. General

Optical rotations were measured with a Perkin Elmer 241 automatic polarimeter in $CHCl_3$ at 20 °C and the $[\alpha]_D$ values are given in 10^{-1} deg cm^2/g. The NMR experiments were recorded using Bruker Avance 500 or 600 spectrometers (Bruker, Wissembourg, France) with the pulse sequences given by Bruker using $CDCl_3$ as a solvent, and the chemical shifts are given in δ (ppm) with TMS as internal reference. Silica gel 60 (particle size 15–40 and 63–200 μm, Macherey-Nagel, Düren, Germany) and Sephadex LH-20 (Pharmacia Biotech, Uppsala, Sweden) were used for column chromatography, and silica gel 60 F254 (Macherey-Nagel) was used for analytical or preparative TLC. The spots were visualized by UV light and heating silica gel plates sprayed with H_2O-H_2SO_4-AcOH (1:4:20). All solvents of analytical grade and the reagents were purchased from Panreac (Barcelona, Spain). PGA culture medium (Potato glucose agar, Sigma-Aldrich, Madrid, Spain) and 9 cm diameter Petri dishes (Sarstedt, Nümbrecht, Germany) were used for the maintenance of the fungal colonies, and to carry out bioassays. Tetracycline (50 mg/L, Sigma-Aldrich, Madrid, Spain) was added to avoid bacterial growth on the Petri dishes. The stock solutions of tested samples were prepared with absolute ethanol (Sigma-Aldrich, Madrid, Spain). Fosbel-Plus (35% Fosetil Al and 35% Mancozeb) from Probelte S.A. (Murcia, Spain) was purchased in a phytosanitary product store, and used as a reference fungicide for tests against *A. alternata*, *B. cinerea*, and *F. oxysporum*. Ortiva PC (Azoxystrobin 250 g/L, 22.8% p/p) from Syngenta España S.A (Madrid, Spain) was also used as a reference fungicide. Ethanol was used as a negative control, using one dish per pathogen and eight discs for each control.

3.2. Plant Material

3.2.1. Plant Collection

Roots of *Argyranthemum frutescens* (L.) Sch. Bip. subsp. frutescens were gathered at Rambla de Castro, in the Los Realejos municipality, Tenerife Province (Canary Islands, Spain) in May 2023. The botanical identification was performed by botanist Ph D Cristina G. Montelongo, and a voucher specimen (TFC 54141) has been deposited in the Herbarium TFC-SEGAI, Universidad de La Laguna (Tenerife, Spain).

3.2.2. Plant Cultivation: Seedling Production

The seed of the species *A. frutescens* were collected from the natural environment and native plants belonging to the garden center of native flora at the Environmental Centre La

Tahonilla (Cabildo Insular de Tenerife, Canary Island, Spain) sited at La Laguna (36°17′25″ N and 59°35′45″ E; 985 m above sea level). The conditions in the greenhouse during the sowing phase were 70–80% humidity and 23 ± 2 °C. This material was cultivated using compost (25%), coconut fiber (50%), and perlite (25%) as substrates and irrigated three times a week. The seedlings with 3–4 leaves were further transplanted into individual pots. In this first phase of growth, plants received only water to maintain optimal conditions, because the application of fungicide can affect the roots' development. In the potting phase, solid fertilizer (8 gr/each pot, 16N-9P-12K) was added once a month to enhance the blooming. Once they were transplanted into a new pot, they were placed in the experimental field in May 2023 (325 plants) to growth until they were collected. Roots of the cultivated *A. frutescens* were gathered in January 2023. The roots were spread on a tray, turned over occasionally, and air-dried at room temperature (22 ± 4 °C) for two weeks. The dried plant materials were ground and stored until extraction.

3.3. Plant Extracts Preparation and Liquid–Liquid Partition Procedure

The air-dried and powdered roots of wild (450.9 g) and cultivated (500.0 g) *A. frutescens* were extracted by maceration with 96% EtOH (5 L, 24 h, three times) at room temperature (22 ± 4 °C). The resulting extracts were filtered, and the solvent removed under reduced pressure at 40 °C. This procedure yielded 17.9 g and 40.3 g of residue (4.0% and 8.1% *w*/*w*) from the wild and cultivated plants, respectively. An aliquot for each one was assayed on phytopathogenic fungi under study, revealing a notable fungal inhibition for both extracts.

Therefore, the ethanolic extracts were subjected to further fractionation through a liquid–liquid partition. Thus, the extracts were suspended in water (500 mL H_2O) and partitioned sequentially with hexane (Hx, 3 × 500 mL) and ethyl acetate (EtOAc, 3 × 500 mL). The organic phases were concentrated under reduced pressure to give the Hx (7.7 g and 8.4 g for wild and cultivated plants, respectively) and EtOAc (3.8 g and 4.2 g for wild and cultivated plants, respectively) fractions. Simultaneously, the aqueous residues were subjected to lyophilization, yielding the aqueous fractions (4.0 and 13.5 g, respectively). Aliquots (approximately 45 mg) of the fractions from the liquid–liquid protocol were assayed for their antifungal activity. Biological evaluation revealed that the organic fractions were active against the phytopathogenic fungi, with the hexane fraction in both wild and cultivated plants showing the highest efficacy. Therefore, those hexane fractions were further investigated.

3.4. Bioactivity-Guided Chromatographic Fractionation and Metabolites Isolation in Wild and Cultivated A. frutescens Roots

The most active fractions from the wild and cultivated *A. frutescens*, the hexane ones (7.7 g and 8.4 g, respectively), were chromatographed on a silica gel column, using a mixture of hexane/EtOAc of increasing polarity (10:0 to 0:10) as eluent to afford subfractions, which were then combined based on their TLC profile into subfractions A1–A8 and B1–B7 (Experimental Part S1 in Supporting Materials). Phytopathogenic fungi evaluation of those fractions revealed that subfractions A2–A6 and B1–B5 were active against the strains of *A. alternata*, *B. cinerea*, and *F. oxysporum*. Therefore, these subfractions were subjected to several chromatography steps until yielding the pure compounds (Figures 1 and 2). Following this procedure, eleven metabolites were isolated and identified by NMR techniques (Figures S1–S10 in Supporting Materials) and compared with those reported in the literature. Thus, compounds identified in the wild plant correspond to: capillinol acetate (**1**, $[\alpha]^{20}_D$ + 42.2°, *c* 1.68 in $CHCl_3$; 80.5 mg) [26], capillinol (**2**, $[\alpha]^{20}_D$ + 3.0°, *c* 0.44 in $CHCl_3$; 19.6 mg) [28], capillin (**3**, 6.4 mg) [28], frutescinol acetate (**4**, $[\alpha]^{20}_D$ + 8.2°, *c* 0.02 in $CHCl_3$; 4.6 mg) [26], frutescinone (**5**, 13.1 mg) [26], frutescin (**6**, 11.4 mg) [26], *N*-isobutyl-6-(2-thienyl)-2*Z*,4*Z*-hexadienamide (**7**, 1.6 mg) [29], and 2*E*,4*E*-tetradecadienoic acid isobutyl amide (**8**, 7.6 mg) [30]. Moreover, the following metabolites were identified in the cultivated plant: capillinol acetate (**1**, 667.4 mg) [26], capillinol (**2**, 69.6 mg) [28], frutescinol acetate (**4**, 78.7 mg) [26], frutescin (**6**, 269.6 mg) [26], 2*E*,4*E*-tetradecadienoic acid isobutyl amide (**8**, 9.4 mg) [30], 3′-*O*-demethyl frutescinol acetate (**9**, $[\alpha]^{20}{}_{D}$ + 14.3°, *c* 0.28 in $CHCl_3$; 163.1 mg) [26], 3′-*O*-demehthyl frutescin (**10**, 8.7 mg) [31], and *N*-isobutyl-6-(2-thienyl)-2*E*,4*E*-hexadienamide (**11**, 18.1 mg) [32].

3.5. Determination of the Absolute Configuration of Capillinol (2)

Capillinol (11.5 mg, 0.067 mmoles) in 1 mL of dry CH_2Cl_2 at room temperature, 10.7 mg (0.064 mmoles) of (*R*)-(-)-α-methoxyphenylacetic acid (*R*-MPA), 14.2 mg (0.069 mmoles) of *N*, *N*′-dicyclohexylcarbodiimide (DCC), and 6.5 mg (0.053 mmoles) of 4-(dimethylamine)pyridine (DMAP) were added. The reaction mixture was stirred for 30 min in an inert nitrogen atmosphere. After this time, the solvent was removed under vacuum, and the reaction mixture was purified by column chromatography (CC) using a hexane/ethyl acetate polarity gradient (9:1 to 8:2) as the mobile phase, yielding 15.2 mg (74.2%) of capillinol *R*-(-)-α-methoxyphenylacetate as a colorless solid. Subsequently, the product was chelated with Ba (II), using a barium perchlorate salt dissolved in CH_3CN, which was added until the solution was saturated to ensure its chelation. The ^1H NMR shift differences between before and after suturing with Ba (II) were analyzed.

NMR data before Ba^{2+} chelation: NMR ^1H ($CDCl_3$, 600 MHz) δ 1.94 (3H, s, H-6), 3.37 (3H, s, OCH_3), 4.89 (1H, s, H-1″), 6.47 (1H, s, H-1), 7.38–7.45 (8H, m, H-3′, H-4′, H-5′, H-4″, H-5″, H-6″, H-7″, H-8″), 7.47 (2H, dd, *J* = 2.1 Hz, *J* = 6.6 Hz, H-2′, H-6′); NMR ^{13}C ($CDCl_3$, 150 MHz) δ 4.15 (q, C-6), 57.8 (q, 2″, OCH_3), 63.3 (s, C-4), 67.4 (s, C-1), 71.5 (s, C-2), 73.1 (s, C-3), 81.0 (s, C-5), 83.0 (s, C-1″), 128.2 (2xd, C-4″, C-8″), 128.4 (2xd, C-2′, C-6′), 129.6 (2xd, C-5″, C-7″), 129.7 (d, C-6″), 129.8 (2xd, C-2′, C-6′), 130.3 (d, C-3′), 137.0 (s, C-1′), 137.2 (s, C-3″).

NMR data after Ba^{2+} chelation: NMR ^1H ($CDCl_3$, 600 MHz) δ 1.94 (3H, s, H-6), 3.34 (3H, s, OCH_3), 5.00 (1H, s, H-1″), 6.51 (1H, s, H-1), 7.38–7.45 (2H, m, H-2′, H-6′), 7.47 (8H, dd, *J* = 2.1 Hz, *J* = 6.6 Hz, H-3′, H-4′, H-5′, H-4″, H-5″, H-6″, H-7″, H-8″); NMR ^{13}C ($CDCl_3$, 150 MHz) δ 4.15 (q, C-6), 57.8 (q, OCH_3), 63.3 (s, C-4), 67.4 (s, C-1), 71.5 (s, C-2), 73.1 (s, C-3), 81.0 (s, C-5), 83.0 (s, C-1″), 128.2 (2xd, C-4″, C-8″), 128.4 (2xd, C-2′, C-6′), 129.6 (2xd, C-5″, C-7″), 129.7 (d, C-6″), 129.8 (2xd, C-2′, C-6′), 130.3 (d, C-3′), 137.0 (s, C-1′), 137.2 (s, C-3″).

3.6. Biological Assays

3.6.1. Fungal Culture

Phytopathogenic fungi, *Alternaria alternata*, *Botrytis cinerea*, and *Fusarium oxysporum*, were maintained at 25 °C in darkness, and periodically replicated in Petri dishes with PGA culture medium and tetracycline to avoid contamination. Strains of *B. cinerea* (B05.10) and *A. alternata* (Aa 100) were isolated from *Vitis vinifera* and *Lycopersicon esculentum*, respectively, both supplied by the Universidad de La Laguna, Tenerife. *F. oxysporum f.* sp. *lycopersici* (2715) strain, isolated from *Lycopersicon esculentum*, was provided by the Colección Española de Cultivos Tipo (CECT) from Valencia, Spain.

3.6.2. In Vitro Test-Assay on Mycelium

The antifungal activity of extracts, fractions, subfractions, and isolated compounds was performed on the mycelium stage of *F. oxysporum*, *B. cinerea*, and *A. alternata* using an agar-dilution protocol as previously described [27]. Briefly, an aliquot of test sample in EtOH (50 mg/mL stock solution) was dissolved in warm PGA (final volume 5 mL) and placed in a 9 cm Petri dish. The medium containing the sample was allowed to solidify. Eight 4 mm diameter discs of each fungus were deposited in a Petri dish with 1 mg/mL final concentration of each tested and control sample and incubated for 48 h (*B. cinerea*) or 72 h (*A. alternata* and *F. oxysporum*). A Petri dish with eight discs was analyzed for each pathogen. Colony growth was measured with an image-processing software ImageJ 1.53e-Wayne Rasband (NIH) (ANOVA). The diameter of each colony was measured twice, making a cross, and the average of both measurements was taken. Percentage growth inhibition (%GI) was calculated using the equation: %GI = $[(C - T)/C] \times 100$, where C and T are the diameters of the control and test treated colonies, respectively. Fractions and subfractions exhibiting %GI higher than 20% at 1 mg/mL were also evaluated at 0.5 and 0.1 mg/mL doses. Pure compounds growth inhibition assays were conducted at 0.1, 0.05 and 0.01 mg/mL. Fosbel-Plus and/or Ortiva PC, and EtOH were used as positive and negative controls, respectively.

3.6.3. Statistical Analysis

All data were shown as a mean ± standard deviation (SD). The percentage of inhibition was analyzed using one-way analysis of variance (ANOVA) [41]. Prior to analysis, data were assessed for normality and homogeneity of variances using the Shapiro–Wilk and Levene's tests, respectively, to ensure the validity of ANOVA assumptions. Tukey's HSD multiple comparisons of means was used to compare concentrations and treatments. Differences of $p < 0.05$ were considered statistically significant. All analyses were performed using Social Science Statistics, 2018.

4. Conclusions

This research pointed out the potential of *A. frutescens* as a sustainable and competitive alternative to conventional pesticides against high-risk phytopathogenic fungi, fostering more ecological and efficient agricultural practices. It overcame significant cultivation challenges and confirmed the preservation of phytochemical and fungicidal profiles in cultivated plants, offering a sustainable alternative source that could mitigate wild-type collection pressure. The bioassay-guided fractionation of the ethanolic extracts successfully identified polyacetylenes, notably capillin, capillinol, and frutescinone exhibiting an outstanding antifungal profile, comparable to that of the commercial pesticides used as positive controls. These results highlight the potential of these compounds as fungicides that could be applied in combination with other biological agents for integrated crop disease management. Furthermore, these compounds could serve as a versatile platform through structural optimization for the development of more effective fungicides. Nonetheless, further investigations are warranted to elucidate their mode of action, and to optimize the experimental extraction process for application in sustainable, environmentally safe, and economically viable integrated crop protection management.

Supplementary Materials: The following supporting information can be downloaded at: https://www.mdpi.com/article/10.3390/plants14070985/s1, Experimental part S1. Bioguided fractionation: Extraction and Isolation; Figures S1–S11: ^{1}H and ^{13}C NMR spectra of metabolites **1–11** isolated from wild and cultivated Argyranthemum frutescens roots; Figure S12: ^{1}H NMR spectra of capillinol (*R*)-α-methoxyphenylacetate before and after saturation with barium (II) salt as the chelating agent; Figure S13: representative photographs of fungal growth inhibition in a dilution agar assay; Table S1: antifungal effects (% growth inhibition) of extract, fractions, and subfractions from roots of wild Argyranthemum frutescens against *Alternaria alternata, Botrytis cinerea,* and *Fusarium oxysporum*; Table S2: antifungal effects (% growth inhibition) of extract, fractions, and subfractions from roots of cultivated *Argyranthemum frutescens* against *Alternaria alternata, Botrytis cinerea,* and *Fusarium oxysporum*.

Author Contributions: Conceptualization, I.L.B. and I.A.J.; methodology, E.H.-Á., S.R.-S., C.P.R. and M.Á.L.-L.; investigation, N.L.-G. and J.H.P.; resources (collection and identification of the plant species), M.Á.L.-L.; writing—original draft preparation, I.A.J., E.H.-Á. and S.R.-S.; writing—reviewing and editing, I.L.B.; funding acquisition, I.L.B. All authors have read and agreed to the published version of the manuscript.

Funding: This research was funded by projects MACBIOPEST (MAC2/1.1a/289), program Interreg MAC 2014–2020 co-funded by the European Regional Development Fund (FEDER), and PID2022-136549OB-100 funded by MCIN/AEI/10.13039/501100011033/FEDER, UE.

Data Availability Statement: Data are contained within the article and Supplementary Materials.

Conflicts of Interest: The authors declare no conflicts of interest.

References

1. Singh, V.K.; Singh, R.; Kumar, A.; Bhadouria, R. Current status of plant diseases and food security. In *Food Security and Plant Disease Management*; Kumar, A., Droby, S., Eds.; Wiley: New York, NY, USA, 2021; pp. 19–35. [CrossRef]
2. Li, P.; Tedersoo, L.; Crowther, T.W.; Wang, B.; Shi, Y.; Kuang, L.; Li, T.; Wu, M.; Liu, M.; Luan, L.; et al. Global diversity andb of potential phytopathogenic fungi in a changing world. *Nat. Commun.* **2023**, *14*, 6482. [CrossRef] [PubMed]
3. John, E.; Singh, K.B.; Oliver, R.P.; Tan, K. Transcription factor control of virulence in phytopathogenic fungi. *Mol. Plant Pathol.* **2021**, *22*, 858–881. [CrossRef] [PubMed]
4. Fernandes, C.; Casadevall, A.; Gonçalves, T. Mechanisms of *Alternaria* pathogenesis in animals and plants. *FEMS Microbiol. Rev.* **2023**, *47*, 1–25. [CrossRef]
5. Wang, H.; Guo, Y.; Luo, Z.; Gao, L.; Li, R.; Zhang, Y.; Kalaji, H.M.; Qiang, S.; Chen, S. Recent advances in *Alternaria* phytotoxins: A review of their occurrence, structure, bioactivity, and biosynthesis. *J. Fungi* **2022**, *8*, 168. [CrossRef]
6. Singh, R.; Caseys, C.; Kliebenstein, D.J. Genetic and molecular landscapes of the generalist phytopathogen *Botrytis cinerea*. *Mol. Plant Pathol.* **2024**, *25*, e13404. [CrossRef]
7. Bi, K.; Liang, Y.; Mengiste, T.; Sharon, A. Killing Softly: A Roadmap of *Botrytis cinerea* pathogenicity. *Trends Plant Sci.* **2023**, *28*, 211–222. [CrossRef]
8. Seepe, H.A.; Nxumalo, W.; Amoo, S.O. Natural products from medicinal plants against phytopathogenic *Fusarium* species: Current research endeavours, challenges and prospects. *Molecules* **2021**, *26*, 6539. [CrossRef]
9. Zhou, W.; Li, M.; Achal, V. A Comprehensive review on environmental and human health impacts of chemical pesticide usage. *Emerg. Contam.* **2025**, *11*, 100410. [CrossRef]
10. Pandian, S.; Ramesh, M. Development of pesticide resistance in pests: A key challenge to the crop protection and environmental safety. In *Pesticides in Crop Production: Physiological and Biochemical Action*; Srivastava, P.K., Singh, V.P., Singh, A., Singh, S., Prasad, S.M., Tripathi, D.K., Chauhan, D.K., Eds.; Wiley: New York, NY, USA, 2020; pp. 1–13.
11. Harte, S.J.; Bray, D.P.; Nash-Woolley, V.; Stevenson, P.C.; Fernández-Grandon, G.M. Antagonistic and additive effect when combining biopesticides against the fall armyworm, *Spodoptera frugiperda*. *Sci. Rep.* **2024**, *14*, 6029. [CrossRef]
12. Aioub, A.A.A.; Ghosh, S.; AL-Farga, A.; Khan, A.N.; Bibi, R.; Elwakeel, A.M.; Nawaz, A.; Sherif, N.T.; Elmasry, S.A.; Ammar, E.E. Back to the origins: Biopesticides as promising alternatives to conventional agrochemicals. *Eur. J. Plant. Pathol.* **2024**, *1*, 697–713. [CrossRef]
13. Fragkouli, R.; Antonopoulou, M.; Asimakis, E.; Spyrou, A.; Kosma, C.; Zotos, A.; Tsiamis, G.; Patakas, A.; Triantafyllidis, V. Mediterranean plants as potential source of biopesticides: An overview of current research and future trends. *Metabolites* **2023**, *13*, 967. [CrossRef] [PubMed]
14. Mateo-Martín, J.; Benítez, G.; Gras, A.; Molina, M.; Reyes-García, V.; Tardío, J.; Verde, A.; Pardo-de-Santayana, M. Cultural importance, availability and conservation status of spanish wild medicinal plants: Implications for sustainability. *People Nat.* **2023**, *5*, 1512–1525. [CrossRef]
15. Dempewolf, H.; Rieseberg, L.H.; Cronk, Q.C. Crop domestication in the *Compositae*: A family-wide trait assessment. *Genet. Resour. Crop Evol.* **2008**, *55*, 1141–1157. [CrossRef]
16. Mandel, J.R.; Barker, M.S.; Bayer, R.J.; Dikow, R.B.; Gao, T.G.; Jones, K.E.; Keeley, S.; Kilian, N.; Ma, H.; Siniscalchi, C.M.; et al. The *Compositae* tree of life in the age of phylogenomics. *J. Syst. Evol.* **2017**, *55*, 405–410. [CrossRef]
17. Rolnik, A.; Olas, B. The Plants of the *Asteraceae* Family as Agents in the Protection of Human Health. *Int. J. Mol. Sci.* **2021**, *22*, 3009. [CrossRef]
18. Matsuda, K. Understanding pyrethrin biosynthesis: Toward and beyond natural pesticide overproduction. *Biochem. Soc. Trans.* **2024**, *52*, 1927–1937. [CrossRef] [PubMed]
19. Jaison, J.P.; Balasubramanian, B.; Gangwar, J.; James, N.; Pappuswamy, M.; Anand, A.V.; Al-Dhabi, N.A.; Arasu, M.V.; Liu, W.-C.; Sebastian, J.K. Green synthesis of bioinspired nanoparticles mediated from plant extracts of *Asteraceae* family for potential biological applications. *Antibiotics* **2023**, *12*, 543. [CrossRef]

20. Petrova, M.; Miladinova-Georgieva, K.; Geneva, M. Influence of abiotic and biotic elicitors on organogenesis, biomass accumulation, and production of key secondary metabolites in *Asteraceae* Plants. *Int. J. Mol. Sci.* **2024**, *25*, 4197. [CrossRef]
21. Bramwell, D. Introduction: Islands and plants. In *The Biology of Island Floras*; Bramwell, D., Caujapé-Castells, J., Eds.; Cambridge University Press: Cambridge, UK, 2011; pp. 297–304.
22. White, O.W.; Reyes-Betancort, J.A.; Chapman, M.A.; Carine, M.A. Geographical isolation, habitat shifts and hybridisation in the diversification of the Macaronesian endemic genus *Argyranthemum* (Asteraceae). *New Phytol.* **2020**, *228*, 1953–1971. [CrossRef]
23. Cruz Suárez, S.J. *Más de 100 Plantas Medicinales*, 1st ed.; Obra Social de la Caja de Canaria: Las Palmas de Gran Canarias, Spain, 2007.
24. Sabotič, J.; Bayram, E.; Ezra, D.; Gaudêncio, S.P.; Haznedaroğlu, B.Z.; Janež, N.; Ktari, L.; Luganini, A.; Mandalakis, M.; Safarik, I.; et al. A guide to the use of bioassays in exploration of natural resources. *Biotechnol. Adv.* **2024**, *71*, 108307. [CrossRef]
25. González, A.G.; Barrera, J.B.; Díaz, J.G.; García, T.Z.; de Paz, P.P. Distribution of acetylenes and sesquiterpene lactones in *Argyranthemum* from Tenerife. *Biochem. Syst. Ecol.* **1988**, *16*, 17–21. [CrossRef]
26. González, A.G.; Estévez-Reyes, R.; Estévez-Braun, A.; Ravelo, A.G.; Jiménez, I.A.; Bazzocchi, I.L.; Aguilar, M.A.; Moujir, L. Biological activities of some *Argyranthemum* species. *Phytochemistry* **1997**, *45*, 963–967. [CrossRef] [PubMed]
27. Cosoveanu, A.; Hernández, M.; Iacomi-Vasilescu, B.; Zhang, X.; Shu, S.; Wang, M.; Cabrera, R. Fungi as endophytes in chinese *Artemisia spp.*: Juxtaposed elements of phylogeny, diversity and bioactivity. *Mycosphere* **2016**, *7*, 102–117. [CrossRef]
28. Zhang, Z.; Guo, S.; Zhang, W.; Geng, Z.; Liang, J.; Du, S.; Wang, C.; Deng, Z. Essential oil and polyacetylenes from *Artemisia ordosica* and their bioactivities against *Tribolium castaneum* Herbst (Coleoptera: Tenebrionidae). *Ind. Crops Prod.* **2017**, *100*, 132–137. [CrossRef]
29. Esmaeili, G.; Fatemi, H.; Avval, M.B.; Azizi, M.; Arouiee, H.; Vaezi, J.; Fujii, Y. Diversity of chemical composition and morphological traits of eight iranian wild *Salvia* species during the first step of domestication. *Agronomy* **2022**, *12*, 2455. [CrossRef]
30. Ee, G.C.L.; Lim, S.K.; Dzulkefly, K. Alkaloids and carboxylic acids from *Piper nigrum*. *Asian J. Chem.* **2008**, *20*, 5931–5940.
31. Bohlmann, F.; Tsankova, E.; Jakupovic, J. Sesquiterpenes and acetylenes from *Argyranthemum adauctum* ssp. *jacobaeifolium*. *Phytochemistry* **1984**, *23*, 1103–1104. [CrossRef]
32. Shahat, A.A.; Apers, S.; Pieters, L.; Vlietinck, A.J. Isolation and Complete NMR assignment of the numbing principle from *Chrysanthemum morifolium*. *Fitoterapia* **2001**, *72*, 89–91. [CrossRef]
33. Seco, J.M.; Quiñoá, E.; Riguera, R. Assignment of the absolute configuration of polyfunctional compounds by NMR using chiral derivatizing agents. *Chem. Rev.* **2012**, *112*, 4603–4641. [CrossRef]
34. Xie, Q.; Wang, C. Polyacetylenes in herbal medicine: A comprehensive review of its occurrence, pharmacology, toxicology, and pharmacokinetics (2014–2021). *Phytochemistry* **2022**, *201*, 113288. [CrossRef]
35. Konovalov, D.A. Medicinal plants polyacetylene compounds of plants of the *Asteraceae* family (review). *Pharm. Chem. J.* **2014**, *48*, 613–631. [CrossRef]
36. Mullins, A.J.; Webster, G.; Kim, H.J.; Zhao, J.; Petrova, Y.D.; Ramming, C.E.; Jenner, M.; Murray, J.A.H.; Connor, T.R.; Hertweck, C.; et al. Discovery of the *Pseudomonas* polyyne protegencin by a phylogeny-guided study of polyyne biosynthetic gene cluster diversity. *mBio* **2021**, *12*, e00715-21. [CrossRef] [PubMed]
37. Chen, H.P.; Zheng, L.S.; Yang, K.; Lei, N.; Geng, Z.F.; Cai, Q.; Du, S.S.; Deng, Z.W. Insecticidal and repellant activities of polyacetylenes and lactones derived from *Atractylodes lancea* rhizomes. *Chem. Biodivers.* **2015**, *12*, 593–598. [CrossRef] [PubMed]
38. Masuda, Y.; Asada, K.; Satoh, R.; Takada, K.; Kitajima, J. Capillin, a major constituent of *Artemisia capillaris Thunb.* flower essential oil, induces apoptosis through the mitochondrial pathway in human leukemia HL-60 cells. *Phytomedicine* **2015**, *22*, 545–552. [CrossRef]
39. Dembitsky, V.M.; Levitsky, D.O. Acetylenic terrestrial anticancer agents. *Nat. Prod. Commun.* **2006**, *1*, 405–429. [CrossRef]
40. Islam, M.N.; Choi, R.J.; Jung, H.A.; Oh, S.H.; Choi, J.S. Promising anti-diabetic potential of capillin and capillinol isolated from *Artemisia capillaris*. *Arch Pharm. Res.* **2016**, *39*, 340–349. [CrossRef]
41. One-Way ANOVA Calculator, Including Tukey HSD. Available online: https://www.socscistatistics.com/tests/chisquare2/default2.aspx (accessed on 3 May 2024).

plants

MDPI

Article

Bioactive Sesquiterpenoids from *Santolina chamaecyparissus* L. Flowers: Chemical Profiling and Antifungal Activity Against *Neocosmospora* Species

Eva Sánchez-Hernández [1], Jesús Martín-Gil [1], Vicente González-García [2,3], José Casanova-Gascón [4] and Pablo Martín-Ramos [1,*]

1 Department of Agricultural and Forestry Engineering, ETSIIAA, Universidad de Valladolid, Avenida de Madrid 44, 34004 Palencia, Spain; eva.sanchez.hernandez@uva.es (E.S.-H.); jesus.martin.gil@uva.es (J.M.-G.)
2 Departamento de Sistemas Agrícolas, Forestales y Medio Ambiente, Centro de Investigación y Tecnología Agroalimentaria de Aragón (CITA), Avda. Montañana 930, 50059 Zaragoza, Spain; vgonzalezg@aragon.es
3 Instituto Agroalimentario de Aragón-IA2 (Universidad de Zaragoza-CITA), Avda. Montañana 930, 50059 Zaragoza, Spain
4 Instituto Agroalimentario de Aragón-IA2 (Universidad de Zaragoza-CITA), EPS, University of Zaragoza, Carretera de Cuarte s/n, 22071 Huesca, Spain; jcasan@unizar.es
* Correspondence: pmr@uva.es

Abstract: *Santolina chamaecyparissus* L. (cotton-lavender) is receiving increasing attention due to its potential for modern medicine and is considered both a functional food and nutraceutical. In this work, the phytochemical profile of its flower hydromethanolic extract was investigated by gas chromatography–mass spectrometry, and its applications as a biorational for crop protection were explored against *Neocosmospora* spp., both in vitro and in planta. The phytochemical profiling analysis identified several terpene groups. Among sesquiterpenoids, which constituted the major fraction (50.4%), compounds featuring cedrane skeleton (8-cedren-13-ol), aromadendrene skeleton (such as (−)-spathulenol, ledol, alloaromadendrene oxide, epiglobulol, and alloaromadendrene), hydroazulene skeleton (ledene oxide, isoledene, and 1,2,3,3a,8,8a-hexahydro-2,2,8-trimethyl-,(3aα,8β,8aα)-5,6-azulenedimethanol), or copaane skeleton (*cis*-α-copaene-8-ol) were predominant. Additional sesquiterpenoids included longiborneol and longifolene. The monoterpenoid fraction (1.51%) was represented by eucalyptol, (+)-4-carene, endoborneol, and 7-norbornenol. In vitro tests against *N. falciformis* and *N. keratoplastica*, two emerging soil phytopathogens, resulted in effective concentration EC_{90} values of 984.4 and 728.6 $\mu g \cdot mL^{-1}$, respectively. A higher dose (3000 $\mu g \cdot mL^{-1}$) was nonetheless required to achieve full protection in the in planta tests conducted on zucchini (*Cucurbita pepo* L.) cv. 'Diamant F1' and tomato (*Solanum lycopersicum* L.) cv. 'Optima F1' plants inoculated with *N. falciformis* by root dipping. The reported data indicate an antimicrobial activity comparable to that of fosetyl-Al and higher than that of azoxystrobin conventional fungicides, thus making the flower extract a promising bioactive product for organic farming and expanding *S. chamaecyparissus* potential applications.

Keywords: aromadendrane skeleton; biorational pesticides; cotton-lavender; hydroazulene framework; tomato; zucchini

1. Introduction

Santolina (*Santolina chamaecyparissus* L., family *Asteraceae*), sometimes called cotton-lavender, is an herbaceous perennial shrub native to the Mediterranean region and parts of

Europe and America [1] (Figure 1a). It has a mounded, round, and dense habit, reaching only 0.5 m in height and width, with vibrant yellow, button-like composite flowerheads (consisting only of disc florets) perched on stems above the foliage (Figure 1b). It has either silvery-grey or green leaves reminiscent of conifers. When bruised, Santolina foliage emits a pungent aroma reminiscent of a mixture of camphor and resin.

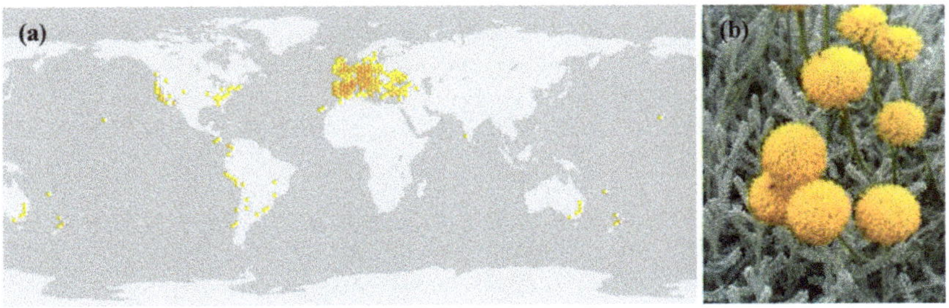

Figure 1. (**a**) *Santolina chamaecyparissus* L. worldwide distribution and (**b**) inflorescences. Source: *Santolina chamaecyparissus* L. in GBIF Secretariat (2023). GBIF Backbone Taxonomy. Checklist dataset https://doi.org/10.15468/39omei [2] accessed via GBIF.org on 11 December 2024.

Santolina chamaecyparissus' phytoconstituents have recognized medicinal properties, with analgesic, anticancer, anti-inflammatory, antimicrobial, antioxidant, and antispasmodic activity, as well as central nervous system depressants [3].

Terpenes are the largest and most diverse plant secondary metabolites in nature, serving as informative and defensive vehicles used by plants for antagonistic and mutualistic interactions. Their biosynthesis occurs through two independent pathways: the mevalonate pathway (MVA) in the cytosols of plants and the 2C-methyl-D-erythritol-4-phosphate (MEP) pathway in plastids. Both pathways lead to the formation of isopentenyl diphosphate (IPP) and its isomer dimethylallyl diphosphate (DMAPP), which serve as the precursors for all terpene classes. Monoterpenes are highly diverse compounds occurring in monocotyledonous and dicotyledonous angiosperms, fungi, bacteria, and gymnosperms. They are odoriferous compounds that partly account for the scent of many flowers and fruits and are vital components of essential oils that give plants aroma and flavor. Sesquiterpenes, which share ring classification with monoterpenes (with the exception of a few tricyclic terpenes), demonstrate significant structural diversity arising from the arrangement of their 15-carbon skeletons, the layering of functional groups, and the substituents on their backbone. Both monoterpenes and sesquiterpenes play crucial roles in plant defense against biotic stresses such as pathogenic microbes and herbivore pests [4].

Phytochemical analyses have revealed that cotton-lavender contains various mono- and sesquiterpenoids alongside other secondary metabolites [5]. The bioactive properties of these terpenes vary depending on their composition and relative concentrations, which are influenced by plant origin, plant part used, and extraction method. For instance, a 1,8-cineole- and β-eudesmol-rich flowerhead essential oil demonstrated potent antibacterial properties against *Pseudomonas aeruginosa* (Schroeter) Migula and *Enterococcus faecalis* (Andrewes and Horder) Schleifer and Kilpper-Balz, with minimum inhibitory concentration (MIC) values as low as 0.625 $\mu g \cdot mL^{-1}$ [6].

Although previous studies have investigated the antimicrobial activity of *S. chamaecyparissus* essential oils and extracts against various human pathogens [3], their potential as biorational agents for crop protection remains unexplored. In particular, no studies have examined the activity of cotton-lavender essential oils or extracts against *Neocosmospora* species. These fungi significantly impact industry, human health, and agriculture [7]. Several species are known pathogens in immunocompromised humans, often resulting in high mortality rates despite antifungal therapy [8], while others serve as opportunistic soil-borne phytopathogens, inducing cankers, stem and root rot, and blight in a variety of plants [9].

This work aimed to investigate the phytoconstituents of the hydromethanolic extract of *S. chamaecyparissus* flowers and to examine its antifungal activity—both in vitro and in planta—against two emerging *Neocosmospora* phytopathogens, namely *Neocosmospora falciformis* (Carrión) Summerb. & Schroers, associated with multiple plant species decline and muskmelon wilting and root rot in Spain [10], and *Neocosmospora keratoplastica* Geiser, O'Donnell, Short & Zhang, which affects cucurbits through rot and root decay [11].

2. Results

2.1. Vibrational Characterization

The Fourier-transform infrared (FTIR) spectrum of the flower extract (Table 1 and Figure S1) displayed characteristic absorption bands assignable to hydroxyl (3292 cm^{-1}) and ester carbonyl (1694 cm^{-1}) functional groups. Additional bands were observed at 2927, 1596, 1515, 1371, 1258, 1158, 1115, 1031, 854, and 811 cm^{-1} (aromatics). Comparison of the spectrum with that of shade-dried flower heads before extraction showed similar functional groups with only minor shifts, indicating successful ultrasonication-assisted extraction. The concurrence of bands at 3300, 1515, and 811 cm^{-1} suggests the presence of sesquiterpenes [12].

Table 1. Main bands in the infrared spectrum of *Santolina chamaecyparissus* L. flowers before and after extraction. Wavenumber values are expressed in cm^{-1}. Band assignments according to Socrates [13].

Flowers	Flower Extract	Assignment
3287	3292	O–H stretching (alcohols)
2919	2927	asymmetric aliphatic C–H stretching vibration (methylene)
1717	1694	C=O stretching of ketones and aldehydes
	1596	aromatic ring (C=C in-plane) stretching symmetric
1516	1515	C=C stretching (aromatic); C=O stretching
1372	1371	symmetric aliphatic C–H bending of CH_3 groups (alkanes)
1239	1258	C–O stretch (in rings); out-of-phase C–C–O stretching
1147	1158	C–O–C group (O–bridge)
-	1115	C–O stretching in alcohols
1023	1031	C–O–H deformation in cellulose; C–O stretching (esters)
896	854	aromatic out-of-plane rings with two neighboring C–H groups
812	811	bending of C–H bonds in out-of-plane deformation in aromatics

2.2. Gas Chromatography–Mass Spectrometry Characterization

Gas chromatography–mass spectrometry (GC-MS) analysis of the extract led to the identification of multiple compounds (Table S1 and Figure S2). Among sesquiterpenoids, which constituted the major fraction (50.4%), compounds featuring the cedrane skeleton, such as 8-cedren-13-ol (27.43%); the aromadendrane skeleton, such as (−)-spathulenol (3.06%), alloaromadendrene and its oxides (2.72%), ledol (2.22%), epiglobulol (1.45%), and aromadendrene (0.65%); the hydroazulene skeleton, including ledene oxide (6.34%), isoledene (1.41%), and 1,2,3,3a,8,8a-hexahydro-2,2,8-trimethyl-,(3aα,8β,8aα)-5,6-azulenedimethanol (0.44%); and the copaane skeleton, represented by *cis*-α-copaene-8-ol (0.77%), were predominant. Additional sesquiterpenoids included perhydrocyclopropa[e]azulene-4,5,6-triol (1.01%), characterized by a fully saturated (perhydro) bicyclo[5.3.0] system with three hydroxyl groups, and spiro[2,4,5,6,7,7a-hexahydro-2-oxo-4,4,7a-trimethylbenzofuran]-7,2′-(oxirane) (1.81%), which is a modified sesquiterpenoid in which the original C15 skeleton has undergone oxidative modifications. The series was completed by longiborneol (0.69%) and longifolene (0.37%), which feature a bicyclo[2.2.1] and a bicyclo[2.2.2] system, respectively (Figure 2). The monoterpenoid fraction (1.51%) was represented by the cyclic ether eucalyptol, the bicyclic (+)-4-carene, and two compounds with bicyclic[2.2.1] systems: endo-borneol (bornane-type) and 7-norbornenol (norbornane-type). Non-terpenoid compounds were also present, with 2,1,3-benzothiadiazole (12%) being the second most abundant component, followed by acetic acid (3.72%), retinol acetate (2.27%), and catechol (1.62%).

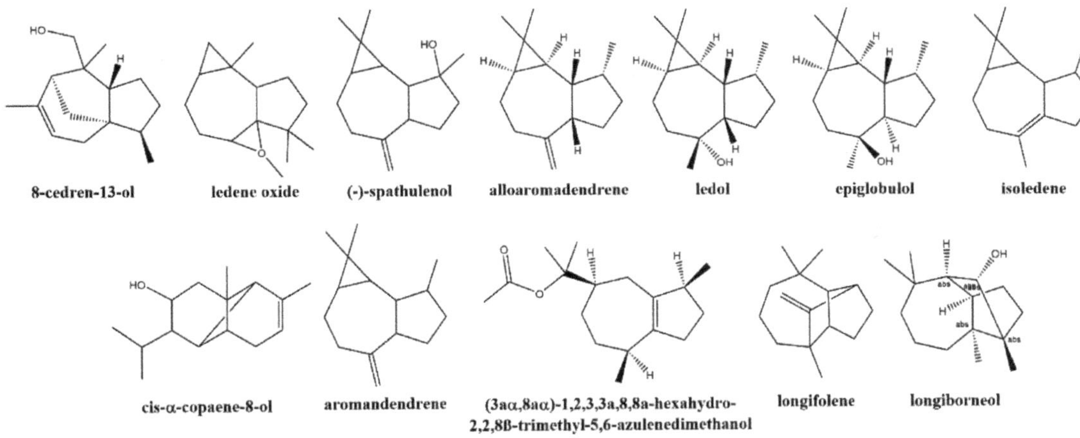

Figure 2. Main sesquiterpenoids identified in *S. chamaecyparissus* hydromethanolic flower extract, ordered from highest to lowest relative abundance (% of total content). Chemical structures were drawn using ChemDraw v.2023 (Revvity Signals Software, Waltham, MA, USA).

2.3. In Vitro Antifungal Activity

Figures 3 and S3 show the antifungal susceptibility test results. Statistically significant differences in mycelial growth were observed with increasing concentrations of *S. chamaecyparissus* extract in all cases. In particular, the MICs were 1000 and 1500 $\mu g \cdot mL^{-1}$ for *N. falciformis* and *N. keratoplastica*, respectively. The 50% and 90% effective concentrations (EC_{50} and EC_{90}, respectively) were 393.4 and 984.4 $\mu g \cdot mL^{-1}$ against *N. falciformis* and 281.0 and 728.6 $\mu g \cdot mL^{-1}$ against *N. keratoplastica*.

2.4. In Planta Activity Against N. falciformis

Given that *N. falciformis* was both highly virulent and the most resistant to the treatment in the in vitro tests, it was chosen for further in planta testing. Protection with the *S. chamaecyparissus* flower extract was assayed both on zucchini (*Cucurbita pepo* L.) cv. 'Diamant F1' (Figure 4a–d) and tomato (*Solanum lycopersicum* L.) cv. 'Optima F1' (Figure 4e–h) plants. In both cases, positive controls (plants artificially inoculated and treated with bi-distilled water only) exhibited the typical symptoms for this pathogen, including yellowing and wilting of leaves, rotting at the stem base and upper root, and—in the case of zucchini plants—the collapse of the entire plant (Figure 4b). Plants treated with the extract at a concentration equal to the MIC (1500 $\mu g \cdot mL^{-1}$) showed disease incidence but without the severity observed in the positive control. Based on plant weight results (Figure 5), zucchini plants were more susceptible than tomato plants, probably because the former is a more specific host for this pathogen. Increasing the treatment concentration to twice the MIC (i.e., 3000 $\mu g \cdot mL^{-1}$) fully prevented symptoms of wilt and root rot without causing phytotoxicity, resulting in treated plants that appeared identical to the negative controls and with no significant statistical differences except in the case of the zucchini weight, with a 5% lower median for the MIC × 2-treated plants (3437.5 mg) compared to the negative control (3640 mg).

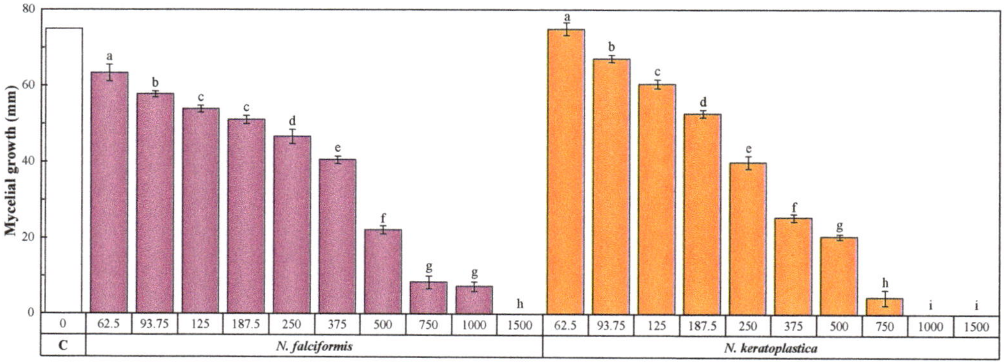

Figure 3. Inhibition of the growth of the mycelium of *N. falciformis* and *N. keratoplastica* in the in vitro tests performed with potato dextrose agar medium amended with different concentrations (in the 15.62–1500 $\mu g \cdot mL^{-1}$ range) of *S. chamaecyparissus* flower extract. C (white bar) represents the controls. The efficacies of the concentrations labeled with the same letters are not statistically different at $p < 0.05$. Standard deviations are represented by error bars.

Figure 4. Symptoms of *N. falciformis* in (**a**–**d**) zucchini (*Cucurbita pepo* L.) cv. 'Diamant F1' and (**e**–**h**) tomato (*Solanum lycopersicum* L.) cv. 'Optima F1' plants 15 days after inoculation. (**a**,**e**) Negative controls with no inoculation or treatment, exhibiting healthy roots and foliage. (**b**,**f**) Positive controls (plants inoculated with *N. falciformis* and treated with bi-distilled water) showing severe symptoms, including leaf yellowing, wilting, and root rot; zucchini plants (**b**) also display stem base collapse. (**c**,**g**) Plants treated with *S. chamaecyparissus* flower extract at 1500 $\mu g \cdot mL^{-1}$, which mitigates disease severity. (**d**,**h**) Plants treated with the *S. chamaecyparissus* flower extract at 3000 $\mu g \cdot mL^{-1}$, showing no disease symptoms and appearing comparable to negative controls, with healthy roots and foliage.

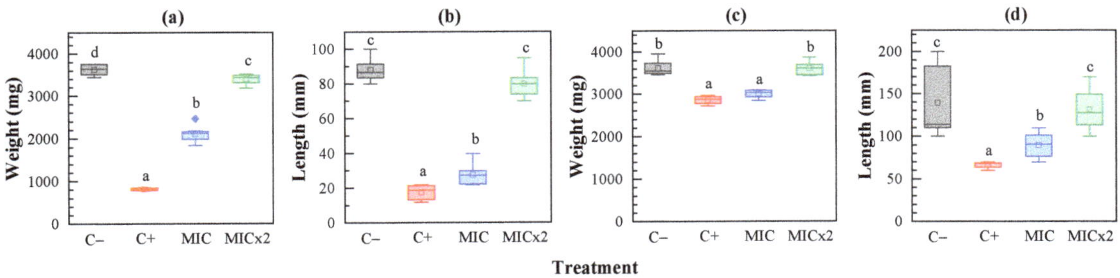

Figure 5. (**a**) Weight and (**b**) root length of zucchini plants; (**c**) weight and (**d**) root length of tomato plants. Different lowercase letters indicate significant differences at the 5% level of significance according to Tukey's test (in the case of zucchini root length) or to Conover–Iman's test (in the case of zucchini weight, tomato weight, and root length). C− and C+ stand for negative control and positive control, respectively. MIC and MIC × 2 represent the minimum inhibitory concentration (1500 $\mu g \cdot mL^{-1}$) and twice that concentration (3000 $\mu g \cdot mL^{-1}$), respectively.

3. Discussion

3.1. Phytochemical Profile and the Antimicrobial Activity of the Phytoconstituents

The characterization of the *S. chamaecyparissus* flower hydromethanolic extract revealed a complex composition that was dominated by sesquiterpenoids. According to the comprehensive review by Tundis and Loizzo [14] on the phytochemistry of the genus *Santolina*, previous studies have reported the presence of terpenoids, particularly eudesmane and germacrane sesquiterpenoids, along with chrysanthemane monoterpenoids. Among the sesquiterpenoids identified herein, spathulenol was previously reported in *S. africana* and *S. pectinata* and aromadendrene only in *S. corsica* [14].

The phytochemical profile of *S. chamaecyparissus* flower extract studied herein shows the closest similarity to two previously reported sesquiterpenoid profiles: that of the essential oil obtained from fresh leaves of *Duguetia glabriuscula* (R.E.Fr.) R.E.Fr. [15], containing ledol, spathulenol, alloaromadendrene, alloaromadendrene oxide, epiglobulol, and ledene oxide, and that of the fruit volatiles from *Eucalyptus camaldulensis* Denham [16], which contain ledene, aromadendrene, alloaromadendrene, globulol, and isolongifolen. Although the three species belong to the class Magnoliopsida, they are classified into different orders and families. Therefore, their shared bioactivity should be attributed to the nature of their phytochemical components rather than to their taxonomic proximity.

The most abundant constituent of *S. chamaecyparissus* flower hydromethanolic extract was 8-cedren-13-ol (27.4%) Previous reports have documented its presence in various aromatic and medicinal plants, including *Artemisia arborescens* L. [17], *Curcuma* species [18], *Juniperus recurva* Buch. [19], *Mentha longifolia* L. [20], *Peucedanum longifolium* Waldst. & Kit. [21], *Plumeria alba* L. [22], and *Zingiber officinale* Roscoe. The antimicrobial activity of cedrene and its derivatives are supported by previous research on *Tetraclinis articulata* (Vahl) Mast. essential oil. This wood-derived oil, containing high levels of cedrene (23%) and cedrenol (9.6%), demonstrated significant antimicrobial activity against several bacterial strains, including *E. faecalis*, *Escherichia coli* (Migula) Castellani & Chalmers, *Klebsiella pneumoniae* (Schroeter) Trevisan, and *Streptococcus* D [23].

The second most abundant compound, 2,1,3-benzothiadiazole (12.0%), has also been found in other plants, including *Armeria maritima* (Mill.) Willd. [24], *Lawsonia inermis* L. [25], and *Sambucus nigra* L. [26], as well as in natural products such as propolis [27]. It serves as a precursor for compounds with fungicidal, herbicidal, and antibacterial properties [28,29].

Concerning the other sesquiterpenoids and their derivatives, they show a widespread distribution across various plant families, in addition to featuring significant biological activities [30], as discussed below.

The presence of ledene has been reported in significant amounts, for instance, in the Australian tea tree oil (*Melaleuca alternifolia* (Maiden & Betche) Cheel) [31] and the *n*-hexane extract of *Barringtonia asiatica* (L.) Kurz seeds (33.6%) [32], with antimicrobial activity against *Staphylococcus aureus* Rosenbach and *E. coli*. That of ledene oxide has been documented —along with spathulenol, discussed below, and caryophyllene oxide— in *Salvia candidissima* Vahl. essential oil [33], with antifungal activity against *Rhizoctonia solani* J.G. Kühn and *Alternaria solani* Sorauer, and in *Shorea robusta* Gaertn F. methanolic extracts [34] together with alloaromadendrene oxide. In turn, high contents of isoledene have been reported in *Cryptomeria japonica* D. Don essential oils (12.4%), with antifungal activity against *Laetiporus sulphureus* (Bull.) Murrill, *Trametes versicolor* (L.) Lloyd, *R. solani*, *Colletotrichum gloeosporioides* (Penz.) Penz. & Sacc., *Fusarium solani* (Mart.) Sacc., and *Ganoderma australe* (Fr.) Pat. [35]. The non-polar twig extract of *Mundulea sericea* (Willd.) A.Chev. contained 29% of isoledene—and 18.2% of spathulenol—and showed activity against *S. aureus*, *E. coli*, *Bacillus subtilis* (Ehrenberg) Cohn, *P. aeruginosa*, and *Candida albicans* (C.P. Robin) Berkhout [36]. *Colocasia esculenta* (L.) Schott leaves' ethanolic extract, rich in D-germacrene and isoledene, showed promising antibacterial activity against *Proteus vulgaris* Hauser, *E. coli*, *P. aeruginosa*, *S. aureus*, and *E. faecalis* as well as antifungal activity against *Cryptococcus neoformans* (San Felice) Vuill., *Aspergillus fumigatus* Fresenius, *Syncephalastrum racemosum* Cohn ex J. Schröt., and *C. albicans* [37].

The concurrent presence of ledol and spathulenol has been reported in several *Eryngium* species [38] and that of ledol, spathulenol, and alloaromadendrene in *Helichrysum petiolare* Hilliard & B.L.Burtt [39]. Ledol-rich *Tabauma gioi* A. Chev wood extract showed antifungal activity against *Trametes ochracea* (Pers.) Gilb. & Ryvarden [40], pure ledol synthesized from natural (+)-aromadendrene-III showed activity against *Cladosporium cucumerinum* Ellis & Arthur [41], while a ledol-enriched (12 %) essential oil showed antimicrobial activity against *S. aureus* (MIC = 50 $\mu g \cdot mL^{-1}$), *Bacillus cereus* Frankland & Frankland (MIC = 100 $\mu g \cdot mL^{-1}$), *Listeria monocytogenes* (Murray et al.) Pirie (MIC = 100 $\mu g \cdot mL^{-1}$), *E. faecalis* (MIC = 400 $\mu g \cdot mL^{-1}$), *Enterobacter aerogenes* Hormaeche & Edwards (MIC = 400 $\mu g \cdot mL^{-1}$), and *Campylobacter jejuni* (Jones et al.) Veron & Chatelain (MIC 30 $\mu g \cdot mL^{-1}$) [42].

Spathulenol is particularly widespread and has been found in *Annonaceae* (*Xylopia aromatica* Mart., *Xylopia emarginata* Mart. [43]), *Lamiaceae* (*Salvia sclarea* L. [44], *Salvia mirzayanii* Rech.f. & Esfand. [45], *Salvia yosgadensis* Freyn & Bornm. [46], and *Salvia rhytidea* Benth. [47]), and *Myrtaceae* (*Eugenia calycina* Cambess. [48], *Eugenia uniflora* L. [49], *Psidium guineense* Sw. [50], *Psidium guajava* L. [51], and *Psidium cattleianum* Afzel.ex Sabine [52]) as well as in other plants that do not belong to those families (e.g., *Barbacenia brasiliensis* Willd. [53]; *Cinnamomum osmophloeum* Kaneh. [54], *Croton arboreus* Millsp. [55], *Croton hirtus* L'Hér. [56], *Perilla frutescens* Britton [57], and *Zanthoxylum nitidum* DC. [58]). It has antiproliferative, anti-inflammatory, antinociceptive, and antimicrobial effects and acts as a mosquito repellent [50,59,60]. In terms of antimicrobial activity, spathulenol was effective against *C. neoformans* and *E. faecalis* (MIC = 200 $\mu g \cdot mL^{-1}$) [61]; *S. aureus* and *Staphylococcus epidermidis* (Winslow and Winslow) Evans (MIC= 1350 and 1500 $\mu g \cdot mL^{-1}$, respectively) [62]; *Micrococcus luteus* (Schroeter) Cohn, *B. subtilis*, and *Xanthomonas campestris* (Pammel) Dowson (MIC = 15 $\mu g \cdot mL^{-1}$) [63]; and *S. aureus*, *Proteus mirabilis* Hauser [44], and *Mycobacterium tuberculosis* (Zopf) Lehmann & Neumann (MIC = 231.9 $\mu g \cdot mL^{-1}$).

Alloaromadendrene, isolated from *Ambrosia peruviana* All., has been previously identified in *Duguetia glabriuscula* (R.E.Fr.) R.E.Fr. [15], *Ledum palustre* Michx. [64], *Monanthotaxis discolor* (Diels) Verdc. [65], *Prostanthera centralis* B.J.Conn [66], *Rhododendron tomentosum*

(Stokes) Harmaja [67], and *Salvia bracteata* Banks & Sol. Alloaromadendrane has been reported to effectively inhibit *Cladosporium herbarum* (Pers.) Link growth [68] and has shown antimycobacterial activity against *M. tuberculosis* [69]. Aromadendrene, as alloaromadendrene, was found in *Eucaliptus* spp. [70], and related compounds such as aromadendrane-4-β,10α-diol, aromadendrane-4α,10α,diol, and 1-epimer-aromadendrane-4β,10α-diol from *Cinnamomum cassia* (L.) D.Don showed significant activity against *S. aureus* [71].

The presence of epiglobulol has been documented in *Achyranthes aspera* L. [72]; *Chamaemelum nobile* (L.) All.; *Matricaria chamomilla* L.; *Ligusticum striatum* DC. [73]; *Conyza dioscoridis* L. [74]; *Croton macrostachyus* Hochst. ex Delile [75]; *Eucalyptus nova-anglica* H.Deane & Maiden [76]; *Eucalyptus citriodora* Hook. [77]; *Jurinea auriculata* (DC.) N.Garcia, Herrando & Susanna [78]; *Ligusticum chuanxiong* S.H.Qiu, Y.Q.Zeng, K.Y.Pan, Y.C.Tang & J.M.Xu [79]; *Moringa oleifera* Lam. [80]; *Psidium guajava* L. [81]; *Salvia officinalis* L. [82]; and *Thottea barberi* Ding Hou [83]. Epiglobulol has been studied extensively for potential therapeutic applications for its anti-inflammatory and anticancer activities and is a natural insecticide [84]. As for its antimicrobial activity, *Croton tricolor* Klotzsch ex Baill. stem essential oil, in which epiglobulol is the main constituent (19%), showed promising antifungal activity against *Candida* spp. [85]

Cis-α-copaene-8-ol has been identified, for instance, in the essential oil from dried fruits of *Xylopia aethiopica* (Dunal) A.Rich. [86], in *Osyris laceolata* Hochst. & Steud. oil [87], in *Otostegia persica* (Burm.f.) Boiss. essential oil [88], and in *Houttuynia cordata* Thunb, with in silico molecular docking studies suggesting strong antibacterial activity [89]. This preconized activity was demonstrated against *B. subtilis* and *E. coli* in the case of essential oils from *Elsholtzia ciliata* (Thunb.) Hyl. [90].

Longiborneol is a constituent of various species including *Juniperus, Pinus, Cupressus, Dacrydium*, as well as *Cedrus deodara* (Lamb.) G.Don [91]. It has also been documented in *Mallotus tetracoccus* (Roxb.) Kurz leaves [92]; in the essential oils of *Cistus salviifolius* L., with antimicrobial activity against Gram-positive and Gram-negative bacteria [93]; in the leaf oil of *Cinnamomum chemungianum* Mohan & Henry, with moderate activity against certain strains of Gram-positive and Gram-negative bacteria [94]; and in the essential oil of *Drimys granadensis* L.fil., with activity against *L. monocytogenes* [95].

Aromadendrene has been detected at concentrations as high as 42.3% in *E. camaldulensis* fruit volatiles [16]. It was also the main constituent (9.1%) of the ether extracts of *Scapania verrucosa* Heeg., which showed antifungal activity against *C. albicans, C. neoformans, Trichophyton rubrum* (Castell.) Sabour., *A. fumigatus*, and *Pyricularia oryzae* Cavara [96].

The essential oil from leaves of *Withania adpressa* Coss. Ex, rich in longifolene (21.4%), showed antibacterial effects against *E. coli, K. pneumoniae, S. aureus*, and *Streptococcus pneumoniae* (Klein) Chester, while its antifungal efficacy was demonstrated against *C. albicans, Aspergillus flavus* Link, *Aspergillus niger* P.E.L. van Tieghem, and *Fusarium oxysporum* Schltdl. [97]. Its autoxidation products also showed antifungal activity against *T. versicolor, Lenzites betulinus* (L.) Fr., *Gloeophyllum trabeum* (Pers.) Murrill, *Trichoderma virens* (J.H. Mill., Giddens & A.A. Foster) Arx, and *Rhizopus oryzae* Went & Prins. Geerl. [98]. Its presence in agarwood essential oil —along with carvacrol— is thought to be responsible for the antifungal activity demonstrated in vitro and in vivo against agricultural and foodborne pathogens [99].

3.2. Structure–Activity Relationship: On the Biological Role of the Skeleton in Sesquiterpenes

As noted above, the characterization of the *S. chamaecyparissus* flower extract phytochemical profile revealed the presence of sesquiterpenoids with distinct skeletal types: compounds with a hydroazulene skeleton (consisting of a bicyclo[5.3.0]decane system, where a seven-membered ring is fused to a five-membered ring) and compounds with an aromadendrane-type.

The biological activity of compounds containing the aforementioned skeletons is influenced by several key structural features:

(i) The degree of unsaturation affects both lipophilicity and molecular flexibility, which in turn influence bacterial membrane penetration. Studies with guaiane-type sesquiterpenes have shown that compounds with higher degrees of unsaturation typically exhibit enhanced antimicrobial activity due to improved membrane interactions [100,101];

(ii) The presence and position of hydroxyl groups also significantly impact biological activity. Hydroxylated derivatives like ledol show enhanced antimicrobial properties through multiple mechanisms: increased water solubility facilitating cellular uptake, potential for hydrogen bonding with cellular targets, formation of reactive oxygen species in bacterial cells, and enhanced interaction with membrane phospholipids [102];

(iii) The presence of a fused cyclopropane ring, as found in ledene and isoledene, introduces structural rigidity that can modify the three-dimensional conformation of the molecule, enhance binding specificity to cellular targets, increase molecular stability, and affect membrane permeability through changes in molecular shape [103,104];

(iv) In alloaromadendrene, the presence of a *gem*-dimethylcyclopropyl unit rather than an isopropyl group confers antimicrobial activity, whereas its corresponding guaiane derivative shows in vitro cytotoxic activity against L-1210 cells [100]. Similarly, spathulenol has demonstrated biological activity, whereas its corresponding isopropyl derivative, nardol, has not shown any biological activity [105,106];

(v) The stereochemistry of the ring junction between the five- and seven-membered rings plays a crucial role in determining the overall molecular shape and consequently its biological activity. The *cis*-fused hydroazulene system is more common in nature and typically shows higher biological activity compared to *trans*-fused analogs [103]

(vi) Additional functional groups and their spatial orientation can modulate activity: methyl groups increase lipophilicity, ketone groups can serve as hydrogen bond acceptors, and epoxide rings can react with cellular nucleophiles [107].

Looking at the key structural characteristics that influence bioactivity, a theoretical hierarchy of potential biological activity, from highest to lowest expected activity, may be proposed:

Ledene and alloaromadendrene oxide > (−)-spathulenol and ledol > 8-cedren-13-ol and *cis*-α-copaene-8-ol > epiglobulol and iso-ledene.

In comparing the basic structure of sesquiterpenoids containing hydroazulene and aromadendrene skeletons with that of the azulene-type (the unsaturated version of hydroazulen) or other sesquiterpenoids with the same biosynthetic origin (i.e., the cyclization of farnesyl pyrophosphate through different folding patterns and rearrangements) as germacrene and guaiane (Figure 6), the former should exhibit higher activity. This is evident when comparing them to the germacrene skeleton (a 10-carbon ring that can adopt pseudobicyclic conformations) [108], which has lower structural rigidity and shows only moderate antimicrobial activity. A similar effect occurs when comparing guaiane-type sesquiterpenoids (compounds with a decahydro-1,4-dimethyl-7-(1-methylethyl)azulene skeleton, a 5/7 system biosynthetically related to germacrene) from *Alisma orientale* (Sam.) Juz., *Enterospermum madagascariense* (Baill.) Homolle, *Amoora rohituka* (Roxb.) Wight & Arn., and *Curcuma phaeocaulis* Valeton [109–112], which exhibit lower activity. When the comparison is extended to sesquiterpenoids with an azulene skeleton, the antimicrobial activity of these is notably reduced due to aromaticity.

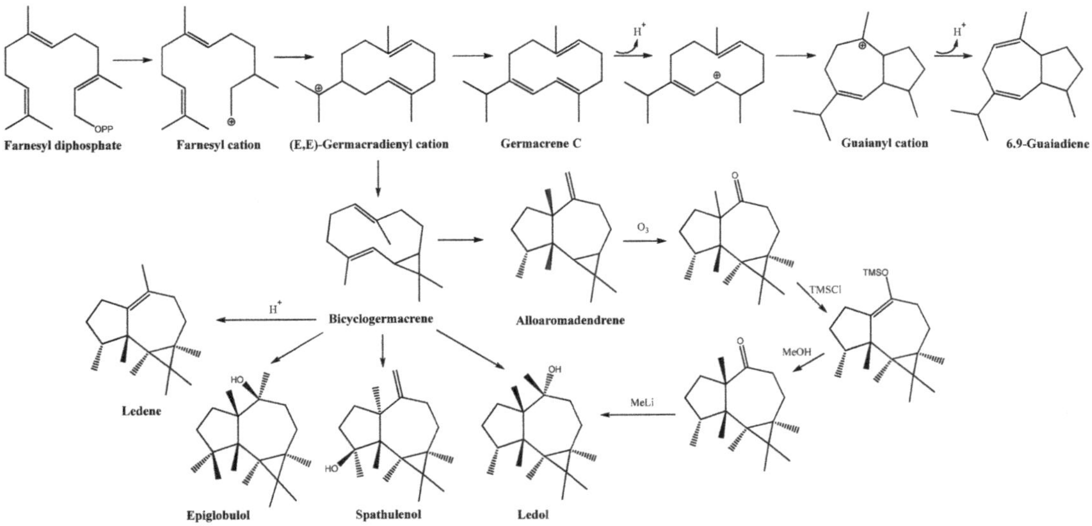

Figure 6. Types of primary cyclization of aromadrendane, bicyclogermacrene, and 6,9-guiadiene (adapted from [113]), along with the conversion of aromadendrene into ledol (adapted from [103]). Chemical structures were drawn using ChemDraw v.2023 (Revvity Signals Software, Waltham, MA, USA).

Briefly, based on structure–activity relationships, the aromadendrene and hydroazulene skeletons typically demonstrate higher antimicrobial activity than related parent compounds, particularly when functionalized with specific oxygenated groups (presence of either epoxide groups, secondary alcohols, and *gem*-dimethylcyclopropyl units) [100,102,107]. This enhanced activity can be attributed to their higher structural rigidity, better hydrophobic/hydrophilic balance, and consequently, improved ability to interact with cell membranes [103,104].

3.3. Comparisons of Efficacy

3.3.1. Comparison Versus Previous Reports on the Antimicrobial Activity of *S. chamaecyparissus*

In this study, the hydromethanolic extract from *S. chamaecyparissus* flowers exhibited antifungal activity against *Neocosmospora* spp. with MICs of 1000–1500 μg/mL. Although there is extensive literature on the antimicrobial activity of *S. chamaecyparissus* (Table S2), most previous studies [6,114–118] have focused on human pathogens and evaluated essential oils rather than hydromethanolic extracts, with reported MICs ranging from 0.625 to 1600 $\mu g{\cdot}mL^{-1}$ depending on the pathogen type, plant part used, and collection site. Studies on plant pathogens are scarce and have shown varying results: an aqueous extract demonstrated limited activity against *Fusarium oxysporum* f.sp. *lentis* W.L. Gordon (MIC > 20,000 μg/mL) [119], while a 30% commercial essential oil [120] showed differential mycelial growth inhibition against various plant pathogenic fungi, ranging from minimal activity against *Alternaria brassicae* (Berk.) Sacc. (10.65% inhibition) to substantial growth suppression of *Phytophthora parasitica* Dastur (72.02%), *Cladobotryum mycophilum* (Oudem.) W. Gams & Hooz (68.69%), *Fusarium oxysporum* Schltdl. (67.65%), and *Sclerotinia sclerotiorum* (Lib.) de Bary (54.50%). Notably, no activity was observed against *Botrytis cinerea* Pers. and *Pythium aphanidermatum* (Edson) Fitzp. at the tested concentration. Our findings extend the known spectrum of *S. chamaecyparissus* antifungal activity against other plant pathogens, representing the first report of its effectiveness against *Neocosmospora* species. The observed MIC values against *Neocosmospora* spp. are particularly promising when compared to the higher concentrations required for activity against other plant pathogenic fungi in previous studies.

3.3.2. Comparison of Efficacy Against *Neocosmospora* spp. Versus Other Natural Compounds

Table S3 [24,121–128] presents a comparison of the reported efficacies of plant extracts and essential oils against the two *Neocosmospora* species studied. It is important to note that sensitivity may vary depending on the isolate, and results are expressed using different parameters (MIC values, percentage of mycelial growth inhibition (PMIG), and inhibition zones (IZ)), which makes direct comparisons challenging.

For *N. falciformis*, the flower extract of *S. chamaecyparissus* showed MIC values of 1500 $\mu g \cdot mL^{-1}$, positioning it in the intermediate-to-high range of efficacy among tested natural products. It was less effective than the hydromethanolic extract of *A. maritima* (MIC = 1000 $\mu g \cdot mL^{-1}$), the hydroethanolic extract of *Syzygium aromaticum* (L.) Merr. & L.M.Perry extract (MIC = 1000 $\mu g \cdot mL^{-1}$) [24,123], and the essential oils from *Cinnamomum aromaticum* Nees (MIC = 625 $\mu g \cdot mL^{-1}$) and *Cymbopogon flexuosus* (Nees ex Steud.) Will.Watson (MIC = 1250 $\mu g \cdot mL^{-1}$)) [122]. Nonetheless, it demonstrated better performance than several other treatments, which included *Citronella* spp. (requiring 20,000 $\mu g \cdot mL^{-1}$ for complete inhibition), *Melaleuca* spp. (achieving only 40.1% inhibition at 25,000 $\mu g \cdot mL^{-1}$) [121], and *Ocimum basilicum* L. essential oils (64.2–72% inhibition at 5000 $\mu g \cdot mL^{-1}$) [122]. Moreover, hydroethanolic extracts of *Hibiscus sabdariffa* L. and *Curcuma longa* L. showed much lower efficacy (MIC = 10,000 $\mu g \cdot mL^{-1}$), while *Cymbopogon citratus* (DC.) Stapf extract showed no significant activity (MIC > 10,000 $\mu g \cdot mL^{-1}$) [123].

Regarding *N. keratoplastica*, the *S. chamaecyparissus* flower extract (MIC = 1000 $\mu g \cdot mL^{-1}$) showed moderately good efficacy among tested products. Although it was less effective than *M. chamomilla* flower essential oil (MIC = 20 $\mu g \cdot mL^{-1}$) [128], it performed better than *A. maritima* flower extract (MIC = 1500 $\mu g \cdot mL^{-1}$) [24] and was comparable to *Origanum vulgare* L. subsp. *hirtum* essential oil (MIC = 800 $\mu g \cdot mL^{-1}$) [127]. Several other essential oils showed lower or no activity, including *Trachyspermum ammi* (L.) Sprague (no activity at the highest concentration tested) [124] and *Pogostemon cablin* (Blanco) Benth. (no activity at 500 $\mu g \cdot mL^{-1}$) [126]. *Kaempferia parviflora* Wall. ex Baker rhizome essential oil produced inhibition zones of 17–18 mm at 500 $\mu g \cdot mL^{-1}$ [125], but a direct comparison is not possible.

3.3.3. Comparison of Efficacy Against *Neocosmospora* spp. Versus Conventional Fungicides

The MIC values obtained for *S. chamaecyparissus* L. flower extract (1500 and 1000 $\mu g \cdot mL^{-1}$ for *N. falciformis* and *N. keratoplastica*, respectively) can be compared with those previously reported by our group for commercial fungicides, as they were tested against the same fungal isolates under identical in vitro experimental conditions [24]. The currently prohibited dithiocarbamate Mancozeb demonstrated superior performance, achieving complete inhibition of both *Neocosmospora* species even at concentrations of 150 $\mu g \cdot mL^{-1}$, ten times lower than those required for the plant extract. The organophosphate fungicide Fosetyl-Al exhibited complete inhibition (100%) of both species at its recommended dose (2 $mg \cdot mL^{-1}$), which is higher than the extract's MICs. The response to the strobilurin fungicide azoxystrobin was species-dependent: while *N. keratoplastica* was completely inhibited at the recommended dose (62.5 $mg \cdot mL^{-1}$), *N. falciformis* showed only partial inhibition (62.2%). Compared to these latter two conventional fungicides, the superior performance of the *S. chamaecyparissus* extract may be attributed to its particular composition, with a family of compounds potentially acting through different non-specific and specific mechanisms, as discussed in the recent review by Wiart et al. [129]. This characteristic could offer advantages for integrated disease management approaches, particularly in rotation schemes aimed at minimizing the risk of resistance development.

4. Materials and Methods

4.1. Reagents and Phytopathogens

Potato dextrose broth (PDB) and potato dextrose agar (PDA) were purchased from Becton Dickinson (Bergen County, NJ, USA). Ortiva® (azoxystrobin 25%; Syngenta, Basel, Switzerland) and Fesil® (fosetyl-Al 80; Bayer, Leverkusen, Germany) commercial fungicides were provided by the Plant Health and Certification Center (CSCV) of the Gobierno de Aragón.

Neocosmospora falciformis (MYC-1345) and *N. keratoplastica* (MYC-1250) isolates were supplied by the Mycology Lab of the Center for Research and Agrifood Technology of Aragón (CITA, Zaragoza, Spain) as living subcultures on PDA.

4.2. Plant Material and Extraction Protocol

Flowers of *S. chamaecyparissus* were collected during full bloom in June near the city of Huesca, NE Spain (42°09′13.6″ N 0°27′46.1″ W). A voucher specimen was authenticated by Prof. J. Ascaso and deposited at the herbarium of the Escuela Politécnica Superior de Huesca, University of Zaragoza. Samples from twenty-five specimens were combined to obtain representative composite samples, which were shade-dried, mechanically ground into powder, homogenized, and sieved through a 1 mm mesh.

For extraction, dried flower powder (34.2 g) was combined with 200 mL of a methanol/water solution (1:17 *v*/*v*). The mixture was heated for 30 min at 50 °C and sonicated using a probe-type ultrasonicator (model UIP1000 hdT; Hielscher Ultrasonics, Teltow, Germany). The mixture was subjected to ultrasonication at 20 kHz and 1000 W in alternating cycles: 10 to 15 min of sonication followed by 5 to 10 min rest periods to maintain the temperature between 30 and 60 °C. The extract was first decanted, then filtered through Whatman No. 1 paper, and subsequently centrifuged at 9000 rpm for 15 min. The resulting solution was freeze-dried to obtain the solid residue. The extraction yield was 18% (*w*/*w*) (6.4 g). For gas chromatography–mass spectrophotometry (GC-MS) analysis, the freeze-dried extract was dissolved in high-performance liquid chromatography (HPLC)-grade methanol to obtain a 5 mg·mL^{-1} solution, followed by filtration.

4.3. Physicochemical Characterization

The infrared vibrational spectra of both the flower heads (pre-extraction) and the freeze-dried extract were recorded using a Nicolet iS50 Fourier-transform infrared (FTIR) spectrometer (Thermo Scientific, Waltham, MA, USA) equipped with a diamond attenuated total reflection (ATR) system. Interferograms were generated by co-adding 64 scans across the 400–4000 cm^{-1} spectral range with a 1 cm^{-1} resolution.

A gas chromatography–mass spectrometry (GC-MS) instrument comprising a system with a model 7890A gas chromatograph coupled to a model 5975C quadrupole mass spectrometer (both from Agilent Technologies, Santa Clara, CA, USA) was used to elucidate the constituents of the flower hydromethanolic extract. This analysis was outsourced to the Research Support Services (STI) at Universidad de Alicante (Alicante, Spain). The chromatographic conditions consisted of an injection volume of 1 μL, an injector temperature of 280 °C in splitless mode, and an initial oven temperature of 60 °C held for 2 min, followed by a ramp of 10 °C·min^{-1} up to a final temperature of 300 °C, held for 15 min. An HP-5MS UI chromatographic column (30 m length, 0.250 mm diameter, and 0.25 μm film), also from Agilent Technologies, was employed for the separation of the compounds. The temperatures of the mass spectrometer's electron impact source and quadrupole were 230 and 150 °C, respectively, with an ionization energy of 70 eV. Components were identified by comparing their mass spectra and retention times with those of authentic compounds and through computer matching with the database of the National Institute of Standards and Technology (NIST11). Test mixture 2 for apolar capillary columns according to Grob (Supelco 86501) and PFTBA tuning standards were used for equipment calibration.

4.4. In Vitro Antimicrobial Activity Assessment

Antifungal activity was evaluated using the agar dilution method following EUCAST standard procedures [130]. Aliquots of the stock solution were incorporated into the PDA medium to achieve final concentrations ranging from 62.5 to 1500 $\mu g \cdot mL^{-1}$. Mycelial plugs (5 mm diameter) were excised from the margin of 7-day-old PDA cultures of *N. falciformis* and *N. keratoplastica* and transferred to the center of the amended PDA plates. For each treatment and concentration, three replicate plates were prepared in duplicate. The control treatment consisted of PDA medium amended with the extraction solvent (methanol/water 1:17 v/v). All plates were incubated at 25 °C in darkness for seven days. Mycelial growth inhibition was calculated using the formula $((d_c - d_t)/d_c) \times 100$, where d_c and d_t represent the mean colony diameters in control and treated plates, respectively. Effective concentrations (EC_{50} and EC_{90}) were determined through PROBIT analysis using SPSS Statistics v. 25 (IBM, New York, NY, USA).

Since the Shapiro–Wilk [131] and Levene [132] tests indicated that the normality and homoscedasticity requirements were fulfilled, the mycelial growth inhibition results were subjected to one-way analysis of variance (ANOVA) and subsequent post hoc comparison of means through Tukey's test at a significance level of $p < 0.05$, also conducted with IBM SPSS Statistics v. 25 software. In the case of in planta weight and root length results, normality and homoscedasticity requirements were only met for zucchini lengths (to which an ANOVA was applied). Kruskal–Wallis test [133] was used for zucchini weight and tomato weight and root length analysis, given that the distribution was not normal, but the groups were homoscedastic, followed by multiple pairwise comparisons using the Conover–Iman procedure.

4.5. In Vivo Tests on Horticultural Crops

Tomato cv. 'Optima F1' and zucchini cv. 'Diamant F1' plants for in vivo experiments were obtained from Agrodepa S.L. (Palencia, Spain) and Fronda Centros de Jardinería S.L. (Madrid, Spain), respectively.

The efficacy of the extract against *N. falciformis* was evaluated through artificial inoculation following modified versions of the methods described by González et al. [10] and Sánchez-Hernández et al. [134]. The fungus was grown in 250 mL flasks containing PDB for 3 days at 25 °C in the dark with constant shaking. Plant roots were first dipped in the treatment (either at MIC or at MIC × 2 concentration, i.e., 1500 and 3000 $\mu g \cdot mL^{-1}$) and then in a suspension of 5×10^6 conidia$\cdot mL^{-1}$ for 2 min, after which they were transferred to the 4 cm × 4 cm plastic pots with sterilized peat substrate. Non-inoculated plants dipped in sterilized water were used as negative controls. Each treatment included twelve plants per replicate, with two independent replicates. Plants were maintained in a growth chamber at 25 °C with a 16/8 h photoperiod for 15 days, after which plant development, wilting, and yellowing were assessed.

5. Conclusions

The characterization and evaluation of bioactive compounds from *S. chamaecyparissus* flowers revealed their significant potential as natural fungicides. The extract's composition was dominated by sesquiterpenoids (50.4% of total compounds), with distinctive structural features including hydroazulene and aromadendrane skeletons. Structure–activity relationship analysis supported that these skeletal types exhibit enhanced antimicrobial activity compared to analogous frameworks (such as azulene, germacrene, or guaiane), primarily due to their unique structural characteristics: higher molecular rigidity, optimal hydrophobic–hydrophilic balance, and specific functional group orientations.

The practical significance of this research lies in the extract's demonstrated effectiveness against two emerging and polyphagous soil phytopathogens. The observed EC_{90} values (984.4 and 728.6 $\mu g \cdot mL^{-1}$ against *N. falciformis* and *N. keratoplastica*, respectively) and successful in planta protection at 3000 $\mu g \cdot mL^{-1}$ position this extract as a promising alternative to conventional fungicides. Its activity was comparable to fosetyl-Al and superior to azoxystrobin, suggesting its potential value in organic farming applications, where natural alternatives to synthetic fungicides are urgently needed.

Future research should focus on optimizing extraction methods, developing stable formulations, conducting field trials broadening the spectrum of plant hosts, utilizing various environmental conditions, and exploring potential applications against insect pests given the volatile nature of the active compounds.

The integration of natural product chemistry with agricultural applications demonstrated in this study contributes to the growing field of biorational pest management, offering a sustainable alternative to conventional synthetic fungicides while expanding the known applications of *S. chamaecyparissus*.

Supplementary Materials: The following supporting information can be downloaded at https://www.mdpi.com/article/10.3390/plants14020235/s1. Figure S1: FTIR spectra of *Santolina chamaecyparissus* L. flowers and flower hydromethanolic extract; Figure S2: GC-MS chromatogram of *S. chamaecyparissus* flower hydromethanolic extract; Figure S3: In vitro antifungal activity assays showing growth inhibition of *Neocosmospora falciformis* and *N. keratoplastica* by *S. chamaecyparissus* extract at different concentrations; Table S1: Main chemical compounds identified by GC-MS in *S. chamaecyparissus* flower extract; Table S2: Antimicrobial activity of *S. chamaecyparissus* extracts and essential oils previously reported in the literature; Table S3: Efficacy of plant extracts and essential oils reported in the literature against the *Neocosmospora* species under study.

Author Contributions: Conceptualization, V.G.-G. and P.M.-R.; methodology, J.M.-G., V.G.-G. and J.C.-G.; validation, J.C.-G.; formal analysis, E.S.-H., J.M.-G. and P.M.-R.; investigation, E.S.-H., J.M.-G., V.G.-G., J.C.-G. and P.M.-R.; resources, J.C.-G. and P.M.-R.; writing—original draft preparation, E.S.-H., J.M.-G., V.G.-G., J.C.-G. and P.M.-R.; writing—review and editing, E.S.-H. and P.M.-R.; visualization, E.S.-H. and P.M.-R.; project administration, P.M.-R.; funding acquisition, P.M.-R. All authors have read and agreed to the published version of the manuscript.

Funding: This research was funded by Junta de Castilla y León under project VA148P23, with FEDER co-funding.

Data Availability Statement: All data supporting the findings of this study are available within the paper and its Supplementary Information. Should any raw data files be needed in another format, they are available from the corresponding author upon reasonable request.

Acknowledgments: The authors acknowledge Pilar Blasco and Pablo Candela from the Technical Research Services of the University of Alicante for conducting the GC-MS analysis.

Conflicts of Interest: The authors declare no conflicts of interest. The funders had no role in the design of the study; in the collection, analyses, or interpretation of data; in the writing of the manuscript; or in the decision to publish the results.

Abbreviations

The following abbreviations are used in this manuscript:

ANOVA	Analysis of variance
ATR	Attenuated total reflection
EC_{50}	50% Effective concentration
EC_{90}	90% Effective concentration
FTIR	Fourier-transform infrared
GC-MS	Gas chromatography–mass spectrometry
HPLC	High-performance liquid chromatography
MIC	Minimum inhibitory concentration
NIST	National Institute of Standards and Technology
PDA	Potato dextrose agar
PDB	Potato dextrose broth

References

1. Calvo, M.I.; Cavero, R.Y. Medicinal plants used for neurological and mental disorders in Navarra and their validation from official sources. *J. Ethnopharmacol.* **2015**, *169*, 263–268. [CrossRef] [PubMed]
2. GBIF Secretariat. GBIF Backbone Taxonomy. Checklist Dataset. 2023. Available online: https://doi.org/10.15468/39omei (accessed on 11 December 2024).
3. Azevedo, T.; Faustino-Rocha, A.I.; Barros, L.; Finimundy, T.C.; Matos, M.; Oliveira, P.A. *Santolina chamaecyparissus* L.: A Brief Overview of Its Medicinal Properties. *Med. Sci. Forum* **2023**, *21*, 8. [CrossRef]
4. Ninkuu, V.; Zhang, L.; Yan, J.; Fu, Z.; Yang, T.; Zeng, H. Biochemistry of terpenes and recent advances in plant protection. *Int. J. Mol. Sci.* **2021**, *22*, 5710. [CrossRef]
5. Barrero, A.F.; Alvarez-Manzaneda, R.; Quilez, J.F.; Herrador, M.M. Sesquiterpenes from *Santolina chamaecyparissus* subsp. *squarrosa*. *Phytochemistry* **1998**, *48*, 807–813. [CrossRef]
6. Bel Hadj Salah-Fatnassi, K.; Hassayoun, F.; Cheraif, I.; Khan, S.; Jannet, H.B.; Hammami, M.; Aouni, M.; Harzallah-Skhiri, F. Chemical composition, antibacterial and antifungal activities of flowerhead and root essential oils of *Santolina chamaecyparissus* L., growing wild in Tunisia. *Saudi J. Biol. Sci.* **2017**, *24*, 875–882. [CrossRef]
7. Sandoval-Denis, M.; Lombard, L.; Crous, P. Back to the roots: A reappraisal of *Neocosmospora*. *Persoonia Mol. Phylogeny Evol. Fungi* **2019**, *43*, 90–185. [CrossRef]
8. Kidd, S.E.; Abdolrasouli, A.; Hagen, F. Fungal nomenclature: Managing change is the name of the game. *Open Forum Infect. Dis.* **2023**, *10*, ofac559. [CrossRef]
9. Grabka, R.; d'Entremont, T.W.; Adams, S.J.; Walker, A.K.; Tanney, J.B.; Abbasi, P.A.; Ali, S. Fungal endophytes and their role in agricultural plant protection against pests and pathogens. *Plants* **2022**, *11*, 384. [CrossRef]
10. González, V.; Ruiz, J.; Picó, B.; García-Martínez, S.; Garcés-Claver, A.; Flores-León, A. First report of *Neocosmospora falciformis* causing wilt and root rot of muskmelon in Spain. *Plant Dis.* **2020**, *104*, 1256. [CrossRef]
11. González, V.; García-Martínez, S.; Flores-León, A.; Ruiz, J.; Picó, B.; Garcés-Claver, A. *Neocosmospora keratoplastica*, a relevant human fusarial pathogen is found to be associated with wilt and root rot of muskmelon and watermelon crops in Spain: Epidemiological and molecular evidences. *Eur. J. Plant Pathol.* **2020**, *156*, 1189–1196. [CrossRef]
12. Sakai, T.; Nishimura, K.; Hirose, Y. The structure and stereochemistry of four new sesquiterpenes isolated from the wood oil of "Kaya" (*Torreya nucifera*). *Bull. Chem. Soc. Jpn.* **1965**, *38*, 381–387. [CrossRef]
13. Socrates, G. *Infrared and Raman Characteristic Group Frequencies: Tables and Charts*, 3rd ed.; Wiley: Chichester, NY, USA, 2001; p. 347.
14. Tundis, R.; Loizzo, M. A review of the traditional uses, phytochemistry and biological activities of the genus *Santolina*. *Planta Med.* **2018**, *84*, 627–637. [CrossRef] [PubMed]
15. De Siqueira, J.M.; MÜLler, L.; Carollo, C.A.; Silva Garcez, W.; Diamantino Boaventura, M.A.; Nascimento, E.A. Aromadendrane sesquiterpenoids from the essential oil of leaves of *Duguetia glabriuscula*—Annonaceae. *J. Chil. Chem. Soc.* **2003**, *48*, 89–93. [CrossRef]
16. Ozel, M.Z.; Yilmaz, S.; Ergin, C. Analysis of *Eucalyptus camaldulensis* volatiles from Turkey using DTD-GC-MS. *J. Essent. Oil Bear. Plants* **2010**, *13*, 297–312. [CrossRef]
17. Polito, F.; Papaianni, M.; Woo, S.L.; Malaspina, P.; Cornara, L.; De Feo, V. *Artemisia arborescens* (Vaill.) L.: Micromorphology, essential oil composition, and its potential as an alternative biocontrol product. *Plants* **2024**, *13*, 825. [CrossRef]
18. Xiang, H.; Zhang, L.; Xi, L.; Yang, Y.; Wang, X.; Lei, D.; Zheng, X.; Liu, X. Phytochemical profiles and bioactivities of essential oils extracted from seven *Curcuma* herbs. *Ind. Crops Prod.* **2018**, *111*, 298–305. [CrossRef]
19. Oda, J.i.; Ando, N.; Nakajima, Y.; Inouye, Y. Studies on insecticidal constituents of *Juniperus recurva* Buch. *Agric. Biol. Chem.* **1977**, *41*, 201–204. [CrossRef]

20. Pirmohammadi, M.; Abai, M.R.; Shayeghi, M.; Vatandoost, H.; Rahimi, S.; Pirmohammadi, M. Influence of agro-climatic conditions on chemical compositions and repellency effect of *Mentha longifolia* plant against malaria vector, *Anopheles stephensi*. *Toxin Rev.* **2022**, *42*, 115–121. [CrossRef]
21. Tepe, B.; Akpulat, H.A.; Sokmen, M. Evaluation of the chemical composition and antioxidant activity of the essential oils of *Peucedanum longifolium* (Waldst. & Kit.) and *P. palimbioides* (Boiss.). *Rec. Nat. Prod.* **2011**, *5*, 108.
22. Sahoo, A.; Dash, B.; Jena, S.; Ray, A.; Panda, P.C.; Nayak, S. Phytochemical composition of flower essential oil of *Plumeria alba* grown in India. *J. Essent. Oil Bear. Plants* **2021**, *24*, 671–676. [CrossRef]
23. El Moussaouiti, M.; Talbaoui, A.; Gmouh, S.; Aberchane, M.; Benjouad, A.; Bakri, Y.; Kamdem, D.P. Chemical composition and bactericidal evaluation of essential oil of *Tetraclinis articulata* burl wood from Morocco. *J. Indian Acad. Wood Sci.* **2010**, *7*, 14–18. [CrossRef]
24. Sánchez-Hernández, E.; Martín-Ramos, P.; Navas Gracia, L.M.; Martín-Gil, J.; Garcés-Claver, A.; Flores-León, A.; González-García, V. *Armeria maritima* (Mill.) Willd. flower hydromethanolic extract for Cucurbitaceae fungal diseases control. *Molecules* **2023**, *28*, 3730. [CrossRef] [PubMed]
25. Kumaran, N.S. Protective effect of *Lawsonia inermis* Linn. on chronic inflammation in rats. *Int. J. Green Pharm.* **2018**, *12*, 549–554.
26. Sánchez-Hernández, E.; Balduque-Gil, J.; González-García, V.; Barriuso-Vargas, J.J.; Casanova-Gascón, J.; Martín-Gil, J.; Martín-Ramos, P. Phytochemical profiling of *Sambucus nigra* L. flower and leaf extracts and their antimicrobial potential against almond tree pathogens. *Int. J. Mol. Sci.* **2023**, *24*, 1154. [CrossRef]
27. Ghavidel, F.; Javadi, A.; Anarjan, N.; Jafarizadeh-Malmiri, H. New approach in process intensification based on subcritical water, as green solvent, in propolis oil in water nanoemulsion preparation. *Green Process. Synth.* **2021**, *10*, 208–218. [CrossRef]
28. Neto, B.A.D.; Lapis, A.A.M.; da Silva Júnior, E.N.; Dupont, J. 2,1,3-benzothiadiazole and derivatives: Synthesis, properties, reactions, and applications in light technology of small molecules. *Eur. J. Org. Chem.* **2012**, *2013*, 228–255. [CrossRef]
29. Saka, E.; Akkoç, S.; Coban, B. Synthesis, DNA interaction and cytotoxic activity studies of a novel 2,1,3-benzothiadiazole-benzimidazole derivative. *Pharm. Chem. J.* **2020**, *7*, 82–89.
30. Perveen, S.; Al-Taweel, A. *Terpenes and Terpenoids*; IntechOpen: London, UK, 2018; p. 152. [CrossRef]
31. Swords, G.; Hunter, G.L.K. Composition of Australian tea tree oil (*Melaleuca alternifolia*). *J. Agric. Food Chem.* **1978**, *26*, 734–737. [CrossRef]
32. Nurhaida, N.; Murniana, M.; Atanta, J.S. Phytochemical screening and antibacterial activity testing of n-hexane extract from *Barringtonia asiatica* seeds. *J. Nat.* **2024**, *24*, 50–54. [CrossRef]
33. Bayar, Y.; Genc, N. Determination of the chemical components, antioxidant and antifungal activities of essential oil and plant extract of *Salvia candidissima* Vahl. *Mediterr. Agric. Sci.* **2018**, *31*, 93–99. [CrossRef]
34. Sushma, V.; Pal, S.M.; Viney, C. GC-MS analysis of phytocomponents in the various extracts of *Shorea robusta* Gaertn F. *Int. J. Pharmacogn. Phytochem. Res.* **2017**, *9*, 783–788. [CrossRef]
35. Cheng, S.-S.; Lin, H.-Y.; Chang, S.-T. Chemical composition and antifungal activity of essential oils from different tissues of Japanese cedar (*Cryptomeria japonica*). *J. Agric. Food. Chem.* **2005**, *53*, 614–619. [CrossRef] [PubMed]
36. Mazimba, O.; Masesane, I.B.; Majinda, R.R.; Muzila, A. GC-MS analysis and antimicrobial activities of the non-polar extracts of *Mundulea sericea*. *S. Afr. J. Chem.* **2012**, *65*, 50–52.
37. Elmosallamy, A.; Eltawil, N.; Ibrahim, S.; Hussein, S. Phenolic profile: Antimicrobial activity and antioxidant capacity of *Colocasia esculenta* (L.) Schott. *Egypt. J. Chem.* **2021**, *64*, 2165–2172. [CrossRef]
38. Landoulsi, A.; Hennebelle, T.; Bero, J.; Rivière, C.; Sahpaz, S.; Quetin-Leclercq, J.; Neut, C.; Benhamida, J.; Roumy, V. Antimicrobial and light-enhanced antimicrobial activities, cytotoxicity and chemical variability of all Tunisian *Eryngium* species. *Chem. Biodivers.* **2020**, *17*, e1900543. [CrossRef]
39. Akinyede, K.A.; Cupido, C.N.; Hughes, G.D.; Oguntibeju, O.O.; Ekpo, O.E. Medicinal properties and in vitro biological activities of selected *Helichrysum* species from South Africa: A review. *Plants* **2021**, *10*, 1566. [CrossRef]
40. Tkhu, D.K.; Roshchin, V.I.; Malysheva, O.N.; Solov'ev, V.A. Fungicidal activity of an extract from the wood of *Tabauma gioi* A. Chev. *Koksnes Kim.* **1987**, *1*, 103–104.
41. Gijsen, H.J.M.; Wijnberg, J.B.P.A.; Stork, G.A.; de Groot, A.; de Waard, M.A.; van Nistelrooy, J.G.M. The synthesis of mono- and dihydroxy aromadendrane sesquiterpenes, starting from natural (+)-aromadendrene-III. *Tetrahedron* **1992**, *48*, 2465–2476. [CrossRef]
42. Guinoiseau, E. *Antibacterial Molecules Isolated from Essential Oils: Fractionation, Identification and Mode of Action*; Université de Corse: Corsica, France, 2010.
43. Lago, J.H.G.; Reis, A.A.; Martins, D.; Cruz, F.G.; Roque, N.F. Composition of the leaf oil of *Xylopia emarginata* Mart. (Annonaceae). *J. Essent. Oil Res.* **2005**, *17*, 622–623. [CrossRef]
44. Ulubelen, A.; Topcu, G.; Eri§, C.; Sönmez, U.; Kartal, M.; Kurucu, S.; Bozok-Johansson, C. Terpenoids from *Salvia sclarea*. *Phytochemistry* **1994**, *36*, 971–974. [CrossRef]

45. Ziaei, A.; Ramezani, M.; Wright, L.; Paetz, C.; Schneider, B.; Amirghofran, Z. Identification of spathulenol in *Salvia mirzayanii* and the immunomodulatory effects. *Phytother. Res.* **2010**, *25*, 557–562. [CrossRef] [PubMed]
46. Topcu, G.; Ulubelen, A.; Tam, T.C.M.; Chun, T.-C. Sesterterpenes and other constituents of *Salvia yosgadensis*. *Phytochemistry* **1996**, *42*, 1089–1092. [CrossRef]
47. Sajjadi, S.-E.; Ghannadi, A. Essential oil of the Persian sage, *Salvia rhytidea* Benth. *Acta Pharm.* **2005**, *55*, 321–326. [PubMed]
48. Sousa, R.M.F.; de Morais, S.A.L.; Vieira, R.B.K.; Napolitano, D.R.; Guzman, V.B.; Moraes, T.S.; Cunha, L.C.S.; Martins, C.H.G.; Chang, R.; de Aquino, F.J.T.; et al. Chemical composition, cytotoxic, and antibacterial activity of the essential oil from *Eugenia calycina* Cambess. leaves against oral bacteria. *Ind. Crops Prod.* **2015**, *65*, 71–78. [CrossRef]
49. Costa, D.P.; Santos, S.C.; Seraphin, J.C.; Ferri, P.H. Seasonal variability of essential oils of *Eugenia uniflora* leaves. *J. Braz. Chem. Soc.* **2009**, *20*, 1287–1293. [CrossRef]
50. do Nascimento, K.F.; Moreira, F.M.F.; Alencar Santos, J.; Kassuya, C.A.L.; Croda, J.H.R.; Cardoso, C.A.L.; Vieira, M.d.C.; Góis Ruiz, A.L.T.; Ann Foglio, M.; de Carvalho, J.E.; et al. Antioxidant, anti-inflammatory, antiproliferative and antimycobacterial activities of the essential oil of *Psidium guineense* Sw. and spathulenol. *J. Ethnopharmacol.* **2018**, *210*, 351–358. [CrossRef]
51. Khadhri, A.; El Mokni, R.; Almeida, C.; Nogueira, J.M.F.; Araújo, M.E.M. Chemical composition of essential oil of *Psidium guajava* L. growing in Tunisia. *Ind. Crops Prod.* **2014**, *52*, 29–31. [CrossRef]
52. Marques, F.A.; Wendler, E.P.; Sales Maia, B.H.L.N.; Coffani-Nunes, J.V.; Campana, J.; Guerrero, P.G. Volatile oil of *Psidium cattleianum* Sabine from the Brazilian Atlantic Forest. *J. Essent. Oil Res.* **2008**, *20*, 519–520. [CrossRef]
53. Martins, F.T.; Doriguetto, A.C.; de Souza, T.C.; de Souza, K.R.D.; dos Santos, M.H.; Moreira, M.E.C.; Barbosa, L.C.A. Composition, and anti-inflammatory and antioxidant activities of the volatile oil from the fruit peel of *Garcinia brasiliensis*. *Chem. Biodivers.* **2008**, *5*, 251–258. [CrossRef]
54. Chao, L.K.; Hua, K.-F.; Hsu, H.-Y.; Cheng, S.-S.; Liu, J.-Y.; Chang, S.-T. Study on the antiinflammatory activity of essential oil from leaves of *Cinnamomum osmophloeum*. *J. Agric. Food Chem.* **2005**, *53*, 7274–7278. [CrossRef]
55. Aguilar-Guadarrama, A.B.; Rios, M.Y. Three new sesquiterpenes from *Croton arboreous*. *J. Nat. Prod.* **2004**, *67*, 914–917. [CrossRef] [PubMed]
56. de Lima, S.G.; Medeiros, L.B.P.; Cunha, C.N.L.C.; da Silva, D.; de Andrade, N.C.; Neto, J.M.M.; Lopes, J.A.D.; Steffen, R.A.; Araújo, B.Q.; Reis, F.d.A.M. Chemical composition of essential oils of *Croton hirtus* L'Her from Piauí (Brazil). *J. Essent. Oil Res.* **2012**, *24*, 371–376. [CrossRef]
57. Ghimire, B.K.; Yoo, J.H.; Yu, C.Y.; Chung, I.-M. GC–MS analysis of volatile compounds of *Perilla frutescens* Britton var. *Japonica accessions: Morphological and seasonal variability*. *Asian Pac. J. Trop. Med.* **2017**, *10*, 643–651. [CrossRef]
58. Yang, C.H.; Cheng, M.J.; Lee, S.J.; Yang, C.W.; Chang, H.S.; Chen, I.S. Secondary metabolites and cytotoxic activities from the stem bark of *Zanthoxylum nitidum*. *Chem. Biodivers.* **2009**, *6*, 846–857. [CrossRef]
59. Dos Santos, E.; Radai, J.A.S.; do Nascimento, K.F.; Formagio, A.S.N.; de Matos Balsalobre, N.; Ziff, E.B.; Castelon Konkiewitz, E.; Kassuya, C.A.L. Contribution of spathulenol to the anti-nociceptive effects of *Psidium guineense*. *Nutr. Neurosci.* **2020**, *25*, 812–822. [CrossRef]
60. Yılar, M.; Bayar, Y.; Bayar, A.; Genc, N. Chemical composition of the essential oil of *Salvia bracteata* Banks and the biological activity of its extracts: Antioxidant, total phenolic, total flavonoid, antifungal and allelopathic effects. *Bot. Serbica* **2020**, *44*, 71–79. [CrossRef]
61. Fernandes, F.H.; da Guterres, Z.d.R.; Violante, I.M.P.; Lopes, T.F.S.; Garcez, W.S.; Garcez, F.R. Evaluation of mutagenic and antimicrobial properties of brown propolis essential oil from the Brazilian Cerrado biome. *Toxicol. Rep.* **2015**, *2*, 1482–1488. [CrossRef]
62. Bougatsos, C.; Ngassapa, O.; Runyoro, D.K.B.; Chinou, I.B. Chemical composition and in vitro antimicrobial activity of the essential oils of two *Helichrysum* species from Tanzania. *Z. Für Naturforschung C* **2004**, *59*, 368–372. [CrossRef]
63. Rahman, A.; Sultana Shanta, Z.; Rashid, M.A.; Parvin, T.; Afrin, S.; Khodeza Khatun, M.; Sattar, M.A. In vitro antibacterial properties of essential oil and organic extracts of *Premna integrifolia* Linn. *Arab. J. Chem.* **2016**, *9*, S475–S479. [CrossRef]
64. Klokova, M.V.; Khan, V.A.; Dubovenko, Z.V.; Pentegova, V.A.; Berezovskaya, T.P.; Serykh, E.A. Terpenoids of the essential oil of *Ledum palustre*. *Chem. Nat. Compd.* **1983**, *19*, 278–281. [CrossRef]
65. Parmena, D.; Mgina, C.; Joseph, C. Composition of non volatile oils and antimicrobial activities of extracts from *Monanthotaxis discolor*, and an undescribed *Uvariondedron* species. *Tanz. J. Sci.* **2012**, *38*, 221–231.
66. Collins, T.; Jones, G.; Sadgrove, N. Volatiles from the rare Australian desert plant *Prostanthera centralis* B.J.Conn (Lamiaceae): Chemical composition and antimicrobial activity. *Agriculture* **2014**, *4*, 308–316. [CrossRef]
67. Jesionek, A.; Poblocka-Olech, L.; Zabiegala, B.; Bucinski, A.; Krauze-Baranowska, M.; Luczkiewicz, M. Validated HPTLC method for determination of ledol and alloaromadendrene in the essential oil fractions of *Rhododendron tomentosum* plants and in vitro cultures and bioautography for their activity screening. *J. Chromatogr. B* **2018**, *1086*, 63–72. [CrossRef]
68. Goldsby, G.; Burke, B.A. Sesquiterpene lactones and a sesquiterpene diol from Jamaican *Ambrosia peruviana*. *Phytochemistry* **1987**, *26*, 1059–1063. [CrossRef]

69. Phongmaykin, J.; Kumamoto, T.; Ishikawa, T.; Suttisri, R.; Saifah, E. A new sesquiterpene and other terpenoid constituents of *Chisocheton penduliflorus*. *Arch. Pharmacal Res.* **2008**, *31*, 21–27. [CrossRef]
70. Smith, H.G. Note on the sesquiterpene of *Eucalyptus* oils. *J. Proc. R. Soc. N. S. W.* **1901**, *35*, 124–126. [CrossRef]
71. Li, H.-Y.; Yang, W.-Q.; Zhou, X.-Z.; Shao, F.; Shen, T.; Guan, H.-Y.; Zheng, J.; Zhang, L.-M. Antibacterial and antifungal sesquiterpenoids: Chemistry, resource, and activity. *Biomolecules* **2022**, *12*, 1271. [CrossRef]
72. Azfal, M.; Zeshan, B.; Ishaq, M.; Batool, S.; Arkmal, Z. Biochemical profiling and HPLC directed antimicrobial and antioxidant activities of *Achyranthes aspera*. *Plant Cell Biotechnol. Mol. Biol.* **2021**, *22*, 19–34.
73. Dai, Y.-L.; Li, Y.; Wang, Q.; Niu, F.-J.; Li, K.-W.; Wang, Y.-Y.; Wang, J.; Zhou, C.-Z.; Gao, L.-N. Chamomile: A review of its traditional uses, chemical constituents, pharmacological activities and quality control studies. *Molecules* **2022**, *28*, 133. [CrossRef]
74. Yousef, H. Insecticidal activity of the wild plant *Conyza dioscoridis* (L.) against the cotton pink bollworm *Pectinophora gossypiella* (Saunders) (Lepidoptera: Gelechiidae). *Egypt. J. Biol. Pest. Control* **2015**, *25*, 389.
75. Endris, Y.A.; Mekonnen, K.D. Formulation of neem leaf and croton seed essential oils as a natural insecticide tested on mosquitoes and cockroaches. *ACS Omega* **2023**, *8*, 15052–15061. [CrossRef] [PubMed]
76. Brophy, J.J.; Lassak, E.V.; Boland, D.J. The leaf essential oils of *Eucalyptus nova-anglica* Deane & Maiden. *J. Essent. Oil Res.* **1992**, *4*, 29–32. [CrossRef]
77. Sahi, N.M. Evaluation of insecticidal activity of bioactive compounds from *Eucalyptus citriodora* against *Tribolium castaneum*. *Int. J. Pharmacogn. Phytochem. Res.* **2016**, *8*, 1256–1270.
78. Arunmathi, C.; Malarvili, T. Analysis of bioactive compounds in methanol extract of *Aplotaxis auriculata* rhizome using GC-MS. *J. Pharmacogn. Phytochem.* **2017**, *6*, 243–247.
79. Zhang, H.; Han, T.; Yu, C.-H.; Jiang, Y.-P.; Peng, C.; Ran, X.; Qin, L.-P. Analysis of the chemical composition, acute toxicity and skin sensitivity of essential oil from rhizomes of *Ligusticum chuanxiong*. *J. Ethnopharmacol.* **2012**, *144*, 791–796. [CrossRef]
80. Vats, S.; Gupta, T. Evaluation of bioactive compounds and antioxidant potential of hydroethanolic extract of *Moringa oleifera* Lam. from Rajasthan, India. *Physiol. Mol. Biol. Plants* **2017**, *23*, 239–248. [CrossRef]
81. Raj, M.S.A.; Santhi, V.P.; Amalraj, S.; Murugan, R.; Gangapriya, P.; Pragadheesh, V.S.; Sundaresan, V.; Gurav, S.S.; Paramaguru, P.; Arulmozhian, R.; et al. A comparative analysis of leaf essential oil profile, in vitro biological properties and in silico studies of four Indian Guava (*Psidium guajava* L.) cultivars, a promising source of functional food. *S. Afr. J. Bot.* **2023**, *153*, 357–369. [CrossRef]
82. Mohamed, A.; Mustafa, A. Gas chromatography-mass spectrometry (GC-MS) analysis of essential oil *Salvia officinalis* in Sudan. *J. Multidis. Res. Rev.* **2019**, *1*, 43–45.
83. Jayaprakash, A.; Johns, A.E.; Haneef, F.K.; Radhamany, P.M. GC-MS analysis and in silico molecular docking studies of anti-inflammatory compounds from *Thottea barberi* (Gamble) Ding Hou root. *Med. Plants Int. J. Phytomed. Relat. Ind.* **2019**, *11*, 286–291. [CrossRef]
84. Khasanah, L.U.; Praseptiangga, D.; Purwanto, E.; Ariviani, S. Bioactive components and bioactivity of essential oils, hydrosol and water steam distillation solvents of lemongrass leaves (*Cymbopogon citratus*). *IOP Conf. Ser. Earth Environ. Sci.* **2024**, *1377*, 012059. [CrossRef]
85. Miranda, F.M.; Braga do Nascimento Junior, B.; Aguiar, R.M.; Pereira, R.S.; De Oliveira Teixeira, A.; De Oliveira, D.M.; Lima, E.d.O.; Oigman, S.S.; de Rezende, C.M.; Froldi, G. Promising antifungal activity of *Croton tricolor* stem essential oil against Candida yeasts. *J. Essent. Oil Res.* **2019**, *31*, 223–227. [CrossRef]
86. Elhassan, I.A.; Elamin, E.E.; Ayoub, S.M.H. Chemical composition of essential oil in dried fruits of *Xylopia aethiopica* from Sudan. *Open Access J. Med. Aromat. Plants* **2010**, *1*, 24–28.
87. Gathara, M.; Kamondo, B.; Mwadalu, R.; Omondi, S.; Kivuti, M.; Wairagu, N.; Oduor, T.; Gachie, P. Influence of soil nutrients, tree age, and sandalwood provenances on sandalwood oil yield and quality. *Int. J. For. Res.* **2022**, *2022*, 1577979. [CrossRef]
88. Tofighi, Z.; Alipour, F.; Yassa, N.; Hadjiakhoondi, A.; Hadavinia, H.; Goodarzy, S.; Golestani, R. Chemical composition and antioxidant activity of *Otostegia persica* essential oil from Iran. *Int. J. Essen. Oil Ther.* **2009**, *3*, 45–48.
89. Biona, T.; Singh, N.B.; Keithellakpam, O.S.; Srinivasan, S.; Patil, S.; Mukherjee, P.K.; Sharma, N. Headspace SPME-GC-MS analysis and in silico molecular docking studies of phytochemical compounds present in *Houttuynia cordata* Thunb. *Indian J. Pharm. Educ. Res.* **2024**, *58*, 870–881. [CrossRef]
90. Tian, G.-h. Chemical constituents in essential oils from *Elsholtzia ciliata* and their antimicrobial activities. *Chin. Herb. Med.* **2013**, *5*, 104–108.
91. Connolly, J.D.; Hill, R.A. *Dictionary of Terpenoids*, 1st ed.; Chapman & Hall: London, UK, 1991.
92. Ramalakshmi, S.; Muthuchelian, K. Analysis of bioactive constituents from the leaves of *Mallotus tetracoccus* (roxb.) Kurz, by gas chromatography-mass spectrometry. *Int. J. Pharm. Sci. Res.* **2011**, *2*, 1449.
93. Demetzos, C.; Angelopoulou, D.; Perdetzoglou, D. A comparative study of the essential oils of *Cistus salviifolius* in several populations of Crete (Greece). *Biochem. Syst. Ecol.* **2002**, *30*, 651–665. [CrossRef]

94. Rameshkumar, K.B.; George, V.; Shiburaj, S. Chemical constituents and antibacterial activity of the leaf oil of *Cinnamomum chemungianum* Mohan et Henry. *J. Essent. Oil Res.* **2007**, *19*, 98–100. [CrossRef]
95. Cartuche, L.; Vallejo, C.; Castillo, E.; Cumbicus, N.; Morocho, V. Chemical profiling of *Drimys granadensis* (Winteraceae) essential oil, and their antimicrobial, antioxidant, and anticholinesterase properties. *Plants* **2024**, *13*, 1806. [CrossRef]
96. Guo, L.; Wu, J.-z.; Han, T.; Cao, T.; Rahman, K.; Qin, L.-p. Chemical composition, antifungal and antitumor properties of ether extracts of *Scapania verrucosa* Heeg. and its endophytic fungus *Chaetomium fusiforme*. *Molecules* **2008**, *13*, 2114–2125. [CrossRef] [PubMed]
97. Bourhia, M.; Alyousef, A.A.; Doumane, G.; Saghrouchni, H.; Giesy, J.P.; Ouahmane, L.; Gueddari, F.E.; Al-Sheikh, Y.A.; Aboul-Soud, M.A.M. Volatile constituents in essential oil from leaves of *Withania adpressa* Coss. Ex exhibit potent antioxidant and antimicrobial properties against clinically-relevant pathogens. *Molecules* **2023**, *28*, 2839. [CrossRef] [PubMed]
98. Mukai, A.; Takahashi, K.; Ashitani, T. Antifungal activity of longifolene and its autoxidation products. *Eur. J. Wood Wood Prod.* **2018**, *76*, 1079–1082. [CrossRef]
99. Liu, J.; Cheng, N.; Li, Y.; Jiang, X.; Luo, Q.; Wu, Z. Protective role of agarwood essential oil and compounds thereof in crop-associated fungal pathogens. *J. Phytopathol.* **2024**, *172*, e13251. [CrossRef]
100. Ma, G.-H.; Chen, K.-X.; Zhang, L.-Q.; Li, Y.-M. Advance in biological activities of natural guaiane-type sesquiterpenes. *Med. Chem. Res.* **2019**, *28*, 1339–1358. [CrossRef]
101. Chakraborty, K.; Lipton, A.P.; Paulraj, R.; Chakraborty, R.D. Guaiane sesquiterpenes from seaweed *Ulva fasciata* Delile and their antibacterial properties. *Eur. J. Med. Chem.* **2010**, *45*, 2237–2244. [CrossRef]
102. Sun, C.; Zeng, R.; Chen, T.; Yang, Y.; Song, Y.; Li, Q.; Cheng, J.; Liu, B. Recent advances and challenges in the production of hydroxylated natural products using microorganisms. *Fermentation* **2024**, *10*, 604. [CrossRef]
103. Gijsen, H.J.M.; Wijnberg, J.B.P.A.; De Groot, A. Structure, occurrence, biosynthesis, biological activity, synthesis, and chemistry of aromadendrane sesquiterpenoids. In *Fortschritte der Chemie Organischer Naturstoffe/Progress in the Chemistry of Organic Natural Products*; Herz, W., Kirby, G.W., Moore, R.E., Steglich, W., Tamm, C., Eds.; Springer Vienna: Vienna, Austria, 1995; pp. 149–193. [CrossRef]
104. Yan, X.; Lin, J.; Liu, Z.; David, S.D.; Liang, D.; Nie, S.; Ge, M.; Xue, Z.; Li, W.; Qiao, J. The recent progress of tricyclic aromadendrene-type sesquiterpenoids: Biological activities and biosynthesis. *Biomolecules* **2024**, *14*, 1133. [CrossRef]
105. Dzul-Beh, A.d.J.; García-Sosa, K.; Uc-Cachón, A.H.; Bórquez, J.; Loyola, L.A.; Barrios-García, H.B.; Peña-Rodríguez, L.M.; Molina-Salinas, G.M. In vitro growth inhibition and bactericidal activity of spathulenol against drug-resistant clinical isolates of *Mycobacterium tuberculosis*. *Rev. Bras. Farmacogn.* **2019**, *29*, 798–800. [CrossRef]
106. Satyal, P.; Chhetri, B.K.; Dosoky, N.S.; Poudel, A.; Setzer, W.N. Chemical composition of *Nardostachys grandiflora* rhizome oil from Nepal—A contribution to the chemotaxonomy and bioactivity of *Nardostachys*. *Nat. Prod. Commun.* **2015**, *10*, 1067–1070. [CrossRef]
107. Kalidindi, S. *Studies Towards Synthesis of Biologically Active Guaianolides: Enantioselective Total Synthesis of (+)-Arglabin*; Universität Regensburg: Regensburg, Germany, 2010.
108. Tran, D.N.; Cramer, N. Biomimetic synthesis of (+)-ledene, (+)-viridiflorol, (−)-palustrol, (+)-spathulenol, and psiguadial A, C, and D via the platform terpene (+)-bicyclogermacrene. *Chem. Eur. J.* **2014**, *20*, 10654–10660. [CrossRef] [PubMed]
109. Peng, G.-P.; Tian, G.; Huang, X.-F.; Lou, F.-C. Guaiane-type sesquiterpenoids from *Alisma orientalis*. *Phytochemistry* **2003**, *63*, 877–881. [CrossRef] [PubMed]
110. Shu, Z.; Pu, J.; Chen, L.; Zhang, Y.; Rahman, K.; Qin, L.; Zheng, C. *Alisma orientale*: Ethnopharmacology, phytochemistry and pharmacology of an important traditional Chinese medicine. *Am. J. Chin. Med.* **2016**, *44*, 227–251. [CrossRef] [PubMed]
111. Li, H.-M.; Fan, M.; Xue, Y.; Peng, L.-Y.; Wu, X.-D.; Liu, D.; Li, R.-T.; Zhao, Q.-S. Guaiane-type sesquiterpenoids from *Alismatis* rhizoma and their anti-inflammatory activity. *Chem. Pharm. Bull.* **2017**, *65*, 403–407. [CrossRef] [PubMed]
112. Feng, L.; Liu, T.T.; Huo, X.K.; Tian, X.G.; Wang, C.; Lv, X.; Ning, J.; Zhao, W.Y.; Zhang, B.J.; Sun, C.P.; et al. *Alisma* genus: Phytochemical constituents, biosynthesis, and biological activities. *Phytother. Res.* **2020**, *35*, 1872–1886. [CrossRef]
113. Barreto, I.C.; de Almeida, A.S.; Sena Filho, J.G. Taxonomic insights and its type cyclization correlation of volatile sesquiterpenes in *Vitex* species and potential source insecticidal compounds: A review. *Molecules* **2021**, *26*, 6405. [CrossRef]
114. Süfer, Ö.; Ceylan, A.; OnbaŞili, D.; Çelik Yuvali, G.; Bozok, F. Chemical compounds and biological activity of Turkish *Santolina chamaecyparissus* L. Essential oil by microwave assisted distillation. *Kastamonu Univ. Orman. Fak. Derg.* **2021**, *21*, 165–175. [CrossRef]
115. Suresh, B.; Sriram, S.; Dhanaraj, S.A.; Elango, K.; Chinnaswamy, K. Anticandidal activity of *Santolina chamaecyparissus* volatile oil. *J. Ethnopharmacol.* **1997**, *55*, 151–159. [CrossRef]
116. Djeddi, S.; Djebile, K.; Hadjbourega, G.; Achour, Z.; Argyropoulou, C.; Skaltsa, H. In vitro antimicrobial properties and chemical composition of *Santolina chamaecyparissus* essential oil from Algeria. *Nat. Prod. Commun.* **2012**, *7*, 937–940. [CrossRef]
117. Chirane, M.S.; Benchabane, O.; Bousbia, N.; Zenia, S. Antioxydant and antimicrobial activities of essential oil and ethanol extract of *Santolina chamaecyparissus* L. *Rev. Agrobiol.* **2019**, *9*, 1660–1668.

118. Khubeiz, M.J.; Mansour, G. In vitro antifungal, antimicrobial properties and chemical composition of *Santolina chamaecyparissus* essential oil in Syria. *Int. J. Toxicol. Pharm. Res.* **2016**, *8*, 11.
119. Aourach, M.; Barbero, G.F.; González de Peredo, A.V.; Diakite, A.; El Boukari, M.; Essalmani, H. Composition and antifungal effects of aqueous extracts of *Cymbopogon citratus, Laurus nobilis* and *Santolina chamaecyparissus* on the growth of *Fusarium oxysporum* f. sp. *lentis*. *Arch. Phytopathol. Pflanzenschutz* **2021**, *54*, 2141–2159. [CrossRef]
120. Diánez, F.; Santos, M.; Parra, C.; Navarro, M.J.; Blanco, R.; Gea, F.J. Screening of antifungal activity of 12 essential oils against eight pathogenic fungi of vegetables and mushroom. *Lett. Appl. Microbiol.* **2018**, *67*, 400–410. [CrossRef] [PubMed]
121. Medeiros Araújo, M.B. *Species of Fusarium Causing Peduncular Rot in Melon in Brazil and Alternative Management Methods*; Universidade Federal Rural do Semi-Árido: Mossoró, Brazil, 2020.
122. de Medeiros, A.S.; Costa Alves, T.R.; dos Santos Silva, J.L.; de Moura, A.P.; de Lima, J.S.S.; de Souza, J.J.F.; de França, M.A.V.; da Costa Fernandes, J.; Bezerra Evangelista, L.F.; de Queiroz Ambrósio, M.M. Thermotherapy combined with alternative products in the management of melon rot caused by *Fusarium falciforme*. *Eur. J. Plant Pathol.* **2023**, *170*, 869–882. [CrossRef]
123. Srihom, C.; Boonyuen, N.; Khewkhom, N.; Leesutthiphonchai, W.; Nuankaew, S.; Suetrong, S.; Chuaseeharonnachai, C.; Piasai, O. Potential of herb crude extracts against Thai isolates of *Fusarium* wilt pathogens. *Curr. Res. Environ. Appl. Mycol.* **2021**, *11*, 570–584. [CrossRef]
124. Dutta, P.; Sarma, N.; Saikia, S.; Gogoi, R.; Begum, T.; Lal, M. Pharmacological activity of *Trachyspermum ammi* L. seeds essential oil grown from Northeast India. *J. Essent. Oil Bear. Plants* **2022**, *24*, 1373–1388. [CrossRef]
125. Begum, T.; Gogoi, R.; Sarma, N.; Pandey, S.K.; Lal, M. Direct sunlight and partial shading alter the quality, quantity, biochemical activities of *Kaempferia parviflora* Wall., ex Baker rhizome essential oil: A high industrially important species. *Ind. Crops Prod.* **2022**, *180*, 114765. [CrossRef]
126. Pandey, S.K.; Gogoi, R.; Bhandari, S.; Sarma, N.; Begum, T.; Munda, S.; Lal, M. A comparative study on chemical composition, pharmacological potential and toxicity of *Pogostemon cablin* Linn., (Patchouli) flower and leaf essential oil. *J. Essent. Oil Bear. Plants* **2022**, *25*, 160–179. [CrossRef]
127. Krumova, E.; Nikolova, M.; Miteva-Staleva, J.; Kostadinova, N.; Abrashev, R.; Dishliyska, V.; Berkov, S.; Mutafova, B.; Angelova, M. Bio-efficacy of the essential oil isolated from *Origanum vulgare* subsp. *Hirtum* against fungal pathogens of potato. *Comptes Rendus Acad. Bulg. Des Sci.* **2021**, *74*, 1571. [CrossRef]
128. Sarma, N.; Gogoi, R.; Begum, T.; Lal, M.; Perveen, K.; Alsahikh, N.A.; Alsulami, J.A. A study on the chemical profile of cultivated chamomile (*Matricaria chamomilla* L.) flower essential oil from North East India with special emphasis on its pharmacological importance. *J. Essent. Oil Bear. Plants* **2023**, *26*, 745–760. [CrossRef]
129. Wiart, C.; Kathirvalu, G.; Raju, C.S.; Nissapatorn, V.; Rahmatullah, M.; Paul, A.K.; Rajagopal, M.; Sathiya Seelan, J.S.; Rusdi, N.A.; Lanting, S.; et al. Antibacterial and antifungal terpenes from the medicinal angiosperms of Asia and the Pacific: Haystacks and gold needles. *Molecules* **2023**, *28*, 3873. [CrossRef] [PubMed]
130. Arendrup, M.C.; Cuenca-Estrella, M.; Lass-Flörl, C.; Hope, W. EUCAST technical note on the EUCAST definitive document EDef 7.2: Method for the determination of broth dilution minimum inhibitory concentrations of antifungal agents for yeasts EDef 7.2 (EUCAST-AFST). *Clin. Microbiol. Infect.* **2012**, *18*, E246–E247. [CrossRef] [PubMed]
131. Shapiro, S.S.; Wilk, M.B. An analysis of variance test for normality (complete samples). *Biometrika* **1965**, *52*, 591–611. [CrossRef]
132. Brown, M.B.; Forsythe, A.B. Robust tests for the equality of variances. *J. Am. Stat. Assoc.* **1974**, *69*, 364–367. [CrossRef]
133. Kruskal, W.H.; Wallis, W.A. Use of ranks in one-criterion variance analysis. *J. Am. Stat. Assoc.* **1952**, *47*, 583–621. [CrossRef]
134. Sánchez-Hernández, E.; González-García, V.; Palacio-Bielsa, A.; Casanova-Gascón, J.; Navas-Gracia, L.M.; Martín-Gil, J.; Martín-Ramos, P. Phytochemical constituents and antimicrobial activity of *Euphorbia serrata* L. extracts for *Borago officinalis* L. crop protection. *Horticulturae* **2023**, *9*, 652. [CrossRef]

plants

MDPI

Article

Antifungal Activity of Genistein Against Phytopathogenic Fungi *Valsa mali* Through ROS-Mediated Lipid Peroxidation

Fangjie Li [1,2,†], Chen Yang [3,†], Maoye Li [3], Su Liu [3], Kuo Xu [1,2,*] and Xianjun Fu [1,2,*]

1 Research Institute for Marine Traditional Chinese Medicine (Qingdao Academy of Chinese Medical Sciences), The SATCM's Key Unit of Discovering and Developing New Marine TCM Drugs, Key Laboratory of Marine Traditional Chinese Medicine in Shandong Universities, Shandong University of Traditional Chinese Medicine, Jinan 250355, China; lifj819007@163.com

2 Shandong University of Traditional Chinese Medicine Qingdao Academy of Chinese Medical Sciences, Qingdao Key Laboratory of Research in Marine Traditional Chinese Medicine, Qingdao Key Technology Innovation Center of Marine Traditional Chinese Medicine's Deep Development and Industrialization, Qingdao 266114, China

3 Anhui Provincial Key Laboratory of Integrated Pest Management on Crops, School of Plant Protection, Anhui Agricultural University, Hefei 230036, China; chenyang199874@163.com (C.Y.); limaoye81@ahau.edu.cn (M.L.); suliu@ahau.edu.cn (S.L.)

* Correspondence: xukuoworld@126.com (K.X.); xianxiu@hotmail.com (X.F.)

† These authors contributed equally to this work.

Abstract: *Valsa mali* (*V. mali*) is a necrotrophic fungus responsible for apple Valsa canker, which significantly diminishes apple production yields and quality in China. Our serendipitous findings revealed that genistein significantly inhibits the mycelial growth of *V. mali*, with an inhibition rate reaching 42.36 ± 3.22% at a concentration of 10 μg/mL. Scanning electron microscopy analysis revealed that genistein caused significant changes in the structure of *V. mali*, including mycelial contraction, distortion, deformity, collapse, and irregular protrusions. Transmission electron microscopy analysis revealed leakage of cellular contents, blurred cell walls, ruptured membranes, and organelle abnormalities. Genistein has been shown to increase reactive oxygen species levels in *V. mali* mycelia, as demonstrated by 2′,7′-dichlorofluorescin diacetate staining. This increase was associated with a decrease in superoxide dismutase activity alongside increases in catalase and peroxidase activities. These changes collectively disrupted the oxidative equilibrium, leading to the induction of oxidative stress. The transcriptomic analysis revealed 13 genes enriched in this process, linked to unsaturated fatty acid biosynthesis (three downregulated DEGs), saturated fatty acid biosynthesis (three upregulated and six downregulated DEGs), and fatty acid metabolism (four upregulated and nine downregulated DEGs). Additionally, the downregulated DEGs VMIG_07417 and VMIG_08675, which are linked to ergosterol biosynthesis, indicate possible changes in membrane composition. In conjunction with the qRT-PCR results, it is hypothesized that genistein exerts an antifungal effect on *V. mali* through ROS-mediated lipid peroxidation. This finding has the potential to contribute to the development of novel biological control agents for industrial crops.

Keywords: *Valsa mali*; phytopathogenic; fungi; pathogenic; genistein

1. Introduction

Phytopathogenic fungi pose a significant threat to global food security, ecosystem services, and human well-being [1]. These fungi typically persist through parasitic and saprophytic mechanisms, often resulting in diseases in their host plants. Currently, crops such as apples, tomatoes, and rice are afflicted by fungal diseases caused by species including *Valsa mali* (*V. mali*) [2], *Botrytis cinerea* [3], and *Ustilaginoidea virens* [4]. Among them, *V. mali* is a necrotrophic fungus responsible for apple Valsa canker (AVC), a disease that adversely affects various parts of the apple tree, including the fruit, primary branches, and trunk. This pathogen not only reduces apple yield and quality but can also result in tree mortality and orchard devastation under severe conditions, thereby inflicting substantial losses on apple production [5]. The disease is characterized by its extensive geographical distribution, severe impact, and considerable challenges in prevention and control, earning it the moniker "cancer" of apple trees.

Several strategies have been developed for the management of *V. mali*, including the application of fungicides, excision of cankers, and pruning of necrotic or weakened branches. Benzimidazole fungicides, such as tebuconazole and thiophanate-methyl, are the most frequently employed agents for controlling *V. mali* [6,7]. Nevertheless, the increased frequency and concentration of fungicide application have inadvertently contributed to the development of pathogen resistance and have raised concerns regarding chemical residues, which pose potential risks to human health and the environment. Furthermore, once the pathogen has extensively infiltrated the phloem and xylem of apple trees, its management becomes exceedingly challenging. In conclusion, there is presently no effective and safe pesticide available for the prevention and control of *V. mali*.

The discovery of highly effective and low-toxicity natural antifungal compounds against the phytopathogenic fungus *V. mali* is of considerable importance. Researchers have conducted extensive and commendable investigations in this domain. For example, in vitro experimental findings demonstrated that 6-methylcoumarin, at a concentration of 400 mg/L, achieved an inhibition rate of 94.6% against *V. mali* [8]. Subsequent experiments revealed that 6-methylcoumarin could impede the growth of *V. mali* mycelium and spore germination, leading to damage of the cell membrane. In the inoculated apple branches, 6-methylcoumarin was observed to effectively inhibit the progression of AVC and decrease its incidence rate. Duan et al. conducted an investigation into the bactericidal activity of various tobacco organ samples, revealing that flowers exhibit greater antibacterial activity against apple rot pathogens compared to leaves [9]. Our research group unexpectedly identified that genistein, a widely occurring natural compound, exhibits a significant inhibitory effect on *V. mali*. This study is the first to document the antifungal activity of genistein against the phytopathogenic fungus *V. mali*, mediated through reactive oxygen species (ROS)-induced lipid peroxidation.

In this study, the micro double-dilution method was employed to determine the minimum inhibitory concentration (MIC) of genistein. Additionally, scanning electron microscopy (SEM) and transmission electron microscopy (TEM) were utilized to examine the morphology and ultrastructure of *V. mali* hyphae. To elucidate the antifungal mechanism of genistein, we employed 2′,7′-dichlorofluorescin diacetate (DCFH-DA) staining, assessed the activities of antioxidant protective enzymes, and utilized RNA-sequencing technology to explore the potential inhibitory mechanisms of genistein against *V. mali*. This study represents the first investigation into the inhibitory effects and molecular mechanisms of genistein on the plant-derived fungus *V. mali*.

2. Results

2.1. Inhibitory Effect of Genistein on the Growth of V. mali

In our study, different concentrations of genistein were prepared using the double-dilution method with anhydrous ethanol as the solvent, and the MIC values of genistein against *V. mali* were determined using the plate method. The results showed that as the concentration of genistein increased, the mycelial inhibition rate of *V. mali* continued to increase. When the concentration of genistein reached 10 μg/mL, no precipitation occurred in the culture medium, and the inhibition rate of genistein on *V. mali* reached 42.36 ± 3.22%, which was close to the inhibition activities of the positive control groups, α-CBT diol (48.03 ± 6.44%) and β-CBT diol (55.99 ± 3.60%) (Figure 1). Therefore, the MIC of genistein against *V. mali* was 10 μg/mL, which was used for subsequent experiments.

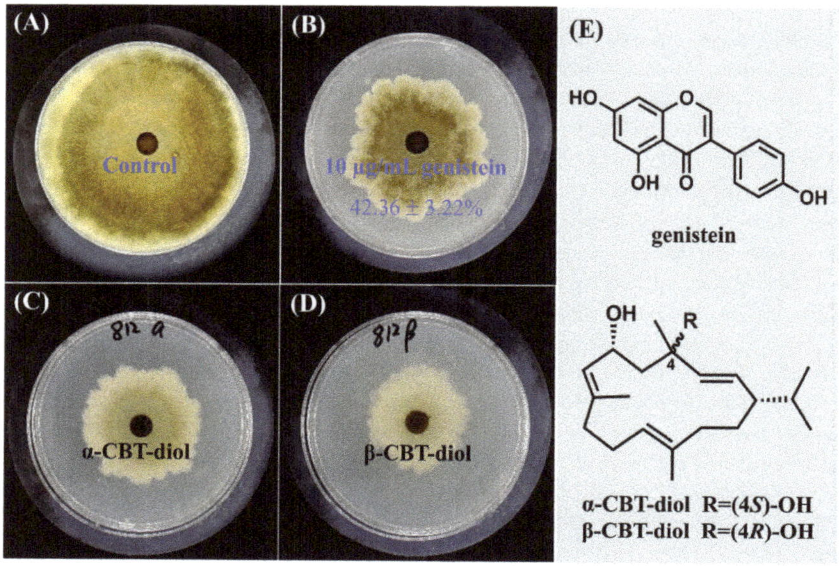

Figure 1. The inhibitory effect of genistein on *V. mali*. (**A**) Blank control; (**B**) Mycelium treated with 10 μg/mL genistein; (**C**) Mycelium treated with 10 μg/mL α-CBT-diol; (**D**) Mycelium treated with 10 μg/mL β-CBT-diol; (**E**) Structures of genistein, α-CBT-diol and β-CBT-diol.

2.2. Effects of Genistein on the Morphology and Ultrastructure of V. mali Mycelia

To further clarify the damage caused by genistein to *V. mali*, SEM and TEM analyses were performed to observe the morphology and ultrastructure of the hyphae. SEM analysis of the preprocessed *V. mali* mycelia revealed that genistein significantly affected the morphology of the mycelia. The mycelia without genistein treatment exhibited normal linear and cylindrical morphologies with a uniform thickness and smooth surface (Figure 2A). Mycelia treated with 10 μg/mL genistein exhibited significant shrinkage, distortion, deformation, and even collapse, losing their cylindrical shape, which was accompanied by irregular protuberances (Figure 2B).

Under TEM, the mycelia of the control group were neat in cross-section, uniform in thickness, and clear and smooth in terms of structure (Figure 2C,E). Organelles, such as the endoplasmic reticulum and mitochondria, were intact and evenly distributed in the cytoplasm. In contrast, the cytoplasm of the hyphae in the treatment group presented pronounced cytoplasmic fluid transformation and an increase in the number of intracellular cavities (Figure 2D,F). Notably, significant extravasation of the cellular contents, a blurry appearance of the cell walls, rupture of the cell membranes, and abnormalities in organelles were observed. These observations suggest that the cellular structure of the hyphae was severely damaged, indicating the significant antifungal effect of genistein on *V. mali* [10].

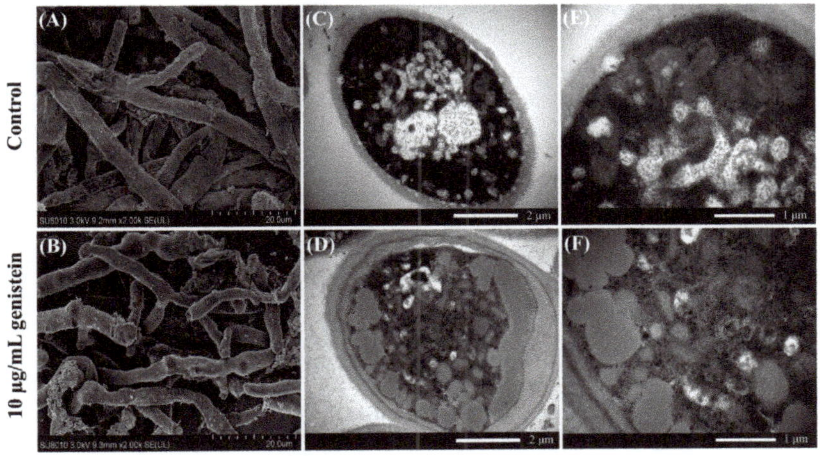

Figure 2. Effects of genistein on hyphal morphology ((**A**,**B**), 2000× magnification) and ultrastructure ((**C**,**D**), 20,000× magnification; (**E**,**F**), 40,000× magnification) of *V. mali*.

2.3. *Effects of Genistein on the Reactive Oxygen Species (ROS) Level of V. mali*

The dynamic balance between oxidation and antioxidation reactions in fungal cells is maintained to ensure normal metabolic processes. Once the balance between the two is disrupted, the permeability of the cell membrane increases, causing cell damage [11]. To further explore the mechanism of genistein against *V. mali*, we determined the content of ROS in the mycelium before and after treatment with genistein. After the mycelium was treated with genistein, the mycelium in the control group was almost completely black in the field of view, making it difficult to detect fluorescence. In contrast, more obvious green fluorescence was detected in the treatment group (Figure 3A,B). The strength of the fluorescence signal reflects the intracellular ROS level of the mycelium. These findings suggest that genistein induces the production of ROS in the mycelium, causing oxidative stress damage to the mycelium. The sudden increase in the reactive oxygen species content within the mycelium is an important indicator of lipid peroxidation in biological membranes. Therefore, in the subsequent transcriptome data analysis, we focused on the suppression of hyphal growth resulting from lipid peroxidation triggered by genistein.

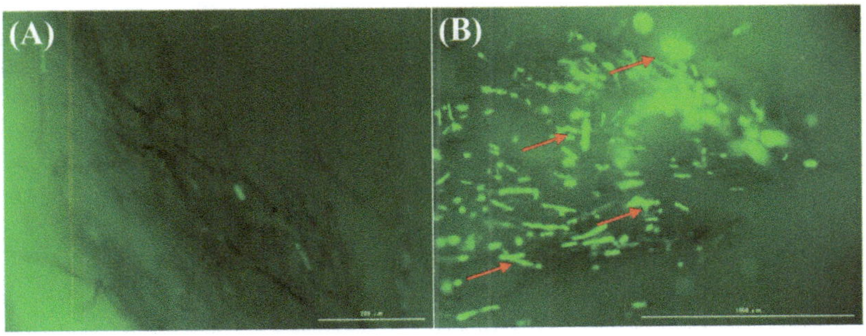

Figure 3. The effect of genistein on the production of reactive oxygen species. (**A**) Blank control; (**B**) Mycelium treated with 10 µg/mL genistein.

2.4. *Determination of Antioxidant Enzyme Activity*

Research has shown that superoxide dismutase (SOD), catalase (CAT), and peroxidase (POD) are the main protective enzymes in fungi, which resist oxidative stress damage to tissues by clearing reactive oxygen species in the fungal body [12]. In this study, after treatment with genistein, the activity of the antioxidant protective enzyme SOD decreased (Figure 4A), whereas the activities of POD (Figure 4B) and CAT increased (Figure 4C), suggesting the activation of the ROS scavenging system in the treatment group [13].

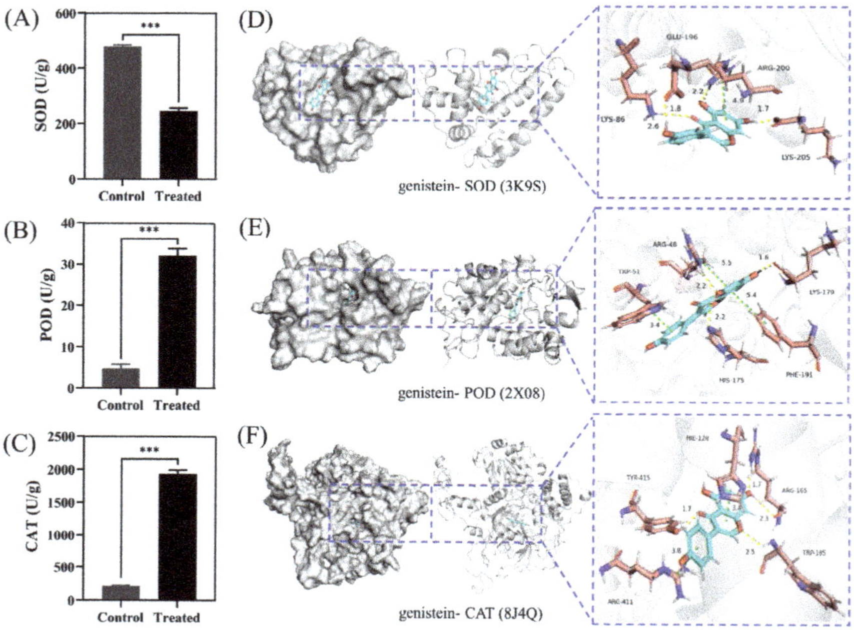

Figure 4. The effect of genistein treatment on the antioxidant enzyme of *V. mali*. (**A–C**) Activity testing of SOD, POD, and CAT enzymes. *** $p < 0.001$. (**D–F**) 3D maps of docking genistein with SOD, POD, and CAT.

Subsequently, we molecularly docked the proteins and genistein to study their binding patterns and interactions. As shown, proteins are represented by a gray surface and cartoon. Genistein is shown as cyan sticks. The key residues are shown as brick red sticks. Through docking, we found that genistein and three proteins had excellent binding energy (docking score: the smaller the value, the stronger the binding energy. Generally, a score less than −5 indicates strong binding energy), which confirms the above results. In addition, we studied the binding patterns and interactions between the compound and proteins. For SOD protein, the binding energy is −9.399 kcal/mol. Genistein could form four hydrogen bonds and one Pi-Pi interaction with amino acids on the protein, respectively. Specifically, the carbonyl oxygen atom, hydroxyl oxygen atom, and hydroxyl H atom on the skeleton core can form three hydrogen bonds with NH on LYS-86, NH on ARG-200, and carboxylic acid oxygen atom on LYS-205 at distances of 2.6, 2.2, and 1.7 Å, respectively. The H atom on the hydroxy group of phenylcyclone can form a hydrogen bond with the carbonyl oxygen atom on GLU-196 at a distance of 1.8 angstroms. In addition, the skeleton core can also form a Pi-Pi interaction with ARG-200 at a distance of 4.9 angstroms (Figure 4D). For POD protein, the binding energy is −8.727 kcal/mol. The compound can form three hydrogen bonds and three Pi-Pi interactions with amino acids on the protein, respectively. Specifically, the two hydroxyl H atoms and hydroxyl oxygen atoms on the skeleton core can form three hydrogen bonds with the carbonyl oxygen atom on LYS-179, the N atom on HIS-175, and the N atom on ARG-48 at distances of 1.6, 2.2, and 2.2 Å, respectively. In addition, the benzene ring of the skeleton core and side chain can also form two Pi-Pi interactions with ARG-48, Ph-191, and TRP-51 at distances of 5.5, 5.4, and 3.4 angstroms, respectively (Figure 4E). For CAT proteins, the binding energy is −8.350 kcal/mol. The compound can form four hydrogen bonds and two Pi-Pi interactions with amino acids on the protein, respectively. Specifically, the hydroxyl H atom, hydroxyl oxygen atom, carbonyl oxygen atom, and the O atom on the opposite side of the carbonyl group can form four hydrogen bonds with the NH and O atom on ARG-165, the hydroxyl H atom on TYR-415, and the oxygen atom on TRP-185, respectively, at distances of 1.7, 2.3, 1.7, and 2.5 angstroms. In addition, the benzene ring of the skeleton core and side chain can also form two Pi-Pi interactions with HIS-128 and ARG-411 at distances of 3.4 and 3.8 angstroms, respectively (Figure 4F).

2.5. Quality Assessment of Sequencing Data

To further investigate the changes in the transcription of genistein in *V. mali*, we collected nine samples, including untreated and treated mycelia, with three biological replicates (Figure 5A). A total of six samples were used to produce RNA-Seq reads using the Illumina NovaSeq platform (Table 1). Among all the sequencing samples, approximately seven Gb of raw reads and (obtained reads) were obtained. To ensure the quality and reliability of the data analysis, we filtered the raw data to obtain clean data, and the sequencing results from each sample had similar Q20 (>98.24%) and Q30 (>94.99%) scores and GC contents (56.59–57.02%). The sequencing error rate was only 0.02%, which was considerably lower than that of the normal condition (1%). The above data quality control results indicate that the high-quality clean data obtained can be used for subsequent analysis.

Table 1. Quality assessment of sequencing data.

Samples	Control-1	Control-2	Control-3	Treated-1	Treated-2	Treated-3
Raw reads	46,768,938	47,448,424	48,968,902	46,731,424	45,432,282	45,201,716
Raw bases	7.02 G	7.12 G	7.35 G	7.01 G	6.81 G	6.78 G
Clean reads	45,370,436	46,266,524	47,631,958	45,241,510	44,104,334	44,120,272
Clean reads/raw reads (%)	97.01%	97.51%	97.27%	96.81%	97.08%	97.61%
Clean bases	6.81 G	6.94 G	7.14 G	6.79 G	6.62 G	6.62 G
Error rate	0.02	0.02	0.02	0.02	0.02	0.02
Q20	98.53	98.34	98.42	98.39	98.36	98.36
Q30	95.79	95.25	95.47	95.44	95.33	95.32
GC pct	56.73	56.87	57.02	56.59	56.62	56.71

Note: Raw reads: the number of reads in the raw data; raw bases: the base number of raw data (raw base = raw reads × 150 bp); clean reads: the number of reads filtered from the raw data; clean bases: the number of bases filtered from the raw data (clean base = clean reads × 150 bp); error rate: overall sequencing error rate of data; Q20: percentage of bases with Phred values greater than 20 in the total number of bases; Q30: percentage of bases with Phred values greater than 30 in the total number of bases; GC pct: the percentage of G and C bases in clean reads.

2.6. Differentially Expressed Genes (DEGs) Expression Analysis

Reference genome and gene model annotation files were downloaded directly from the genome website. Using HISAT2 v2.0.5, the clean reads were aligned to the reference genome (PRJNA268126) based on the downloaded files. The results of read mapping demonstrated that all the sequencing data were of high quality because the proportion of reads (≥95.86%) aligned to the unique site of the reference genome exceeded 70%, and the number of reads (≤0.54%) aligned to numerous positions did not exceed 10% (Table 2). As a result, reads that were aligned to the reference genome's specific location were employed for subsequent quantitative data analysis.

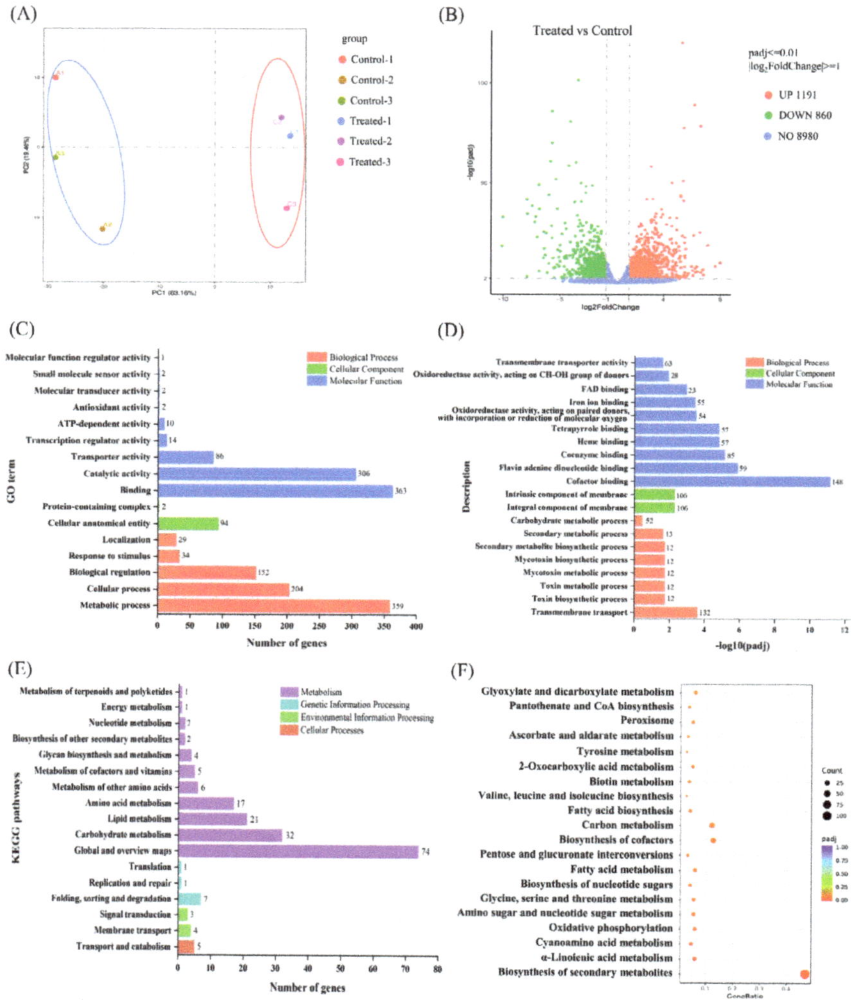

Figure 5. Transcriptome analysis of *V. mali* with genistein treatment. (**A**) PCA assays of control and treated group samples; (**B**) DEGs obtained by transcriptome assays; (**C**,**D**) GO pathway enrichment analysis based on DEGs; (**E**,**F**) KEGG pathway enrichment analysis based on DEGs.

Before the quantitative analysis of gene expression levels, the number of reads covered by each gene (including newly predicted genes) from beginning to end was calculated based on the positional information of the gene alignment on the reference genome. The majority of the filtered reads were noncontrasting pairs, reads with alignment quality values less than 10, and reads that were matched to several different genomic regions. The gene expression level of an RNA sequence is typically reported using the FPKM value rather than the read count because of the impact of sequencing depth and gene length. FPKM mapping resulted in several corrections to the gene length and sequencing depth. We then displayed the distribution of gene expression levels in various samples after determining the FPKM value of each gene in each sample. Moreover, the correlation coefficients of the intragroup and intergroup samples were calculated based on the FPKM values of all the genes in each sample, and a heatmap was drawn to visually display the differences and repetitions of the intergroup samples. In addition, all of the intragroup biological duplicate

R^2 values in this investigation were greater than 0.8 and greater than the R^2 values of the intergroup biological samples. These results set the stage for later differential gene analysis to produce more trustworthy data. After the quantification of gene expression levels, we screened for genes whose expression levels significantly differed among the three groups of samples.

Table 2. Reads mapping to the reference genome.

Samples	Total Reads	Total Map	Unique Map	Multi Map
Control-1	45,370,436	43,541,348 (95.97%)	43,327,458 (95.5%)	213,890 (0.47%)
Control-2	46,266,524	44,370,961 (95.9%)	44,121,004 (95.36%)	249,957 (0.54%)
Control-3	47,631,958	45,658,834 (95.86%)	45,424,635 (95.37%)	234,199 (0.49%)
Treated-1	45,241,510	43,476,439 (96.1%)	43,277,906 (95.66%)	198,533 (0.44%)
Treated-2	44,104,334	42,428,393 (96.2%)	42,239,746 (95.77%)	188,647 (0.43%)
Treated-3	44,120,272	42,441,929 (96.2%)	42,227,638 (95.71%)	214,291 (0.49%)

Note: Total reads: the number of clean reads of sequencing data after quality control; total map: the number and percentage of reads aligned to the genome; unique map: the number and percentage of reads aligned to the unique position of the reference genome (used for subsequent quantitative data analysis of reads); multi map: the number and percentage of reads aligned to multiple positions in the reference genome.

2.7. Gene Ontology (GO) Term and Kyoto Encyclopedia of Genes and Genomes (KEGG) Functional Enrichment Analyses of DEGs

To further investigate the variations in transcript abundance and expression patterns of genistein compared with those of *V. mali*, the Illumina RNA-Seq reads were mapped to the reference genome to determine the expression levels of the transcripts using the RPKM value. DESeq2 R package (1.20.0) was used to carry out DEG analysis based on the RNA-Seq data. A total of 923 DEGs (treated group vs. control group), including 618 upregulated genes and 305 downregulated genes in the treatment group (Figure 5B), were subsequently subjected to functional gene analysis for further exploration.

All DEGs were subjected to BLAST searches against the GO and KEGG databases using the clusterProfiler R package (3.8.1) to define their molecular functions. For the GO classification analysis, all DEGs were assigned to three main GO functional categories, namely, biological process (BP), cellular component (CC), and molecular function (MF), and were categorized into 20 main terms. As depicted in Figure 5C, the most prevalent subcategories within each of these three primary functional categories are distinctly outlined and presented separately. Among them, "binding", "metabolic process", and "catalytic activity" were the dominant terms. In the MF category, notable enrichment was observed among DEGs involved in "binding" (363 genes) and "catalytic activity" (306 genes). A total of 94 genes were enriched in the CC category "cellular anatomical entity". "Metabolic process" (359 genes), "cellular processes" (204 genes), and "biological regulation" (152 genes) were the main terms within the BP category. We subsequently analyzed all the DEGs and displayed the GO terms with the most significant enrichment in each category, as shown in Figure 5D. In the MF category, the enrichment of DEGs involved in "cofactor binding", "flavin adenine dinucleotide binding", "coenzyme binding", "heme binding", and "tetrapyrrole binding" was most significant. Moreover, 54 and 28 genes were involved in "oxidoreductase activity, acting on paired donors, with incorporation or reduction of molecular oxygen" and "oxidoreductase activity, acting on the CH-OH group of donors", respectively, which may be related to the oxidation and antioxidant processes within the mycelium. In addition, 63 genes related to "transmembrane transporter activity" were significantly enriched, suggesting that the integrity of *V. mali* cell membranes may be compromised. This speculation was also supported in the other two categories, which was consistent with the TEM results regarding cell membrane damage. For the CC category, 106 genes were actively involved in the "integral component of the membrane", whereas 106 genes participated in the "intrinsic component of the membrane". In the context of BP function, notable enrichment was observed among all DEGs involved in "transmembrane transport". Therefore, we focused on the mechanism of cell membrane damage caused by genistein in subsequent research. Moreover, DEGs involved in the "toxin biosynthetic process", "toxin metabolism process", "mycotoxin metabolic process", "mycotoxin biosynthetic process", "secondary metabolite biosynthetic process", and "secondary metabolic process" were significantly enriched, suggesting that genistein may exert its antifungal mechanism by modulating these genes, thereby mitigating the production of detrimental substances or metabolites by *V. mali*. The above findings indicate that genistein may affect transmembrane transport, redox function, and metabolic processes, which affect normal substance exchange, energy conversion within biological membranes, and metabolic activities.

KEGG pathway enrichment analysis of DEGs is a way to indicate whether there are significant differences in certain pathways. According to the functional distribution of the DEGs, they were distributed into three main groups: metabolism, genetic information processing, environmental information processing, and cellular processes. (Figure 5E). All DEGs were annotated to 74 KEGG pathways, and metabolic processes were dominant. As shown in Figure 5F, KEGG enrichment analysis revealed that the DEGs were enriched in 20 pathways. "Biosynthesis of secondary metabolites" stands as the preeminent enriched pathway, with a pivotal role in the biological process category, involving a total of 104 genes. Concurrently, "fatty acid metabolism", "fatty acid biosynthesis", and "oxidative phosphorylation" were significantly enriched, incorporating a substantial number of genes into their respective pathways.

2.8. Effects of Genistein on the Cell Membrane System of *V. mali*

Based on the results of the transcriptome analysis, our treatment with genistein resulted in a decrease in the expression levels of multiple genes involved in fungal cell membrane stability. Among them, two genes were involved in the biosynthesis of ergosterol, three genes were involved in the biosynthesis of unsaturated fatty acids, six genes were involved in the biosynthesis of saturated fatty acids, and nine genes were involved in the biosynthesis of lipid metabolism (Figure 6A). The quantitative real-time reverse transcription PCR (qRT-PCR) validation results revealed that the expression levels of 15 genes, including VM1G_04618, VM1G_04926, VM1G_05255, VM1G_07069, VM1G_07417, VM1G_07605, VM1G_08675, VM1G_09419, VM1G_09989, VM1G_11144, VM1G_07635, VM1G_05401, VM1G_10990, VM1G_02847, and VM1G_01285 were consistent with the transcriptome sequencing results (Figure 6B,E). Among them, a total of 13 DEGs were associated with lipid peroxidation, with nine genes downregulated and four genes upregulated. Moreover, genes related to the biosynthesis of unsaturated fatty acids and saturated fatty acids were also differentially regulated. A decrease in the ratio of unsaturated fatty acids to saturated fatty acids affects the fluidity and structural stability of the plasma membrane [14]. These results further indicate that the structure and function of the cell membrane may be abnormal. Therefore, we speculate that changes in the proportion of fatty acids lead to lipid peroxidation, which in turn inhibits the normal growth of hyphae. In addition, ergosterol is the main component of the cell membrane. Research has demonstrated that citral inhibits *Penicillium digitatum* by disrupting the biosynthesis of ergosterol [15]. In our study, two genes involved in ergosterol biosynthesis were significantly downregulated in the treatment group. In summary, we speculate that genistein may exert antifungal activity by disrupting the cell membrane system of *V. mali*. However, TEM analysis revealed that the hyphal cell wall incurred damage. Future research should investigate the molecular mechanisms by which genistein inhibits the plant-derived fungus *V. mali*, with particular attention paid to aspects such as cell wall damage.

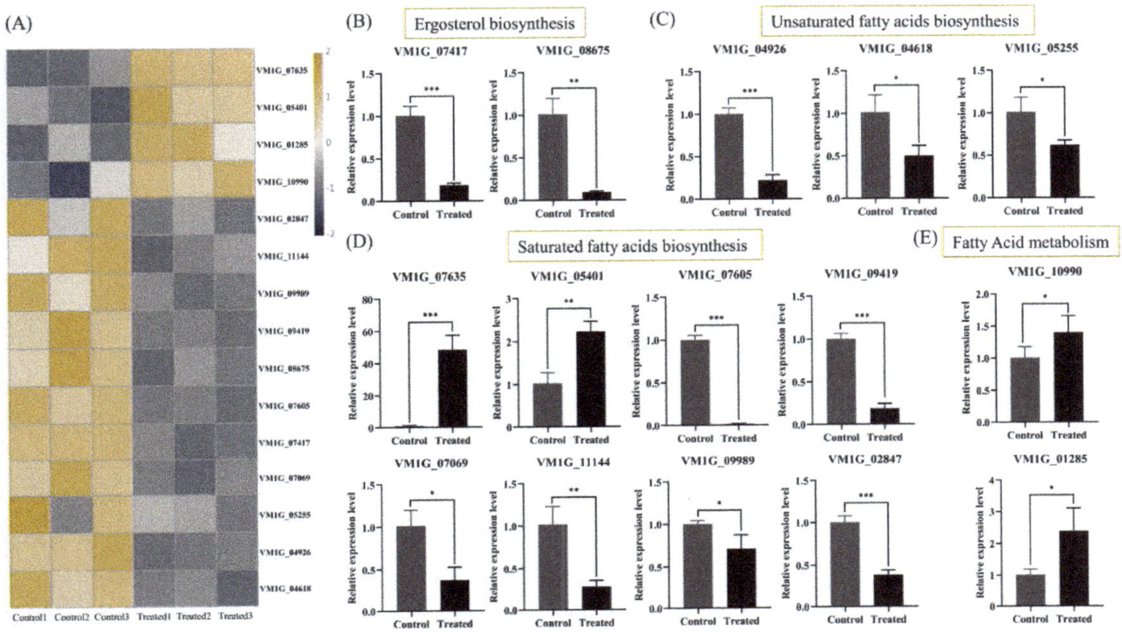

Figure 6. The regulation of the cell membrane system of *V. mali* by genistein. (**A**) A total of 15 DEGs were detected in the regulation cell membrane system by transcriptome. (**B–E**) Gene expression levels of 15 DEGs verified by qRT-PCR. * $p < 0.05$, ** $p < 0.01$, *** $p < 0.001$.

3. Discussion

V. mali is a necrotrophic fungus causing AVC, which limits apple production in China. This weakly parasitic fungus is characterized by its ability to establish a latent infection. According to reports, after infecting host tissue, *V. mali* mycelium can remain dormant at the wound site and survive in the xylem for many years [16]. In general, the various wounds caused by frostbite, sun exposure, trimming of the ends, and other mechanical injuries are the main pathways leading to *V. mali* infection in apple trees. This pathogen primarily damages apple trunks, branches, scaffolds, and fruits, significantly impacting fruit yield and quality, thereby inflicting substantial economic losses on the apple industry.

At present, the prevention and control of this disease mainly rely on measures such as scraping off the scar and applying chemical agents or spraying chemical agents across the entire tree [17]. However, *V. mali* may infect the woody parts of apple branches from cuts and other wounds and remain dormant for a long time, making it difficult for chemical agents to directly kill the pathogens [18]. In addition, long-term use of chemical pesticides may cause pathogenic fungi to develop drug resistance, leading to the deterioration of orchard ecological environments [19]. At the same time, chemical pesticides may cause a decline in fruit quality and even lead to food safety issues.

Plant-derived pesticides are safe, efficient, and residue-free "green pesticides" extracted from plants. Compared with chemical pesticides, they are easily degradable, safe for nontarget organisms, have low resistance development in harmful organisms, and exhibit a special mode of action [20]. In our previous research, we accidentally discovered that genistein has the potential to inhibit *V. mali* [21]. Genistein (4′,5,7-trihydroxyisoflavone) is a secondary metabolite of isoflavone predominantly sourced from leguminous plants. It has been identified in a variety of plant species [22–30], with a notable presence in *Glycine max* [31]. Based on data from the Web of Science (WOS) database, over 2200 studies in the past five years have documented various biological activities of genistein. The majority of these investigations have concentrated on its potential roles in the prevention and treatment of cancer [32], cardiovascular diseases [33], neurological disorders [34], and diabetes [35].

As a typical representative of plant estrogens, genistein is essentially an endocrine disruptor. At present, leguminous ingredients and soy products are considered the main sources of human intake of genistein. However, improper intake of genistein may disrupt hormone balance in the human body, especially estrogen receptor signaling, leading to a range of health problems such as reproductive dysfunction and breast cancer. Hu carried out a risk assessment on the potential harm of genistein intake to human breast health and found that improper intake of genistein had a certain negative impact on breast tissue development, breast cancer course development, and chemosensitivity [36]. At the same time, the authors also estimated that genistein intake in postmenopausal breast cancer patients poses a greater potential risk of harm to disease progression and chemotherapy treatment process; additionally, the mean risk quotient for minors on the Chinese mainland affected by dietary genistein intake on breast development was 45.56, with a 90% confidence interval range of 7.80–126.70. This shows that genistein is very harmful to the health of postmenopausal breast cancer patients and young people. In another study, the authors found that treatment with a dose of 100 mg/kg genistein in neonatal mice severely disrupted the development of ovarian and uterine structures, which is related to its inhibition of cell proliferation activity in ovarian and uterine tissues [37]. In view of the current consumption trends of legumes and the rising incidence of breast cancer and other related diseases in recent years, greater attention should be paid to the safety of legumes, related products, and even healthcare products for at-risk populations. In addition, Rocha also studied the impact of endocrine disruptor compounds on the environment. Due to signs of pollution at the mouth of the Mondego River, Rocha et al. measured the levels of endocrine disruptors in the water samples [38]. However, the results showed that more biological and toxicological analysis is still needed to confirm the estrogenic contributions of these plant estrogens and other compounds in the estuarine environment in order to fully understand their effects on local animal populations and humans.

To the best of our knowledge, there have been no reports on the inhibitory effects and mechanisms of genistein on plant pathogens. In the present study, the growth inhibition rate of *V. mali* treated with 10 μg/mL genistein reached 42.36 $\pm$ 3.22%, indicating that genistein has significant antifungal activity. SEM analysis demonstrated that genistein induced notable alterations in the structure of *V. mali*, encompassing mycelial contraction, distortion, deformity, collapse, as well as irregular protrusions. Furthermore, TEM analysis unveiled the leakage of cellular contents, blurred cell walls, membrane ruptures, and abnormalities in organelles. Through GO enrichment analysis of differentially expressed genes, we found that multiple genes related to cell membrane function were enriched in "integral component of membrane", "intrinsic component of membrane", "transmembrane transport", and "transmembrane transporter activity". Therefore, we focused on the mechanism of cell membrane damage by genistein in subsequent research. The dynamic balance between lipid peroxidation and oxygen-free radical reactions in fungal cells is maintained to support normal metabolic processes. Once the equilibrium between the two is disturbed, the permeability of the cell membrane rises, leading to cellular damage [11]. After treatment with 10 μg/mL genistein, green fluorescence in mycelia was significantly higher than that in the untreated group. This suggested that genistein caused ROS accumulation in mycelia.

Research has shown that CAT, POD, and SOD are the main protective enzymes in fungi, which resist oxidative stress damage to tissues by clearing reactive oxygen species in the fungal body. In this study, after treatment with genistein, the activity of the antioxidant protective enzyme SOD decreased. In contrast, the activities of CAT and POD increased, leading to an imbalance between oxidation and antioxidant processes in the mycelium, resulting in oxidative stress and promoting the accumulation of reactive oxygen species in the mycelium. Meanwhile, molecular docking results showed that genistein exhibited strong binding affinity with SOD(3K9S), POD(2 × 08), and CAT(8J4Q), which supported the above results. More importantly, based on the results of molecular docking, we analyzed and speculated that genistein might adapt and stably bind to the active pockets of the above proteins. This would reduce the flexibility and root mean square deviation of the proteins, alter their dynamic and thermodynamic conformations, and thereby increase the stability of the proteins [39,40]. These interactions are crucial for the stability of proteins and compounds and can significantly affect their biological functions [41,42].

The sudden increase in ROS content within the mycelium is an important indicator of lipid peroxidation in biological membranes [43]. Therefore, we focused on differentially expressed genes related to lipid peroxidation and found that 13 differentially expressed genes were associated with lipid peroxidation, with nine genes downregulated and four genes upregulated. Meanwhile, genes related to unsaturated fatty acids and saturated fatty acids are also regulated. The decrease in the ratio of unsaturated fatty acids to saturated fatty acids will affect the fluidity and structural stability of the plasma membrane. These results further indicate that the structure and function of the cell membrane may be abnormal. Therefore, we speculate that changes in the proportion of fatty acids lead to lipid peroxidation, which in turn inhibits the normal growth of hyphae. In addition, ergosterol, as the main component of the cell membrane, has significantly downregulated two genes involved in its biosynthesis. Therefore, we speculate that genistein may exert antifungal activity by triggering ROS-mediated lipid peroxidation, further disrupting the cell membrane system of *V. mali*. However, the 15 candidate DEGs we selected did not receive further experimental validation. Future research will aim to confirm these mechanisms at the molecular level.

4. Materials and Methods

4.1. Media, Pathogens and Reagents

Genistein (batch number: G1719012) was purchased from Aladdin Biochemical Technology Co., Ltd. (Shanghai, China). *V. mali* was provided by the Tobacco Research Institute of the Chinese Academy of Agricultural Sciences (Qingdao, China) [44]. Potato dextrose agar (PDA, batch number: G1719012) from Beijing Land Bridge Technology Co., Ltd. (Beijing, China) and DCFH-DA detection kits (batch number: D06IR2025SA) from Solarbio (Beijing, China) were used in this study. SOD (batch number: A001-3-1), CAT (batch number: A007-1-1), and POD (batch number: A084-3-1) kits were obtained from Nanjing Jiancheng Bioengineering Institute (Nanjing, China).

4.2. Growth Inhibition Assay of Genistein Against V. mali

The micro double-dilution method was used to determine the MIC of genistein that inhibited *V. mali*. The sterilized culture medium was added to a 96-well plate, and different concentrations of genistein (40, 20, 10, 5, 2.5, 1.25, 0.62, and 0.31 μg/mL) were added, followed by the addition of the fungal mixture [45]. Each group was analyzed in triplicate. Then, the 96-well plates were placed in a constant-temperature incubator at 25 °C for 20 h. The maximum concentration of medium without turbidity in a 96 well plate was recorded as the MIC value.

Using *V. mali* as the test strain, the effect of genistein on fungal hyphal growth was detected using the growth rate method [12]. Under sterile conditions, a 1:100 mixture of 10 μg/mL genistein and autoclaved PDA medium was poured into plates. An equal volume of 95% aqueous ethanol without genistein or other reagents was added to the control group. Each group was analyzed in triplicate. After the agar plates solidified, 9 mm long fungal cakes were obtained from the edges of single colonies, and the plates were inoculated and cultured in a 25 °C incubator. When the mycelia of the control group spread to the edge of the culture dish, the diameters of the fungal colonies were measured using the cross method. The growth inhibition rate of *V. mali* by genistein was calculated according to the following formula:

$$\text{Inhibition (\%)} = [(\text{average diameter of the control group} - \text{average diameter of the treatment group})/(\text{average diameter of the control} - 9)] \times 100\%$$

4.3. Preparation and Examination of Samples for SEM and TEM

For SEM analysis, *V. mali* was cultured on PDA medium supplemented with 10 μg/mL genistein in the dark at 25 °C for 3 days. Mycelia were selected, cut, and fixed in 3% aqueous glutaraldehyde for 4 h. Subsequently, they were washed six times with phosphate-buffered saline (0.1 M, pH = 6.8) for 20 min to remove the glutaraldehyde as thoroughly as possible and then dehydrated using graded aqueous ethanol solutions (30%, 50%, 70%, 80%, 90%, 95%, and 100%) for 30 min at each concentration. The mycelia were subsequently transferred to isoamyl acetate for 3 h, which made the mycelia fuller and prevented deformation or wrinkling. After the samples were dried at the critical point of carbon dioxide and sprayed with gold using an ion-sputtering instrument, they were observed and recorded using a SU8010 scanning electron microscope (HITACHI, Tokyo, Japan).

TEM analysis of the samples was performed using the sample processing and dehydration methods described above. After dehydration, the mycelial samples were embedded in epoxy resin and then cut into mycelial slices of approximately 70 nm by ultrathin sectioning. After double staining with uranyl acetate and lead citrate, the ultrastructure of untreated and treated mycelia was observed using an H-7650 transmission electron microscope (HITACHI, Tokyo, Japan).

4.4. Determination of the Intracellular ROS Content

To further verify the antifungal activity of genistein, which is based on the oxidative stress mechanism, the ROS level in mycelia was detected [46]. The ROS levels of the samples were measured using a reactive oxygen species assay kit (CA1410; Beijing Solarbio Science & Technology Co., Ltd., Beijing, China). The obtained *V. mali* mycelia were processed with DCFH-DA for 20 min at 28 °C. Mycelia were subsequently observed with a TE2000U fluorescence microscope (Nikon Ltd., Tokyo, Japan).

4.5. Determination of the Activities of Antioxidant Enzymes

V. mali mycelium was treated according to a previously described method. The SOD activity, POD activity, and CAT activity were measured using different commercial kits (A001-1, A084-3, A007-1, Nanjing Jiancheng Biotechnology Research Institute, China), respectively [12]. The structure of the compound was downloaded from the PubChem database and imported into ChemBio3D 14.0 software to adjust the spatial conformation of active ingredients and calculate the optimization of energy. After AutoDockTools 1.5.6 processing, the files were saved in pdbqt format [47]. The three-dimensional crystal structure of the target proteins, including SOD (PDB code: 3K9S, resolution: 1.55 Å), POD (PDB code: 2X08, resolution: 2.01 Å), and CAT (PDB code: 8J4Q, resolution: 2.10 Å), was downloaded from the PDB database (https://www.rcsb.org/ accessed on 23

September 2024). The water molecule and organic matter in the target protein were removed by Notepad2, and then the target protein was imported into AutoDockTools 1.5.6 for hydrogenation, charge distribution, and atomic type addition [48]. AutoDockVina was used for molecular docking, and the docking results were plotted using Pymol 2.6.1 [49].

4.6. RNA Extraction and Sequencing

For RNA-Seq analysis, *V. mali* was treated with genistein using the same methods as those described above but with slight modifications. Briefly, mycelia (0.3 g) on genistein-treated (treated group) and control plates (control group) were also collected and transferred to sterile centrifuge tubes, quickly frozen in liquid nitrogen, and stored at −80 °C for subsequent total RNA extraction. Both the treatment and control experiments were performed in triplicate. The total RNA of each sample was extracted using TRIzol reagent (Life Technologies, Carlsbad, CA, USA) on dry ice and processed following the protocol provided by the manufacturer. The integrity of the RNA was assessed using an RNA Nano 6000 Assay Kit on a Bioanalyzer 2100 system (Agilent Technologies, California, CA, USA). mRNA was purified from total RNA using poly-T oligo-attached magnetic beads. An AMPure XP system (Beckman Coulter, Beverly, CA, USA) was used to purify the library fragments to select cDNA fragments that were preferentially 370–420 bp in length. This system was also used to purify the PCR products. The Agilent Bioanalyzer 2100 system was subsequently used for quality assessment of the cDNA library. Clustering of the index-coded samples was performed on a cBot Cluster Generation System using the TruSeg PE Cluster Kit v3-cBot-HS (Illumina) according to the manufacturer's instructions. After cluster generation, the library preparations were sequenced on the Illumina NovaSeq platform by Beijing Nuohe Zhiyuan Technology Co., Ltd. (Beijing, China).

4.7. Analysis of Sequencing Data

After processing raw data (raw reads) in fastq format through in-house Perl scripts, clean data (clean reads) were obtained by removing reads containing adapters, reads containing poly-N runs, and low-quality reads from the raw data. Moreover, the Q20 and Q30 scores and the GC content of the clean data were calculated. All the downstream analyses were based on high-quality, clean data. HISAT2 v2.0.5 was used to construct an index of the reference genome and align paired-end clean reads to it. Then, the HISAT2 mapping tool was used to generate a database of splice junctions with better mapping results [50].

Feature Counts vl.5.0-p3 was used to count the number of reads mapped to each gene. The gene expression levels were subsequently calculated and normalized to the FPKM values based on the lengths of the genes and reads [51–53]. Principal component analysis is also commonly used to evaluate intergroup differences and intragroup sample duplication. PCA uses linear algebra to reduce dimensionality and extract principal components from tens of thousands of gene variables. Differential gene expression analysis of the three groups was performed using the DESeq2R package (1.20.0). To control the false discovery rate, Benjamini and Hochberg's approach was used to adjust the resulting p values [54–57]. Genes with p values < 0.05 and absolute fold changes ≥ 1 were considered differentially expressed. GO enrichment analysis of DEGs was implemented by the clusterProfiler R package, in which gene length bias was corrected. We used the clusterProfiler R package to test the statistical enrichment of DEGs in KEGG pathways (https://www.genome.jp/kegg/ (accessed on 2 March 2023)).

4.8. qRT-PCR Analysis

The expression profiles of ten randomly selected DEGs were analyzed by qRT-PCR to confirm the accuracy of the Illumina RNA-Seq data. The SPARKeasy Improved Tissue/Cell

RNA Kit, SPARKscriptIIRT Plus Kit (With gDNAEraser), and 2 × SYBR Green qPCR Mix Kit (Shandong Sparkjade Biotechnology Co., Ltd., Jinan, China) were used for total RNA preparation, reverse transcription and qRT-PCR analysis of the DEGs. The expression of the α-actin gene was used as the internal standard. The primers were designed, and their sequences are listed in Table 3. The PCR program was as follows: initial denaturation at 95 °C for 3 min, followed by 40 cycles of denaturation at 95 °C for 10 s, annealing at 55 °C for 30 s, and extension at 72 °C for 30 s (Jena Analytical Instruments GmbH, Jena, Germany). The relative expression levels of selected genes were analyzed using the $2^{-\Delta\Delta Ct}$ method.

Table 3. Primers for qRT-PCR.

Genes Name	Primers (5′ to 3′)
G6PDH [58]	F: TCAGAACAAGTTCGAGGGCGACAA R: TGAGGGCAATAGAGGGCTTGTTCA
VM1G_04618	F: AGCTTTCCGGTCATTATCACA R: GTCCGAAGTCAATGATCCACA
VM1G_04926	F: AGCTCCTCTCGGTCGTGCTCA R: CACACCCACGGCCACGTCGTC
VM1G_05255	F: AAAGGCTCTTCTTGTCGTCCA R: TTTGCTGTCCTGCTCCTCGTC
VM1G_07069	F: ACACAACTCGCCATATCCCA R: TCCTCTTTGGATTCGACGATG
VM1G_07417	F: TGATTTCATGGACCTGCCGTT R: TAGATCTGGCTGTACACGCTCT
VM1G_07065	F: ACCATCCTGACATTGAGCCTA R: AACCTGCCCCTTTTGTACCAC
VM1G_08675	F: CTCACGATACCCGTATTTGGC R: AATCAAGGGAACATGCGACT
VM1G_09419	F: CGTGTTGGTCTTCCCATTGACA R: AGAACCTTGCCATTGATCCAC
VM1G_09989	F: CCGAGACACATCTACATGCAC R: ATATCATGCGGCCAATTCCC
VM1G_11144	F: GATACGTCTTGATGCACCGAT R: CTTTCCCAGTTGTCCGCTTG
VM1G_05401	F: AATGTATCACCGCAAAATCCG R: GTACTCTTGACTCAGCGACCA
VM1G_07635	F: GCGGTCCTTCCCTACCTCC R: CTACTCCCCTGGATAGCGTCT

4.9. Statistical Analysis

To compare the differences between groups, an independent sample *t*-test was conducted. Each group consisted of three replicate experiments to ensure the reliability of the results. The following symbols are used to indicate significance levels: * represents $p < 0.05$; ** represents $p < 0.01$; and *** represents $p < 0.001$.

5. Conclusions

This study provides a preliminary investigation into the inhibitory activity and mechanism of genistein on *V. mali*. This study revealed that genistein induced cell membrane damage in *V. mali*, as evidenced by TEM of the hyphae and corroborated by the transcriptome sequencing data. Additionally, genistein promoted the accumulation of ROS, leading to oxidative stress and the activation of the antioxidant defense system. This was specifically observed as a decrease in SOD activity and an increase in CAT and POD activities. Moreover, genistein was implicated in the regulation of the biosynthesis of unsaturated (three DEGs) and saturated fatty acids (nine DEGs), potentially altering the fatty acid composition and ultimately resulting in lipid peroxidation. Furthermore, thirteen genes associated with fatty acid metabolism were identified that may be linked to membrane lipid peroxidation, resulting in oxidative stress damage to the mycelia and ultimately inhibiting their growth (Figure 7). In conclusion, this study is the first to demonstrate that genistein may exert antifungal activity by disrupting the cell membrane system of *V. mali.*

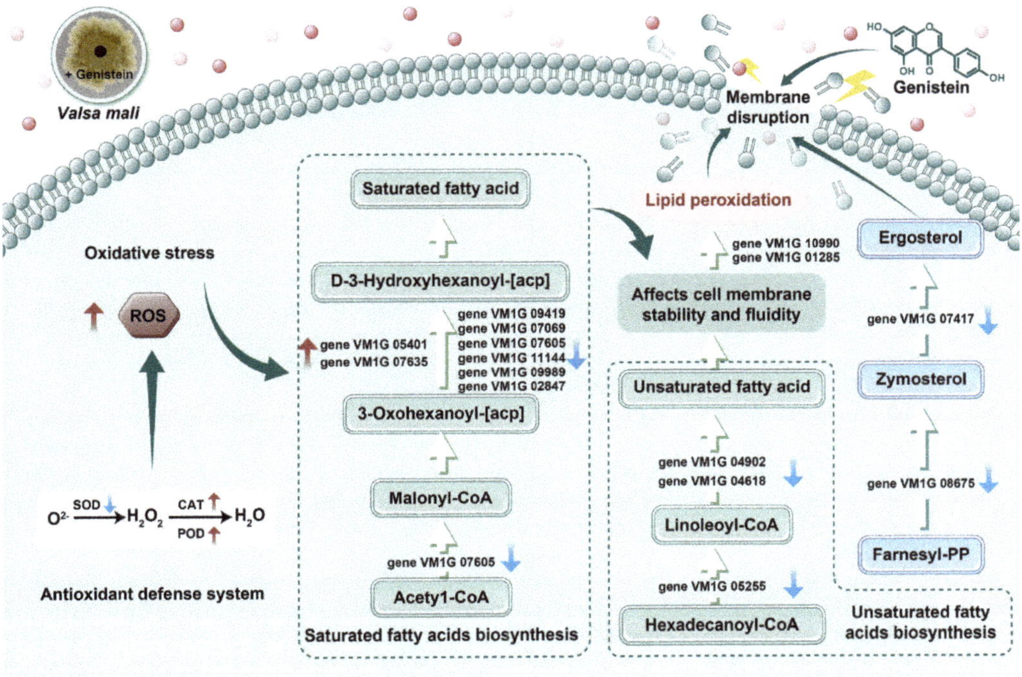

Figure 7. The mechanism of genistein inhibiting *V. mali* growth through ROS-mediated lipid peroxidation.

Author Contributions: Investigation, F.L., C.Y. and S.L.; methodology, F.L., C.Y., M.L., S.L., K.X. and X.F.; software, F.L. and C.Y.; visualization, F.L., C.Y. and M.L.; writing—original draft, F.L.; data curation, M.L. and S.L.; conceptualization, K.X.; supervision, K.X. and X.F.; writing—review and editing, K.X. and X.F. All authors have read and agreed to the published version of the manuscript.

Funding: This research was funded by the Shandong Provincial Natural Science Foundation (No. ZR2022YQ71), NATCM's Project of High-level Construction of Key TCM Disciplines (Marine Traditional Chinese Medicine; No. zyyzdxk-2023124), Taishan Scholars Program (No. tstp20240825), the China Postdoctoral Science Foundation (No. 2023M742171), the Shandong Postdoctoral Science Foundation (SDCX-ZG-202400074), the Shandong Province Key Discipline Construction Project of Traditional Chinese Medicine (Marine Traditional Chinese Medicine), and the Youth Innovation Team of Higher Education Institutions in Shandong (No. 2022KJ256).

Data Availability Statement: The data will be made available upon request.

Conflicts of Interest: The authors declare no conflicts of interest.

References

1. Li, P.F.; Tedersoo, L.; Crowther, T.W.; Wang, B.Z.; Shi, Y.; Kuang, L.; Li, T.; Wu, M.; Liu, M.; Luan, L.; et al. Global diversity and biogeography of potential phytopathogenic fungi in a changing world. *Nat. Commun.* **2023**, *14*, 6482. [CrossRef] [PubMed]

2. Xu, L.S.; Meng, Y.G.; Liu, R.H.; Xiao, Y.Z.; Wang, Y.H.; Huang, L.L. Inhibitory effects of *Bacillus vallismortis* T27 against apple Valsa canker caused by *Valsa mali*. *Pestic. Biochem. Physiol.* **2023**, *195*, 105564. [CrossRef]
3. Niu, X.D.; Wang, Z.Y.; Wang, C.Y.; Wang, H.S. Dibenzylideneacetone overcomes *Botrytis cinerea* infection in cherry tomatoes by inhibiting chitinase activity. *J. Agric. Food Chem.* **2023**, *71*, 19422–19433. [CrossRef]
4. Fisher, M.C.; Henk, D.A.; Briggs, C.J.; Brownstein, J.S.; Madoff, L.C.; McCraw, S.L.; Gurr, S.J. Emerging fungal threats to animal, plant and ecosystem health. *Nature* **2012**, *484*, 186–194. [CrossRef] [PubMed]
5. Li, J.Z.; Wang, Y.; Li, Y.X.; Zhuo, X.Y.; Wang, Z.X.; Zhao, X.L.; Wang, M.C.; Hou, Y.L.; Li, X.Y. Research progress and prospect on *Valsa mali* Miyabe & Yamada. *Mol. Plant Breed.* **2019**, *17*, 7222–7227.
6. Zhao, H.F.; Feng, H.L.; Sun, G.P.; Zheng, Y.Q.; Yan, J.L. Effects of different agents on the prevention and control effect of apple tree rot disease. *North. Fruits* **2024**, 18–20. [CrossRef]
7. Wang, M.; Duan, H.M.; Li, Z.S.; Meng, X.T.; Lu, W.J. Inhibitory activity of seven fungicides combined with fermentation. *J. Xichang Univ. (Nat. Sci. Ed.)* **2023**, *37*, 15–20.
8. Chen, Y.Z.; Wang, S.R.; Li, T.; Zhang, G.C.; Yang, J. Antifungal activity of 6-methylcoumarin against *Valsa mali* and its possible mechanism of action. *J. Fungi.* **2023**, *9*, 5. [CrossRef]
9. Duan, C.B.; Du, Y.; Hou, X.; Yan, N.; Dong, W.; Mao, X.; Zhang, Z. Chemical basis of the fungicidal activity of tobacco extracts against *Valsa mali*. *Molecules* **2016**, *21*, 1743. [CrossRef] [PubMed]
10. Jia, Y.N.; Tang, J.R.; Dang, Y.F.; Yang, Y.Z.; He, S.; Jiang, Y.; Wang, Y.; Feng, J.T.; Ma, Z.Q.; Liu, X.L.; et al. Lansiumamide B induced the degradation of polysaccharides and oligosaccharides showing antifungal activity against *Rhizoctonia solani*. *Ind. Crops Prod.* **2024**, *214*, 118529. [CrossRef]
11. Zhang, J.X.; Wang, X.L.; Vikash, V.; Ye, Q.; Wu, D.D.; Liu, Y.L.; Dong, W.G. ROS and ROS-mediated cellular signaling. *Oxid. Med. Cell. Longev.* **2016**, *2016*, 4350965. [CrossRef] [PubMed]
12. Yang, Q.; Wang, J.; Zhang, P.; Xie, S.N.; Yuan, X.L.; Hou, X.D.; Yan, N.; Fang, Y.D.; Du, Y.M. In vitro and in vivo antifungal activity and preliminary mechanism of cembratrien-diols against *Botrytis cinerea*. *Ind. Crops Prod.* **2020**, *154*, 112745. [CrossRef]
13. Song, C.G.; Shi, W.L.; Yang, J.; Ding, G.; Chen, W.; Jia, C.G.; Zhang, M.Z.; Qin, J.C. Mechanism study on *Monarda didyma* essential oil inhibiting *Aspergillus flavus* infection and aflatoxins accumulation in peanuts. *Ind. Crops Prod.* **2024**, *212*, 118311. [CrossRef]
14. Yang, Q. Antifungal Activity of CBT-diols Against *Botrytis cinerea*. Master's Thesis, Chinese Academy of Agricultural Sciences, Beijing, China, June 2020.
15. OuYang, Q.L.; Tao, N.G.; Jing, G.X. Transcriptional profiling analysis of *Penicillium digitatum*, the causal agent of citrus green mold, unravels an inhibited ergosterol biosynthesis pathway in response to citral. *BMC Genom.* **2016**, *17*, 599. [CrossRef]
16. Chen, C.; Li, B.H.; Dong, X.L.; Wang, C.X.; Lian, S.; Liang, W.X. Effects of temperature, humidity, and wound age on *Valsa mali* infection of apple shoot pruning. *Plant Dis.* **2016**, *100*, 2394–2401. [CrossRef]
17. Zhai, H.Z.; Hu, T.L.; Chen, Q.; Cao, K.Q. Control effect of 10 fungicides against apple valsa canker. *Plant Prot.* **2012**, *38*, 151–154+171.
18. Wang, X.H.; Pan, T.T.; Lian, S.; Wang, C.X.; Li, B.H. Effects of environmental factors on the growth and extension of *Valsa mali* in the xylem of apple branches. *Agric. Sci. China* **2018**, *51*, 3291–3301.
19. Liu, Z.Y.; Wang, S.; Gao, Y.Q.; Huang, L.L.; Feng, H. The cross-resistance of two geographically different *Valsa mali* strains to SBIs inhibitors and their biotic fitness. *J. Northwest For. Univ.* **2020**, *35*, 119–124.
20. Ma, H. Study on the synergistic technology of *Psoralea corylifolia* seed extract against apple valsa canker based on Box-Behnken response methodology. *Chin. Fruit* **2023**, 96–99. [CrossRef]
21. Yang, C.; Xie, S.N.; Ni, L.; Du, Y.M.; Liu, S.; Li, M.Y.; Xu, K. Chemical constituents from *Nicotiana tabacum* L. and their antifungal activity. *Nat. Prod. Commun.* **2021**, *16*, 1–5. [CrossRef]
22. Chen, X.; Zheng, G.P.; Yu, Z.G. Simultaneous assay of isoflavones in *Fructus Sophorae* by QAMS. *Chin. J. Pharm. Anal.* **2013**, *33*, 1904–1909.
23. Chen, X.X.; Luan, F.; Liu, H.T. HPLC determination of daidzein, genistein and formononetin in *Trifolium pratense*. *J. Lanzhou Univ.* **2014**, *40*, 35–37.
24. Jin, X.; Qin, X.H.; Xu, L.; Su, B.; Mu, X.T.; Liu, H.H.; Li, Y.L.; Zhu, Z.; Zhang, D.F. Simultaneous determination of genistein and kaempferol from *Kummerowia striata*. *Pract. Pharm. Clin. Rem.* **2015**, *18*, 1476–1478.
25. Xing, J.H.; Sun, X.L.; Zhou, J. HPLC determination of genistein, formononetin and biochanin A in *Cicer arietinum* L. *Chin. J. Pharm. Anal.* **2009**, *29*, 73–75.
26. Xu, J.B. Simultaneous determination of quercetin-7-*O*-β-D-rhamnoside and genistein in extract of Persimmon leaves by HPLC. *J. Chang. Med. Coll.* **2016**, *30*, 264–266.
27. Yang, Q.M.; Zhang, C.F.; Yang, Z.L. Simultaneous determination of genistin and genistein in *Moghania Philippinens* (Merr. et Rolfe) Li. from different places by HPLC. *Chin. J. Mod. Appl. Pharm.* **2012**, *29*, 744–748.
28. Yao, L.W.; Chen, X.H.; Shi, Y.; Wei, F.; Ma, S.C. Determination of quercetin, genistein and kaempferol in *Cudrania tricuspidata*. *Chin. J. Pharm. Anal.* **2014**, *34*, 415–417.

29. Yu, Y.Y.; Zhang, L.H.; Cao, S.W. Determination of daidzein, genistein, formononetin, and biochanin A in *Millettla nitida* var. *hirsutissima* extract by RP-HPLC. *Chin. Tradit. Herb. Drugs* **2006**, *37*, 1655–1656.
30. Zheng, L.; Long, F. Determination of genistein in root of *Pueraria omeiensis* by RP HPLC. *China Pract. Med.* **2006**, *1*, 32–33.
31. Zhang, L. RP-HPLC determination of daidzein, genistein, daidzin and genistin in soybean extract. *Chin. Tradit. Herb. Drugs* **2001**, *32*, 118–120.
32. Konstantinou, E.K.; Gioxari, A.; Dimitriou, M.; Panoutsopoulos, G.I.; Panagiotopoulos, A.A. Molecular pathways of genistein activity in breast cancer cells. *Int. J. Mol. Sci.* **2024**, *25*, 5556. [CrossRef] [PubMed]
33. Wei, T.T.; Chandy, M.; Nishiga, M.; Zhang, A.; Kumar, K.K.; Thomas, D.; Manhas, A.; Rhee, S.; Justesen, J.M.; Chen, L.Y.; et al. Cannabinoid receptor 1 antagonist genistein attenuates marijuana-induced vascular inflammation. *Cell* **2022**, *185*, 1676–1693. [CrossRef]
34. Viña, J.; Escudero, J.; Baquero, M.; Cebrián, M.; Carbonell-Asíns, J.A.; Muñoz, J.E.; Satorres, E.; Meléndez, J.C.; Ferrer-Rebolleda, J.; Cózar-Santiago, M.D.P.; et al. Genistein effect on cognition in prodromal Alzheimer's disease patients. The GENIAL clinical trial. *Alzheimers Res. Ther.* **2022**, *14*, 164. [CrossRef] [PubMed]
35. Jiang, T.; Dong, Y.H.; Zhu, W.Y.; Wu, T.; Chen, L.Y.; Cao, Y.T.; Yu, X.; Peng, Y.; Wang, L.; Xiao, Y.; et al. Underlying mechanisms and molecular targets of genistein in the management of type 2 diabetes mellitus and related complications. *Crit. Rev. Food Sci. Nutr.* **2023**, *64*, 11543–11555. [CrossRef]
36. Hu, X.J. The Assessment of Potential Breast Health Hazard due to Dietary Genistein Exposure. Ph.D. Thesis, Nanchang University, Nanchang, China, June 2014.
37. Shi, F.X. Effect of Genistein and Bisphenol-A on Female Reproduction. Ph.D. Thesis, Nanjing Agricultural University, Nanjing, China, June 2019.
38. Rocha, M.J.; Cruzeiro, C.; Reis, M.; Pardal, M.; Rocha, E. Spatial and seasonal distribution of 17 endocrine disruptor compounds in an urban estuary (Mondego River, Portugal): Evaluation of the estrogenic load of the area. *Environ. Monit. Assess.* **2014**, *186*, 3337–3350. [CrossRef]
39. Teilum, K.; Olsen, J.G.; Kragelund, B.B. Protein stability, flexibility and function. *Biochim. Biophys. Acta* **2011**, *1814*, 969–976. [CrossRef] [PubMed]
40. Naqvi, A.A.T.; Mohammad, T.; Hasan, G.M.; Hassan, M.I. Advancements in docking and molecular dynamics simulations towards ligand-receptor interactions and structure-function relationships. *Curr. Top. Med. Chem.* **2018**, *18*, 1755–1768. [CrossRef]
41. Porta, J.; Vahedi-Faridi, A.; Borgstahl, G.E. Structural analysis of peroxide-soaked MnSOD crystals reveals side-on binding of peroxide to active-site manganese. *J. Mol. Biol.* **2010**, *399*, 377–384. [CrossRef] [PubMed]
42. Murphy, E.J.; Metcalfe, C.L.; Basran, J.; Moody, P.C.; Raven, E.L. Engineering the substrate specificity and reactivity of a heme protein: Creation of an ascorbate binding site in cytochrome c peroxidase. *Biochemistry* **2008**, *47*, 13933–13941. [CrossRef] [PubMed]
43. Dutta, R.K.; Nenavathu, B.P.; Gangishetty, M.K.; Reddy, A.V. Studies on antibacterial activity of ZnO nanoparticles by ROS induced lipid peroxidation. *Colloids Surf. B Biointerfaces* **2012**, *94*, 143–150. [CrossRef] [PubMed]
44. Yan, N.; Du, Y.M.; Liu, X.M.; Zhang, H.B.; Liu, Y.H.; Shi, J.; Xue, S.J.; Zhang, Z.F. Analyses of effects of α-cembratrien-diol on cell morphology and transcriptome of *Valsa mali* var. *mali*. *Food Chem.* **2017**, *214*, 110–118. [CrossRef]
45. Ke, Y.; Ding, B.B.; Zhang, M.M.; Dong, T.J.; Fu, Y.; Lv, Q.Y.; Ding, W.P.; Wang, X.D. Study on inhibitory activity and mechanism of chitosan oligosaccharides on *Aspergillus Flavus* and *Aspergillus Fumigatus*. *Carbohydr. Polym.* **2022**, *275*, 118673. [CrossRef] [PubMed]
46. Jian, Y.Q.; Chen, X.; Ahmed, T.; Shang, Q.H.; Zhang, S.; Ma, Z.H.; Yin, Y.N. Toxicity and action mechanisms of silver nanoparticles against the mycotoxin-producing fungus *Fusarium graminearum*. *J. Adv. Res.* **2021**, *38*, 1–12. [CrossRef] [PubMed]
47. Morris, G.M.; Huey, R.; Lindstrom, W.; Sanner, M.F.; Belew, R.K.; Goodsell, D.S.; Olson, A.J. AutoDock4 and AutoDockTools4: Automated docking with selective receptor flexibility. *J. Comput. Chem.* **2009**, *30*, 2785–2791. [CrossRef] [PubMed]
48. Trott, O.; Olson, A.J. AutoDock Vina: Improving the speed and accuracy of docking with a new scoring function, efficient optimization, and multithreading. *J. Comput. Chem.* **2010**, *31*, 455–461. [CrossRef]
49. Eberhardt, J.; Santos-Martins, D.; Tillack, A.F.; Forli, S. AutoDock Vina 1.2.0: New docking methods, expanded force field, and python bindings. *J. Chem. Inf. Model.* **2021**, *61*, 3891–3898. [CrossRef]
50. Mortazavi, A.; Williams, B.A.; McCue, K.; Schaeffer, L.; Wold, B. Mapping and quantifying mammalian transcriptomes by RNA-Seq. *Nat. Methods* **2008**, *5*, 621–628. [CrossRef]
51. Bray, N.L.; Pimentel, H.; Melsted, P.; Pachter, L. Near-optimal probabilistic RNA-seq quantification. *Nat. Biotechnol.* **2016**, *34*, 525–527. [CrossRef] [PubMed]
52. Garber, M.; Grabherr, M.G.; Guttman, M.; Trapnell, C. Computational methods for transcriptome annotation and quantification using RNA-seq. *Nat. Methods* **2011**, *8*, 469–477. [CrossRef]
53. Liao, Y.; Smyth, G.K.; Shi, W. featureCounts: An efficient general purpose program for assigning sequence reads to genomic features. *Bioinformatics* **2014**, *30*, 923–930. [CrossRef]

54. Anders, S.; Huber, W. Differential expression analysis for sequence count data. *Genome Biol.* **2010**, *11*, R106. [CrossRef]
55. Love, M.I.; Huber, W.; Anders, S. Moderated estimation of fold change and dispersion for RNA-seq data with DESeq2. *Genome Biol.* **2014**, *15*, 550. [CrossRef]
56. Robinson, M.D.; McCarthy, D.J.; Smyth, G.K. edgeR: A Bioconductor package for differential expression analysis of digital gene expression data. *Bioinformatics* **2010**, *26*, 139–140. [CrossRef]
57. Young, M.D.; Wakefield, M.J.; Smyth, G.K.; Oshlack, A. Gene ontology analysis for RNA-seq: Accounting for selection bias. *Genome Biol.* **2010**, *11*, R14. Available online: http://genomebiology.com/2010/11/2/R14 (accessed on 1 March 2023). [CrossRef] [PubMed]
58. Yin, Z.Y.; Ke, X.W.; Huang, D.X.; Gao, X.N.; Voegele, R.T.; Kang, Z.S.; Huang, L.L. Validation of reference genes for gene expression analysis in *Valsa mali* var. *mali* using real-time quantitative PCR. *World J. Microbiol. Biotechnol.* **2013**, *29*, 1563–1571. [CrossRef] [PubMed]

MDPI AG
Grosspeteranlage 5
4052 Basel
Switzerland
Tel.: +41 61 683 77 34

Plants Editorial Office
E-mail: plants@mdpi.com
www.mdpi.com/journal/plants

www.ingramcontent.com/pod-product-compliance
Lightning Source LLC
Chambersburg PA
CBHW061623210926
51850CB00009B/2421
* 9 7 8 3 7 2 5 8 7 6 0 9 9 *